NATURAL
MAGIC

NATURAL MAGIC

*Emily Dickinson,
Charles Darwin,
and the
Dawn of
Modern Science*

RENÉE BERGLAND

PRINCETON UNIVERSITY PRESS
PRINCETON & OXFORD

Copyright © 2024 by Renée Bergland

Princeton University Press is committed to the protection of copyright and the intellectual property our authors entrust to us. Copyright promotes the progress and integrity of knowledge. Thank you for supporting free speech and the global exchange of ideas by purchasing an authorized edition of this book. If you wish to reproduce or distribute any part of it in any form, please obtain permission.

Requests for permission to reproduce material from this work should be sent to permissions@press.princeton.edu

Published by Princeton University Press
41 William Street, Princeton, New Jersey 08540
99 Banbury Road, Oxford OX2 6JX

press.princeton.edu

All Rights Reserved

Library of Congress Cataloging-in-Publication Data

Names: Bergland, Renée L., 1963– author.
Title: Natural magic : Emily Dickinson, Charles Darwin, and the dawn of modern science / Renée Bergland.
Description: Princeton : Princeton University Press, 2024. | Includes bibliographical references and index.
Identifiers: LCCN 2023035363 (print) | LCCN 2023035364 (ebook) | ISBN 9780691235288 (hardback) | ISBN 9780691235295 (ebook)
Subjects: LCSH: Dickinson, Emily, 1830–1886—Criticism and interpretation. | Darwin, Charles, 1809–1882. | American literature—19th century—History and criticism. | Literature and science—United States—History—19th century. | Nature in literature. | Philosophy of nature. | BISAC: LITERARY CRITICISM / Modern / 19th Century | SCIENCE / Philosophy & Social Aspects
Classification: LCC PS1541.Z5 B443 2024 (print) | LCC PS1541.Z5 (ebook) | DDC 811.4—dc23/eng/20231030
LC record available at https://lccn.loc.gov/2023035363
LC ebook record available at https://lccn.loc.gov/2023035364

British Library Cataloging-in-Publication Data is available

Editorial: Anne Savarese, James Collier
Jacket: Heather Hansen
Production: Erin Suydam
Publicity: Alyssa Sanford (US), Carmen Jimenez (UK)
Copyeditor: Wendy Lawrence

Jacket Credit: Yevheniia Lytvynovych / Shutterstock

This book has been composed in Arno Pro and Garamond Premier Pro.

Printed in the United States of America

10 9 8 7 6 5 4 3 2 1

For Kim and Annelise
and all our Shaggy Allies
(especially Ola, Stjerne, and Bamse)

CONTENTS

Preface: An Orchis' Heart xi

Introduction: An Enchanted World
1

CHAPTER 1
Darwin and Dickinson, Childhood Portraits
18

CHAPTER 2
Darwin the Naturalist
*Shropshire, Edinburgh, Cambridge,
1809–1831; Darwin, to Age 22*
46

CHAPTER 3
Nature's People: Scientific Amherst
Amherst, 1830–1836; Dickinson, to Age 6
72

CHAPTER 4
Juggler, Geologist, Dark Horse
Aboard the Beagle, *1832–1836;
Darwin, Age 23–27*
93

CHAPTER 5
Dickinson the Bold
Amherst, 1836–1847;
Dickinson, Age 6–16
115

CHAPTER 6
The Leading Scientific Men
London and Amherst, 1836–1845
133

CHAPTER 7
Religion of Geology
South Hadley, Amherst, 1847–1851;
Dickinson, Age 16–20
153

CHAPTER 8
A Slow-Sailing Ship
Downe, Great Malvern, 1842–1851;
Darwin, Age 33–42
178

CHAPTER 9
Excitement in the Village
Amherst, 1851–1857;
Dickinson, Age 20–26
201

CHAPTER 10
On the Origin of Species
Downe, 1858–1860; Darwin, Age 49–51
221

CHAPTER 11
If You Saw a Bullet
Amherst, 1857–1861;
Dickinson, Age 26–31
237

CHAPTER 12
Wild Experiment
Downe and Amherst, 1860–1862
260

CHAPTER 13
Melody or Witchcraft?
Amherst, 1862–1866
277

CHAPTER 14
Mutual Friends
Downe and Amherst, 1866–1882
299

CHAPTER 15
Perfectly Disinterested: Darwin's Last Days
313

CHAPTER 16
Nature Is a Haunted House: Dickinson Faces Death
330

Afterword: Hope Is a Strange Invention
*Darwin and Dickinson in the
Twenty-First Century*
351

Acknowledgments 359

Notes 363

Bibliographic Note 389

Bibliography 393

Index 405

A page featuring a pressed wild orchid, *Cypripedium acaule* (pink lady's slipper), from Emily Dickinson's *Herbarium* (1839–1846). MS AM 11181.11, Houghton Library, Harvard University.

PREFACE

An Orchis' Heart

THE SUMMER of 1858 was a hard one. Epidemics of diphtheria and scarlet fever spread across Britain and the United States. In Massachusetts, Emily Dickinson's mother was ill and needed constant care. Dickinson spent hours nursing her, keeping her company, worrying. Late at night while the household rested, she wrote with surprising focus. The pressures of the sickbed gave new intensity to her literary experiments. She started two projects: a series of "Master" documents that discussed love and literature using the vocabularies of natural philosophy, history, and theology and another project that included careful copies of her poems sewn together into booklets. Both were private, closely guarded secrets. She did not intend to publish.

Thousands of miles away in England, three of Charles Darwin's children were also sick. As his youngest son's fever worsened, an unexpected letter arrived from the East Indies. Until that summer, Darwin had kept his theory of natural selection private as he worked painstakingly to substantiate it. If he continued to avoid making a public declaration, another man—Alfred Russel Wallace, the man who had sent the letter—would get sole credit for the theory. Darwin, distracted by the illness in his household, found it almost impossible to think. His friends advised him to publish the ideas he had long held close. They arranged the simultaneous release of two brief papers outlining Darwin's and Wallace's ideas. Darwin did not participate much in the publication.

His son, Charles Waring Darwin, died of scarlet fever in late June. Despite his grief, when the summer ended Darwin started writing *On the Origin of Species*.

Darwin and Dickinson never met. When they embarked on their greatest works, neither was aware of the other. Though they vaguely knew that many families around the world were going through similar trials during that season of epidemics, neither had any idea of the parallels between their ideas.

Dickinson had studied the natural sciences in college. In the years afterward, she would become familiar with Darwin's work. But in 1858 she had no thoughts of Darwin. At night she listened to the music of crickets and peeping frogs. When she could get away, she walked out into the meadows, forests, and wetlands of Amherst for comfort. More often, she worked in the garden, staying within earshot of her mother. She tended the flowers expertly, as her mother had taught her.

Dickinson's booklet of poems—her *fascicle*—started with botanical observation. The poet fit her verses to the paper, arranging them as she had arranged dried plant specimens in an album—her *Herbarium*—as a student. She wrote the first line carefully, in rusty black ink: "The Gentian weaves her fringes –."[1] Dickinson was describing a brilliant blue wildflower that opens in sunlight and clasps its finely lashed petals tight when cold weather threatens. Her fascicle blossomed with flowers and buzzed with pollinators. The first sheet mentioned six different species of plants and described them with a naturalist's precision. A rose was "A sepal – petal – and a thorn –" interacting with a pollinating bee.[2]

Now in her late twenties, Dickinson was more serious about her fascicle than about any of her school projects. This was a more ambitious endeavor. An herbarium could never have contained the plants and animals she wanted to describe, much less the philosophical speculations and moments of surprising emotional intensity she hoped to bring together. Though her poems were metaphorical "blossoms," she would not display them. She wanted to hold her wild ideas close. The gentian poem concluded, "My departing blossoms / Obviate parade."[3] For now, her work required privacy. Though the circumstances were ideal for creating her experimental mix of poetry, empirical observation, and philosophy, the climate was not right for exposing her work to the world. As far as we know, she did not show her fascicle to anyone.

In private she could explore her own longings and beliefs. On the first page she wrote a brief prayer, or parody of prayer:

In the name of the Bee –
And of the Butterfly –
And of the Breeze – Amen![4]

In these lines the poet invoked the material world—physically real animals and air currents—in place of religious figures. "In the name of the Bee –" demonstrated that Dickinson did not separate poetry from philosophy or religion, much less from the natural sciences. The bee, the butterfly, and the breeze were not mere metaphors for her. Dickinson believed that the nonhuman natural world was rich with meaning.

At the bottom of the first sheet of paper, the poet was left with just an inch or two of blank space. She added two lines that brightened the page with imagined pink:

To him who keeps an Orchis' heart –
The swamps are pink with June.[5]

Many years afterward, Dickinson's family would place a pink orchid, *Cypripedium acaule* (a pink lady's slipper), above her heart before she was buried. But in 1858 Dickinson was not worried about dying. Instead she feared that she would be forever mired in caring for her mother and doing the household labor that her mother had always done. She found housework unbearably dull. In the dismal swamp of chores, her late-night literary experiments helped her keep "an Orchis' heart."

After her mother recovered in the autumn, Dickinson no longer needed to stay so close to home. But the course of her life had changed over the summer. Dickinson had learned that she could thrive in isolation, keeping odd hours, dividing her time between the garden and her manuscripts. Putting pen to paper filled her with excitement. Her vocation had become insistently clear: she would be a poet.

That summer marked a similar turning point for Charles Darwin. In 1858 and 1859, in the face of his grief over his son and his fear of public controversy, he worked furiously to complete his book on natural selection. Like Dickinson, Darwin, beset by care and family illness, looked to flowers—and to the natural world—for comfort.

After their son's death, Charles and his wife, Emma, tried to console each other. They often went out to a spot near their house that they called the Orchis Bank. Now known as the Downe Bank Reserve, it still hosts eleven species of orchids. Scholars believe that the "entangled bank" Darwin described in the conclusion to *On the Origin of Species* was probably the Orchis Bank where he and his family liked to walk.[6]

Of course, Darwin had no idea that a woman in Massachusetts had written a poem about a man who managed to keep an "Orchis' heart." Yet he filled his pages with the bright colors and remarkably vibrant, intricately interwoven relationships of the flowers, insects, worms, and other species he had found entangled together on the Orchis Bank.

Darwin published *On the Origin of Species* in December 1859. In the years that followed, he would often be at the center of public debate, subjected to bitter denunciations and passionate advocacy. At times his allies' misguided defenses of his ideas would be as excruciating as the condemnations. Darwin would try again and again to explain that the adaptation and change at the heart of natural selection could offer consolation and inspire hope.

Darwin had hoped his ideas would be important in scientific circles, but he had never dreamed they would have such a widespread impact. Indeed, he was startled by his sudden scientific fame. As it turned out, few aspects of life or death remained untouched by Darwin's thought. His ideas were quickly woven into the fabric of the modern world. Across the Atlantic, Dickinson's response to Darwin helped her to write the poetry that speaks to us today.

Meanwhile, Darwin took refuge in his greenhouse, working with orchids. When he asked an Irish botanist to send him a very large shipment of orchids, he explained, "I fear I am very unreasonable; but this subject is a passion with me."[7] Darwin showed the world his "Orchis' heart" in 1862 with the publication of his book *On the Various Contrivances by Which British and Foreign Orchids Are Fertilised by Insects, and on the Good Effects of Intercrossing*. In a private letter to his good friend, the botanist Joseph Hooker, he begged for orchid pods and the mosses that hosted them: "Remember Orchid pods. – I have a passion to grow the seeds . . . *for love of Heaven favour my madness* & have some [lichens or mosses] scraped off & sent me. I am like a gambler, & love a wild experiment."[8]

Dickinson loved a wild experiment just as much as Darwin, who was one generation ahead of her. Darwin's remarkable ideas about the natural world would influence her thought profoundly. She did not have the same impact on Darwin as he had on her—in fact, since she published almost nothing during her lifetime, her circle of influence was very small until after her death. But being in the next generation also conferred some benefits: able to read Darwin, consider his ideas at leisure, and record her responses for posterity, Dickinson usually got the last word. In this account I have taken the liberty of giving her the first word, too.

For me, this book started with the puzzling realization that many of Dickinson's poems seemed profoundly Darwinian. Although she never mentioned Darwin by name in her poems, she rarely mentioned anyone by name in her poetry, so this absence did not rule him out of her important influences. She did name Darwin in two letters, which confirmed that she knew about his work. Still, there was not much to go on. She returned again and again to the topics that fascinated Darwin, but was that enough to demonstrate that she was responding to his thought?

"In the name of the Bee –" could serve as an invocation for *On the Origin of Species*. Yet Dickinson wrote these early poems before Darwin published *On the Origin of Species*. I wondered if it was just a "striking coincidence," like the parallels between Darwin's theory of natural selection and the very similar theory developed by Alfred Russel Wallace at the same time.[9] Did Dickinson write about Darwinian ideas simply because she was his contemporary? Did her writings seem to apply to Darwin merely because she was a great poet whose writings were almost universally applicable? Both explanations are valid, as far as they go. But neither goes far enough.

There is a stronger connection between Dickinson and Darwin than the proximity of history or the universality of literature. They both understood natural science and the natural world in ways that seem strange and somewhat surprising in the twenty-first century. Their nineteenth-century attitudes to nature and the study of it are so different from ours that when we trace their stories, a vanished world begins to emerge. The more I consider these figures together, the more I feel their world—and my world—come alive. Darwin and Dickinson illuminate each other. By reading them together, we can start to understand the interconnected relationships that animated nineteenth-century poetry and science.

Dickinson's works would not become public until long after both Darwin's and Dickinson's deaths. Yet although she kept her writing private, her lines form a fitting epigraph for a book about two people who loved botany and natural philosophy. Today, we remember them as towering figures, among the greatest of the nineteenth century's poets and scientists. In 1858, however, these professional identities were still emerging. Dickinson and Darwin had not yet become cultural giants. They divided their time between caring for their families and writing, and they shared a curious passion for orchids. For them, science was full of magic. Unknown to each other, they both believed in the bee, the butterfly, and the breeze, and they trusted that

> To him who keeps an Orchis' heart –
> The swamps are pink with June.[10]

NATURAL
MAGIC

NATURAL MAGICK

BY

John Baptista Porta,

A NEAPOLITANE:

IN

TWENTY BOOKS:

1 Of the Causes of Wonderful things.
2 Of the Generation of Animals.
3 Of the Production of new Plants.
4 Of increasing Houshold-Stuff.
5 Of changing Metals.
6 Of counterfeiting Gold.
7 Of the Wonders of the Load-stone.
8 Of strange Cures.
9 Of Beautifying Women.
10 Of Distillation.
11 Of Perfuming.
12 Of Artificial Fires.
13 Of Tempering Steel.
14 Of Cookery.
15 Of Fishing, Fowling, Hunting, &c.
16 Of Invisible Writing.
17 Of Strange Glasses.
18 Of Statick Experiments.
19 Of Pneumatick Experiments.
20 Of the Chaos.

Wherein are set forth
All the RICHES and DELIGHTS
Of the
NATURAL SCIENCES.

LONDON,
Printed for *Thomas Young*, and *Samuel Speed*; and are to be
sold at the three Pigeons, and at the Angel in St.
Paul's Church-yard. 1658.

Title page of the English translation of Giambattista della Porta's *Natural Magick* (1658). Smithsonian Libraries and Archives.

INTRODUCTION

An Enchanted World

CHARLES DARWIN did not call himself a scientist. When he was young, it was not possible. The word *scientist* was not invented until 1833, when Darwin reached his twenties. Before then, people who studied the natural world were known as *naturalists* and *natural philosophers*. Darwin's shipmates on HMS *Beagle* thought of him as the ship's philosopher. They called him Philos. The nickname did not imply that Darwin was a lofty thinker, prone to idle reveries or flights of fancy. At the time, natural philosophy was a surprisingly active, physical pursuit.

When the word "scientist" replaced "natural philosopher," the shift marked a change in the way people thought about studying the natural world. It was more than mere semantics. In the decades before "scientist" was coined, there was no clear separation between the arts and the sciences; after *scientist* was proposed as a parallel to *artist*, these realms began to divide. Until then, Romantic poets and philosophers (who were often one and the same) tended to think of pursuits that we would now call artistic (writing poems, sketching landscapes, making botanical drawings) as empirical investigations into the natural world. At the same time, their idea of philosophy was often very close to what we now think of as natural science. Natural philosophers conducted experiments. They collected specimens of animals, plants, and minerals and exchanged them with others. They anatomized, classified, and dissected. Breaking rocks open, pulling flowers apart, cutting frogs into pieces—philosophers were people who got their hands dirty.

In *Frankenstein*, her 1818 novel about a deranged natural philosopher, Mary Shelley described philosophers as men "whose hands seem only made to dabble in dirt, and their eyes to pore over the microscope or crucible."[1] Although the fictional Victor Frankenstein would become the prototype for mad scientists in thousands of subsequent novels and films, Shelley never described him as a scientist. At the time, years before the word existed, such crucible-wielding, microscope-peering, dirty-handed dabblers were known instead as philosophers.

Decades after the word "scientist" was coined, Charles Darwin and Victor Frankenstein would come to stand for two sides of the nineteenth-century scientist in the popular imagination. The fictional Frankenstein would represent the horrifying danger of interfering with the natural order of things, while Darwin's systematic, secular approach to the natural world would be associated with the dry logic of disenchantment. On the one hand, supernatural horror. On the other, a wholly rationalist world so drained of emotional or spiritual significance that it inspired another type of horror. Scientists would become disturbing cultural figures either way, whether they were too close to magic or too far removed from it.

Both extremes point toward a second remarkable shift that happened around the same time. Just as natural philosophers were transformed into scientists, natural magic was banished from serious conversation. In the course of the nineteenth century, scientific objectivity replaced more subjective emotional approaches to the natural world. Science distanced itself from wonder. The German sociologist Max Weber would describe this process as the "disenchanting of the world."[2] As Weber defined it, the central principle of disenchantment was that "there are no mysterious incalculable forces intervening in our lives, but instead all things, in theory, can be *mastered* through *calculation*."[3] Weber held that modern thinkers had replaced the old sense of mystery with a surprisingly unsubstantiated belief in the principle that humans could— theoretically—have mastery over every aspect of nature.

Charles Darwin's approach to the natural world was unquestionably secular and systematic. He was not a mad scientist like Victor Frankenstein, much less an evil wizard. His attitude was closer to the "rationalization and intellectualization" Max Weber would later describe.[4] However, it would be wrongheaded to imagine him as a great disenchanter who wanted to master the universe. Mastery was alien to him. Darwin challenged human supremacy and stressed kinship among living beings. The humility of his works shocked some of his contemporaries. Throughout his life Darwin remained fascinated and energized by the mysteries of the natural world. He never lost his sense of enchantment. In fact, the scholar George Levine has argued that Darwin is best understood as a secular enchanter.[5]

Though Darwin witnessed—and even participated in—the separation of science from philosophy and poetry, he regretted the divide. In his *Autobiography*, Darwin recalled, "Up to the age of thirty, or beyond it, poetry of many kinds, such as the works of Milton, Gray, Byron, Wordsworth, Coleridge, and Shelley, gave me great pleasure, and even as a schoolboy I took intense delight in Shakespeare, especially in the historical plays."[6] But in 1839, his thirtieth year, Darwin published his first book, *The Voyage of the Beagle*, an account of his five-year journey around the world as a naturalist. In honor of his work, Darwin was made a Fellow of the Royal Society. After he was inducted into the highest echelon of science, he narrowed his scope and renounced other pursuits.

In later years, he would come to feel that his turn away from poetry was a "curious and lamentable loss."[7] In a letter to the botanist Joseph Hooker, written when he was nearly 60, Darwin remarked, "I am a withered leaf for every subject except Science."[8]

Darwin was not as much of a withered leaf as he feared or as others imagined. Yet to his chagrin, *Darwin* became a shorthand way of referring to a version of materialism that saw the word as wholly disenchanted, meaningless, and dispiriting. Darwin's own view was richer. *On the Origin of Species* celebrated the grandeur of the natural world, with its "endless forms most beautiful and most wonderful."[9] That sense of aesthetic appreciation and open-ended wonder imbued all his works—including the books on corals, barnacles, and worms. In *The Descent of Man* (1874), Darwin concluded that "appreciation of the beautiful" shaped sexual selection for many species.[10] In Darwin's thought, beauty was central. He believed that evolution hinged on aesthetics. The impact of his work, however, was beyond his control. Darwin would watch helplessly as poetry, beauty, and magic drained away from science.

The story of how science and religion became opposed to each other during the nineteenth century has often been told. In many accounts, Charles Darwin plays the role of the great Victorian scientist whose work opened the floodgates of secularization. In these pages, however, I will unfold a different narrative. Darwin will begin as an aspiring natural philosopher, born in an age when "poets were philosophers, and philosophers poets" (as Darwin's biographer Janet Browne describes it).[11] Together, poetry and philosophy played the role that science plays today, and magic had a place in serious thought. During Darwin's lifetime, science—and scientists—would come to be imagined in opposition to poetry, philosophy, and magic as well as religion, but Darwin would remain skeptical of the hardening boundaries between science and other ways of understanding the natural world. My account will explore how magic invisibly persisted for Darwin and his closest intellectual kin, infusing their sense of nineteenth-century science with infinite possibility.

We will begin in the years before magic seemed to disappear—the years when poetry and natural philosophy were inseparable. To gain a sense of the approaches to nature that nourished Darwin's thinking, we will consider the poet-philosophers who shaped his world (his grandfather Erasmus Darwin and the circle of Romantics and radicals that included Samuel Taylor Coleridge and Mary Shelley). But although Charles Darwin's Romantic roots reached deep, he was no Romantic. By 1859, when he published *On the Origin of Species*, the world of his grandfathers had vanished. In his own time, when poetry was often separated from science, the poet who best captured the relationships between science, religion, and magic was Emily Dickinson.

Born in Amherst, Massachusetts, where girls were expected to study the natural sciences, Dickinson was extraordinarily well positioned to respond to Darwin's ideas. After studying at Mount Holyoke, Dickinson moved back to Amherst and lived in her family home on the edge of the

Amherst College campus. She and her family were among the intellectual and cultural elite of the United States. Her father was a U.S. congressman. The family was instrumental in establishing Amherst College.

Dickinson was 29 years old when *On the Origin of Species* was published. She was an accomplished poet, but she usually kept her writing private. One of the few people who had the chance to read her poetry was her mentor, Thomas Wentworth Higginson, who shared Dickinson's knowledge of—and interest in—the natural world. This common interest bonded them together. In one early letter to him, she proclaimed, "I know the butterfly, and the lizard, and the orchis. Are not those your countrymen?"[12] She knew that Higginson was as interested in natural science as he was in poetry.

When Dickinson started writing to him, Higginson was one of the most prominent literary Darwinians in the United States. He frequently quoted Darwin. His 1860 lecture at the Concord Lyceum a few weeks after *On the Origin of Species* was published used quotations from Darwin's newest book to argue against slavery.[13] He framed his 1862 essay on the "Life of Birds" with a quotation from *The Voyage of the Beagle*: "'We do not steadily bear in mind,' says Darwin, with a noble scientific humility, 'how profoundly ignorant we are of the condition of existence of every animal.'"[14] Although it was somewhat unusual to cite Darwin in the mid-nineteenth century, what is most striking about this mention is that Higginson praised him for "noble scientific humility." He understood that Darwin did not pretend to be a master of nature. Few others were as admiring or as perceptive.

Higginson would eventually provide the closest personal link between Darwin and Dickinson. They had other social connections, but Higginson was the strongest tie. He was one of the only friends invited to visit Dickinson at home in Western Massachusetts in the poet's later years. He was also acquainted with Darwin and visited him twice in Kent. On his second visit, in 1878, Higginson would stay overnight at Darwin's home.

The social ties between Darwin and Dickinson are not particularly surprising. In cultural terms, nineteenth-century Massachusetts was practically a British colony. Britain and New England were closely connected at the time, though everything in New England tended to be on a smaller scale. New England's yeoman farmers were very different from Britain's aristocratic landowners, but the differences between the emerging professional classes were much less significant. Darwin and Dickinson were born into similar social echelons. Darwin's father was a doctor; Dickinson's was a lawyer. They both moved in highly educated, socially progressive circles. While Dickinson's family had been instrumental in the founding of Amherst College, Darwin's family patronized a wide variety of institutions (Darwin's grandfathers were both *Lunar Men*, founders of the Lunar Society of Birmingham. Ten members of the Darwin-Wedgwood clan were Fellows of the Royal Society). Darwin's fortune came partly from the industrial pottery established by his Wedgwood grandfather and partly from his family investments in railroad companies and other ventures. Dickinson's Norcross grandfather was a canny investor who left substantial property to the poet's mother. Her father was an enthusiastic proponent of the railroad and the telegraph who made significant investments to bring the new technologies to Amherst. Thus, Darwin and Dickinson were heirs of industrialization, fortunate scions of nineteenth-century capitalism.

They were both wealthy enough to be able to avoid the usual obligations. Darwin did not need to find a paying job; Dickinson did not need to marry. Instead, both could stay home and spend their time focused diligently on observing the natural world and writing about it. Their financial independence allowed them to concentrate full time on writing. Darwin cared deeply about literature; he was a writer as well as a naturalist. Dickinson's interest in nature was equally profound; she was a naturalist as well as a poet.

Darwin sailed around the world but never visited North America. After he returned from his great voyage and established himself in the village of Downe, he became somewhat reclusive. Dickinson never left the United States. In her twenties she traveled south along the Eastern Seaboard to Washington, DC, but as she grew older she stayed closer and closer to home. From the 1850s to the 1880s, Darwin and Dickinson lived surprisingly similar lives in similar places.

The connections between Darwin and Dickinson went much deeper than their overlapping friendships and their similar circumstances. Caught in the middle of the great disenchantment that Weber would document, both Darwin and Dickinson grappled with the massive cultural changes wrought by modern science. Over their careers, each would wrestle with the implications of dividing the study of nature from philosophy and poetry. Was it possible that the material world was entirely separate from any higher ideals? Was the universe random—neither good nor bad but somehow outside of ethics? Was the cosmos meaningless? Through their influential works, both addressed these questions. They advanced scientific ways of thinking while continuing to insist that the natural world was rich with mystery.

Today, Dickinson and Darwin are remembered very differently, most obviously in relation to gender and nationality. We tend to picture Dickinson as a feminine character: a young lady in a white dress in her bedroom in Massachusetts, her days spent writing or picking wildflowers at the edge of her family's hayfield, far from the center of the British Empire. In contrast, we imagine Darwin sailing around the world aboard HMS *Beagle* or striding across London with a wild Victorian beard.

Of course, the differences between these two figures go deeper than their popular images. During her lifetime, Dickinson was not a public figure. Although she wrote thousands of poems, she published only ten. In contrast, Darwin was a celebrity who published more than a dozen books and played many public roles. Dickinson never married. Darwin and his wife, Emma, had ten children, seven of whom survived to adulthood. While Darwin was a Fellow of the Royal Society, Dickinson wondered if the pine tree outside her window was a "'Fellow of the Royal' Infinity."[15]

The professionalization of science in the mid-nineteenth century had unexpected consequences for both figures. Darwin had set out to be a naturalist and natural philosopher—a well-rounded man of science—before British universities offered degrees in the natural sciences. His formal education focused on the classics, and he trained to be a clergyman. Later in his career, he would embrace the developing scientific method, but he mourned—and never entirely relinquished—a deeply Romantic love of nature. His work helped the sciences to become culturally central, yet a year before he died, he complained, "My mind seems to have become a kind of machine for grinding general laws out of large collections of facts."[16]

Dickinson was born in 1830, during a brief window of time when people in Massachusetts thought the physical sciences were the most appropriate topic for girls to study. Indeed, one of the reasons for coining the word "scientist" was to make room for the accomplished women who could not readily be described as men of science. The historian Kim Tolley has documented that in the United States in the first decades of the nineteenth century "female higher schools placed a greater emphasis on scientific subjects than did similar, contemporary institutions for males."[17] Dickinson's formal education included botany, geology, astronomy, and chemistry. She wrote most of her poems after Darwin published *On the Origin of Species* in 1859. Yet although Dickinson found ways to breathe natural magic back into scientific thought, few readers were able to recognize the science in her poetry. In 1890, when the first edition of her poems was published, it was almost impossible for readers to see that this very private Massachusetts woman could have been a profound philosophical, theological, and scientific thinker as well as a major poet. By the time readers caught on to her importance as a poet, poetry had come to be viewed as largely irrelevant to science.

Many of Dickinson's readers have overlooked her scientific acuity and her conviction that poetry and magic were valid and useful approaches to nature. Her beloved sister-in-law, Susan Gilbert Dickinson, did not make that mistake. After Dickinson's death, Susan described her as a "magician." According to Susan, Dickinson's magic was not supernatural but a new, modern version of natural magic. She was "quick as the electric spark in her intuitions and analyses," Susan wrote. Her conservatory bloomed with "rare flowers," and she "knew her subtle chemistries." Though her imagination was shaped by her study of botany, physiology, geology, and astronomy, Dickinson rejected the approach to the scientific method that Darwin described as "grinding general laws out of large collections of facts."[18] Instead, Susan reported, Dickinson thought life was "all aglow."[19] Dickinson never allowed her close and careful observation of the natural world to diminish her sense of enchantment.

Dickinson was not alone in her effort to resist disenchantment. In *The Myth of Disenchantment*, the historian of religion Jason Ā. Josephson Storm explains that "for all the polemical attacks against superstition and magic, disenchanting efforts were only sporadically enforced.... Notions of magic and spirits keep resurfacing as redemptive possibilities."[20] Many nineteenth-century thinkers regretted the retreat of magical thinking. The pragmatist philosopher William James described it as "a very sad loss" "for certain poetic constitutions" "that the naturalistic superstition, the worship of the God of nature simply taken as such, should have begun to loosen its hold upon the educated mind."[21] Without enchantment, James worried, life would not be worth living. In his view, disenchantment was not absolute or inevitable. It was not even desirable.[22] Even so, by the end of the nineteenth century, universities and schools were full of dogmatic adherents of disenchantment. Gradually, the way that educated people in Europe, Britain, and America looked at the world changed.

As higher education grew hostile to magic, the ideas of earlier scholars became somewhat embarrassing. University libraries and museums reorganized, deemphasizing works by figures like Isaac Newton and Cotton Mather that focused on such topics as alchemy and witchcraft. Magic was so discredited among scholars that its terms became unfamiliar, even for historians. Everything from alchemy to Zoroastrianism was labeled *superstition*.

The new hostility to superstition was a by-product of secularization. The growing consensus among educated people was that scientific approaches had more validity than both religion and magic. As religion and science came into conflict, schools and universities separated religion from science and moved explicitly theological thought into specialized divinity schools. No one knew quite what to do with magical thought. Although the shift in the credibility accorded to different approaches was dramatic, the more radical change was that magic, religion, and science came to be viewed as not only separated from but even opposed to one another.

If natural philosophy and its bygone companions natural history and natural theology seem strange to twenty-first-century readers, natural magic might seem almost unimaginable. During Charles Darwin's and Emily Dickinson's childhood years, however, "magic" was not necessarily dismissed as superstition or associated with the supernatural. The study of nature embraced many overlapping approaches. Collecting and classifying objects was called natural history. Deducing general laws was the work of natural philosophers. The search for God—the Christian God—in the natural world was the purview of natural theology. Along similar lines, mysterious natural forces and transformations—changes related to life and death, electricity and magnetism, the formation of crystals and gases—had long been understood as natural magic.

As early as 1496, the Renaissance humanist Giovanni Pico della Mirandola had argued that natural magic "when well-researched" was "nothing more than the final realization of natural philosophy."[23] For hundreds of years afterwards, natural magic and natural philosophy were interchangeable. In 1605, Francis Bacon explained that natural magic was characterized by "good and fruitful inventions and experiments."[24] When the English translation of Giambattista della Porta's book *Natural Magick* was published in 1658, its title page promised "all the riches and delights of the natural sciences."[25]

For early modern thinkers, "natural magic" was an attempt to understand and explain the hidden properties of things. For the most part, these thinkers tended to use *occult* to describe mysterious or invisible phenomena—such as the phenomena we know today as gravity, electromagnetism, and thermodynamics. In later years, "occult" began to imply intentional secrecy on the part of the practitioner. The association of the occult with supernatural rituals came even later. Even after the occult and the supernatural became entwined with each other, natural magic was imagined in opposition to the supernatural: by the early nineteenth century, it encompassed the work of chemists, horticulturalists, animal breeders and trainers, and medical practitioners (from midwives to mesmerists).

Magic was not the only English word whose meaning shifted over time. *Science* changed meaning as well. When Francis Bacon and his contemporaries used the word "science," they meant to describe any coherent body of knowledge, what an English speaker of the twenty-first century might describe as an academic discipline. For Bacon the sciences included grammar, logic, and rhetoric as well as arithmetic, geometry, music, and astronomy—the trivium and quadrivium of medieval universities.

When Darwin and Dickinson were young, science and magic were still closely associated. Popular demonstrations of new technologies and scientific discoveries were often framed as displays of natural magic. Although many people found the advances in physics and chemistry mysterious, few saw them as diabolical. In fact, it was the other way around: increasingly, witches, demons, and evil spirits were dismissed as hoaxes or delusions.

Emily Dickinson was born in 1830, just as modern science emerged from disparate strands of natural history, natural theology, natural philosophy, and natural magic. In Britain, one turning point, when "science" became more like science as we know it today, can be pinpointed in 1831, when the British Association for the Advancement of Science (BAAS) was launched. The men who started the BAAS hoped their new organization would become a more professional alternative to the clubby Royal Society, which was full of aristocratic amateurs.

The founders of the BAAS, John Herschel, Charles Babbage, and David Brewster, all published influential books on scientific method around this time. In 1830, Herschel published *Preliminary Discourse on Natural Philosophy*, which described the experimental methods of natural philosophy as rooted in alchemy and natural magic.

His colleagues urged their contemporaries to turn away from the old ways. Babbage's *Reflections on the Decline of Science in England and Some of Its Causes* (1830) argued for the professionalization and intellectual division of labor. Meanwhile, Brewster hoped to demystify science completely. His book *Letters on Natural Magic* (1832) carefully separated science from magic. In Brewster's mind, mystery and magic were the enemies of the modern scientific methods that the BAAS wanted to promote. Brewster emphatically rejected the old notion of equivalency between physical science and natural magic and starkly distinguished natural philosophy from natural magic.

According to Brewster, the central distinction between natural philosophy and natural magic lay in how knowledge was used. Natural philosophers shared their knowledge widely for the good of all humanity, while practitioners of natural magic tried to keep scientific knowledge hidden (or occult) so that science and technology could be used to manipulate the credulous. Brewster explained, "The subject of Natural Magic is one of great extent as well as of deep interest. In its widest range, it embraces the history of the governments and the superstitions of ancient times, – of means by which they maintained their influence over the human mind, – of the assistance which they derived from the arts and the sciences, and from a knowledge of the powers and phenomena of nature."[26]

Brewster defined natural magic as science in the service of tyranny. As he put it, "The prince, the priest, and the sage, were leagued in a dark conspiracy to deceive and enslave their species."[27] Brewster wanted to fight against the "dark conspiracy" by explaining the science behind marvels and illusions.

The new emphasis on making ideas public brought new opportunities and new pressures. One strange side effect of the shift toward a scientific method that required publication was that the privacy long accorded to amateurs and independent scholars was practically demonized—keeping ideas private came to be seen as somewhat occult, while marvels and wonders were greeted with new skepticism. Brewster's *Letters on Natural Magic* embraced this cultural shift. As he put it, "The science of chemistry has from its infancy been pre-eminently the science of wonders. In her laboratory the alchemist and the magician have revelled uncontrolled, and from her treasures was forged the sceptre which was so long and so fatally wielded over human reason."²⁸ In Brewster's mind, any "science of wonders" was dubious, even dangerous.

Interest in scientific explanations of magic tricks extended from Edinburgh and London to Dickinson's hometown. In 1834, in Amherst, Massachusetts, an anonymous author published *Ventriloquism Explained: And Juggler's Tricks, or Legerdemain Exposed: With Remarks on Vulgar Superstitions. In a Series of Letters to an Instructor*. Dickinson's neighbor Edward Hitchcock, presumably the "instructor" mentioned in the title, wrote the preface.²⁹ Like Brewster, the Amherst author tried to explain magical effects by revealing the trickery and deception behind them. At the same time, despite the author's claim that the book was intended to compel "wandering jugglers to live by honest labor rather than by infamous deception,"³⁰ *Ventriloquism Explained* aimed to profit from the popularity of the magic shows and magicians it purported to debunk.

At times these works seemed to be attempts to popularize scientific principles by piggybacking on interest in the paranormal. Sometimes they even functioned as instruction manuals. In *Natural Magic*, Brewster offered detailed diagrams of magic lanterns and other illusion-producing devices. Similarly, *Ventriloquism Explained* gave careful instructions on how to produce a variety of vocal effects and ended by encouraging pupils "to try experiments, in hours of amusement, with their vocal organs" while cautioning them against carrying "these imitations so far as to diminish their own self-respect."³¹

Something about these attitudes toward magic was inherently contradictory. On the one hand, magic was fraudulent, if not diabolical. On the other, it was entertaining, even fascinating. Books about magic sold very well—often better than works of natural philosophy or science. According to Janet Browne, popular displays of scientific materials in Britain fell "indiscriminately into a miscellaneous spectrum of stage shows, art exhibitions, pageants, theatres, circuses, painted panoramic displays, fireworks, magic lanterns, freak shows, funfairs, and the crammed glass cases of civic museums."³² The author of *Ventriloquism Explained* remarked, "I can give no better epithet than *mountebank* or *juggler*, to those penny-seeking idlers who impose upon the public by their petty tricks of legerdemain,"[33] declaring, "Lectures on scientific subjects have done much good throughout our country, but the dark deeds of mountebanks cannot be too severely reprobated."[34] Paradoxically, readers would not necessarily have purchased a volume of scientific lectures by the anonymous author. The "dark deeds of mountebanks" were the draw—they made the book attractive to readers. But how dark were such deeds? Were they diabolical, or were they just unscientific? The book described "a painful uncertainty in the minds of many as to the various phenomena of Legerdemain."[35] Such uncertainty was not painful to everyone, but there is no question that the boundaries between science and magic were often unclear.

Darwin's and Dickinson's understanding of magic grew from these tangled roots. For them, science and magic—and disenchantment and enchantment—were intertwined. Jason Ā. Josephson Storm describes the early nineteenth-century conception of the relationship between enchantment and disenchantment as a "Romantic Spiral" in which the effect of disenchantment is to create more powerful enchantments. As he puts it, "*Magic had to be eliminated so that we could make it real.*"[36]

Shortly after Darwin and Dickinson died, *The Golden Bough* by J. G. Frazer argued against circular accounts of history. Frazer thought that history moved in steps, like a staircase. He claimed that as societies developed, belief systems progressed in distinct, separate stages from magic to religion to science. Frazer's influential account made it hard to imagine magic, religion, and science coexisting. More recently, many scholars have pushed back against *The Golden Bough*. Twenty-first-century approaches to the history of magic tend to stress the coexistence of many ways of understanding the world and continuity across centuries. Unlike the scholars of the late nineteenth and early twentieth centuries, current scholars avoid belittling magical thought.[37]

I am interested in natural magic because I understand it as a way of thinking about nature that is interactive and participatory. When viewed from a magical perspective, the world appears to be alive with relationships. Other ways of thinking—those that we imagine as nonmagical or even antimagical—render the world as either a hierarchy ordained by a superior being or a random grouping of unrelated material objects.

The twenty-first-century archaeologist Chris Gosden's definition of magic harks back to the old tradition of natural magic: "Magic sees spirits in the land, considers how people and animals are related, and tries to understand transformations around birth and death."[38] This explanation has three parts. First, there is the attempt to see the generally unseen "spirits in the land." Next, there is a sense that "people and animals are related." Finally, there is a focus on how birth and death do transformative work in the world. Gosden claims that magic and modern science resonate with each other, not only because of their intertwined history but also because of their shared interest in unseen forces, complex interconnections, and transformative change. As he puts it, "The forces defined by science find echoes in magic's insistence that spirits animate the world."[39] In *Magic: A History*, Gosden argues that magic, science, and religion continue to coexist and that magical thinking persists in the twenty-first century.

In addition to these historians, scholars in other fields—philosophers, physicists, and biologists for the most part—have reengaged with the concepts of natural magic in recent years. These figures do not see enchantment as confined to the past. In *A Thousand Plateaus*, the philosophers Gilles Deleuze and Felix Guittari turned toward botany, describing the tangled relationships between human and nonhuman beings in terms of rhizomes, the buried network of root systems that form underground, invisible connections between different plants. In *The Enchantment of Modern Life* and *Vibrant Matter*, the philosopher Jane Bennett has celebrated enchantment and worked to reanimate the early nineteenth-century concept of vitalism. In *Meeting the Universe Halfway: Quantum Physics and the Entanglement of Matter and Meaning*, the physicist Karen Barad has argued that matter has agency. Donna Haraway, the biologist turned feminist philosopher, started with simians and cyborgs, then focused on the companionship between human and nonhuman animals. Most recently, in *Staying with the Trouble*, Haraway has turned her attention to the mysterious transformations that take place deep in the compost pile. These scholars and others have pushed back against nineteenth- and twentieth-century discourses of disenchantment, but they do not usually describe their approaches as magical. Even critics of disenchantment are somewhat wary of mentioning magic.

Magic continues to have a place in the popular imagination. However, although many people enjoy fantastical works of fiction about witchcraft and wizardry, few educated people like to admit that they believe in magic. Even fewer would be able to separate the idea of magic from the supernatural. In fact, the concept of magic generally goes unexamined. For most, it works as a metaphor. Some religious believers take the idea more seriously. Evangelical Christians tend to see magic as supernatural and anti-Christian. In contrast, Wiccans and other modern pagans make magical rituals cornerstones of their religious practice. These groups have very different views, but Wiccans and Evangelicals are both drawn to supernatural phenomena. They are often skeptical of science and scientific approaches to understanding the world, and they do not tend to express as much interest in natural magic—magic entwined with natural science—as they do in supernatural magic.

Darwin and Dickinson both saw nature as enchanted, but their magical way of knowing—their interest in natural magic rather than supernatural magic—can be hard to access in the contemporary world. To imagine enchantment as an epistemology—an approach to understanding the world—we need to turn our focus away from magical practices such as rituals and incantations. At the same time, if we want to focus on natural magic as Darwin and Dickinson understood it, we need to imagine a kind of enchantment that is not necessarily supernatural. Jane Bennett is helpful here. She describes enchantment as a feeling that "the marvelous vitality of bodies human and nonhuman" can inspire "deep attachment" and "a mood of fullness, plenitude, or liveliness."[40] Darwin and Dickinson described such experiences as moments of "wonder."

Darwin never lost his sense of wonder or his conviction that all kinds of beings were mysteriously entangled. His endless experiments with plants and animals were always framed in terms of learning about connections. He was a conscientious specific naturalist who tried to learn as much as he could about particular species, but he was not limited to finding the dividing lines between species. Darwin was always more interested in the mystery of mysteries at the heart of interspecies relationships. He tried to persuade his readers of their kinship with seaweeds, coral reefs, and barnacles; mammoths, mastodons, and elephants; dogs, finches, and primates; beetles, orchids, and earthworms. Darwin's great project was to show how the hardest things that individuals face—war, famine, struggle, even death—can give rise to new forms of beauty and new kinds of love.

A few of Darwin's contemporaries saw that Darwinian science could create new forms of enchantment as it built new connections and delved down into the mysterious roots of the great green tree of life and death. Dickinson was not the only person (nor even the only poet) to explore the magical possibilities of Darwinian science. But she was among the greatest of them. Her poetry sings with the strange green magic of Darwinian science. By putting Emily Dickinson and Charles Darwin on the page together, I hope to open a window into a time before thinkers worked in atomized disciplinary silos, a time when scientists, philosophers, theologians, poets, and political activists were in constant conversation. Their lives and works invite us home to an enchanted world.

CHAPTER 1

Darwin and Dickinson, Childhood Portraits

IN THE WINTER OF 1816, Ellen Sharples visited the Darwin family home in Shropshire, England, to make portraits of Robert Darwin and his children. Sharples was an artist with a quick eye, a professional portraitist with decades of experience and a long list of wealthy and well-known clients. Her fees were high. Her pastels and miniatures of American political figures, including George Washington and Alexander Hamilton, were sought after in Britain and the United States. She had made portraits of a few celebrated natural philosophers: Joseph Priestley, the discoverer of oxygen, and Sir Joseph Banks, the botanist who had sailed around the world with Captain Cook.

Her hosts, Susanna Wedgwood Darwin and Robert Darwin,[1] were in the same milieu. Each was descended from a well-known scientific figure of the eighteenth century. Susanna's father, Josiah Wedgwood, and Robert's father, Erasmus Darwin, had both been "Lunar Men." With Priestley and a few others, they founded the Lunar Society of Birmingham, an extraordinary coterie of experimenters, industrialists, and philosophers whose radical ideas powered the industrial revolution in Britain.

In the next generation, Susanna and Robert lived quietly, out of the public eye. Sharples had been hired to make private family portraits. There were no plans to sell copies. At the time, there were few hints that one of the Darwin children would eventually become more of a cultural giant than either of his grandfathers.

The Darwin family was obviously prosperous. Their house, known as the Mount, had an air of ease and comfort different from the anxious grandeur of more ancient aristocratic places. It was a large, new brick house set a comfortable distance from the town of Shrewsbury, looking out across well-tended fields that stretched down to the banks of the Severn River. The soft hills of Wales were almost visible in the distance. A warm and sunny atmosphere pervaded the rooms despite the damp chill outside. A glass-paned conservatory overflowed with exotic plants. Thriving orange trees perfumed the air. The children seemed just as carefully nurtured.

Charles Robert Darwin was a pleasant, rather ordinary 6-year-old boy, the fifth of six children. His family called him Bobby. The portraits suggest that Sharples liked the boy and his brother and sisters. Her pastels glow with affection. A gifted artist with keen insight into her subjects, Sharples tried to capture something of the children's potential. As she set up their poses, she looked for props that would capture Charles and his younger sister Catherine's attention and perhaps show hints of the future.

Painters often give models something to hold in their hands keep them still. Shiny toys or bright flowers are common for children. In Sharples's portrait of Charles and Catherine Darwin, the young girl's pose is typical. She is seated with a bunch of flowers trailing across her white dress. Her brother Charles, however, is placed in an unusual position. He kneels on one knee, with a large pot of rare African opal flowers balanced on an outstretched leg. Only a child with great patience, strength, and flexibility could hold the awkward pose for more than a few seconds.

The boy's hands wrap around the glossy red planter as if it is a treasure. In some respects, it is. The spiky bulbous perennial would have been impossible to cultivate in Shropshire without a greenhouse and skilled gardeners.[2] Thanks to their family connections, the Darwins had both. The neatly turned flowerpot looks as if it could be the product of his grandfather Josiah Wedgwood's industrial pottery. The Wedgwood

Charles and Catherine Darwin by Ellen Sharples (1816). Wikimedia Commons.

fortune bankrolled the family and made their large hothouse possible. The plant itself, a rare African opal flower native to Cape Town, South Africa, gestures toward the Darwin side of the family.[3] African opal flowers (*Lachenalia aloides*) were very rare in Britain at the time. The Darwins had access to such exotic plants because Charles's other grandfather, Erasmus Darwin, was a leading British botanist. The plant balanced on the boy's knee signaled wealth, the expanding British Empire, and extraordinary exposure to botany. It was a fitting emblem for a grandson of Josiah Wedgwood and Erasmus Darwin and a perfect

Emily, Austin, and Lavinia Dickinson by Otis Allen Bullard (ca. 1840). HNA41, Houghton Library, Harvard University.

symbol of this particular child's future. Charles Darwin would eventually become one of the world's great naturalists.

Almost twenty-five years after Ellen Sharples made her portrait of the youngest Darwins, Otis Bullard painted the Dickinson children in Amherst, Massachusetts. Bullard was a young man who had been painting portraits for only a year or two, after getting his start as a sign painter. In January 1840 he opened a temporary studio in Amherst, in the building on Main Street that also housed Edward Dickinson's law office.[4] An article in the *Amherst Gazette* in February 1840 praised "the well-executed portraits, just from his life-giving pencil."[5] In March a neighbor commented that Bullard had been particularly "successful in getting the expression of the countenance" of the Dickinson children and their parents.[6] Despite this success, Bullard's time as a portrait painter would be relatively brief. By 1846 he had set up as an impresario, traveling around the United States to display a gigantic painted panorama of New York City. The portrait of the Dickinson children is his best-known surviving work. As was customary for itinerant portraitists, Bullard started with a partially completed canvas, the three headless figures already painted. Later he added the children's faces and emblematic props. This piecework method would result in a slightly off-kilter painting.

Although the Dickinson portrait is a little wooden compared to Ellen Sharples's pastels, Bullard's experience as a sign painter enabled him to capture telling details that signal the children's personalities.[7] Emily's older brother, Austin, and younger sister, Lavinia, look calm, obedient to the painter's instructions. The 10-year-old Emily Dickinson comes across as fiercely intelligent but not necessarily patient. She has regular features and light-brown eyes. Her glossy, chestnut-red hair is very short, cut close to her head. She crushes a rose into an open book, its pages illustrated with colorful diagrams of similar flowers. Neither the book nor the flower is particularly rare, but the way the child holds them together creates a strange, wild energy. Emily Dickinson glares from the canvas with preternatural intensity. She makes ordinary things disturbing.

Some striking parallels between the childhood portraits of Darwin and Dickinson jump out. Both images show carefully tended, wealthy white children with beautiful clothes and lovely playthings. In spite of the different times and places, the children look surprisingly alike. The resemblance is striking. Charles Darwin and Emily Dickinson have similar coloring, with fair, reddish hair cut short and combed across the forehead. Even their faces are similar. They could be brother and sister.

The differences between Darwin's and Dickinson's childhoods also become clear in the portraits. Compared to the elegant Darwin portrait, the Dickinson childhood portrait looks a bit awkward, even provincial. The Darwin pastels are from 1816, twenty-four years before the 1840 Dickinson painting. The Darwin children wear airy Regency fashions. Their portraits could be illustrations from Jane Austen's *Pride and Prejudice*. In contrast, the Dickinson children wear darker, heavier Victorian clothes. Even with its American flavor, the Dickinson painting would not be completely out of place in a dark corner of Charles Dickens's *Old Curiosity Shop*.

Neither the Darwin portrait nor the Dickinson portrait offers many clues about their homes, though the commissioned portraits indicate that their families were relatively prosperous. The white clapboard house by the town cemetery in Amherst where the Dickinsons lived in 1840 was similar to the Darwin's house in Shropshire, though not quite as grand. Like the Mount, Emily Dickinson's childhood home was filled with greenery and surrounded by flourishing gardens. Her mother, Emily Norcross Dickinson, was a skilled gardener with formal training in botany. Her Amherst neighbors were impressed with her ability to coax her fig tree to produce fruit despite the cold climate. The Dickinsons' Massachusetts figs were at least as remarkable as the Darwin's Shropshire oranges.[8]

As the props in the Dickinson children's portrait indicate, their family's highest priority was education. Edward Dickinson, the poet's father, was a prosperous lawyer who served as treasurer of Amherst College. Her grandfather Samuel Fowler Dickinson had helped to establish the college. They were a reading family, with books and periodicals in easy reach. The children were well supplied with sheet music, pens, pencils, bright crayons, and an abundance of paper.

The portraits of the Darwin and Dickinson children signal alike to viewers that their young subjects are being carefully cultivated and appropriately educated. Both images are relatively conventional. After examining them, viewers will not be surprised that both the Darwin and the Dickinson families supported their local churches, sometimes more out of a sense of social obligation than religious fervor. The paintings depict bourgeois children who are well groomed, well organized, and neatly slotted into the social order. Both portraits seem to promise children who will eventually become upstanding members of the social and religious establishment. However, with the benefit of hindsight, the depiction of Charles and Emily as young botanists points toward their unconventional futures. Though interest in botany was widespread in the early nineteenth century, the implications of the subject were complicated. *Botanizing*—collecting and classifying plants to study how they are interrelated—was not always as conventional as it may have seemed on the surface. To the contrary, botany had the potential to be socially disruptive.

By associating Charles and Emily with botany, the portraitists singled them out from the others. None of their siblings hold botanical props. In the Darwin family portraits by Sharples, Charles's brother, Erasmus, toys with a glossy top hat while his sister Caroline holds an open book. The Sharples portrait of their father, Robert, shows him as a man of business beside a desk with a sealed letter and an inkwell. Pictured together, Charles and Catherine both hold plants, but Catherine's bouquet of flowers is not botanically interesting. Charles is the only Darwin who displays a rare plant. Along similar lines, in Bullard's portrait of the Dickinson children, Austin extends his arms protectively around his younger sisters. Lavinia holds a picture of a cat. Emily clasps the flower and the botany book.

As it turned out, Dickinson and Darwin would study botany in school. The science would be foundational for both of their careers. Botany's links to religion, magic, and science as well as poetry and philosophy—and the increasingly contradictory social and political implications of these ways of thinking about the world—make Darwin's African opal flower and Dickinson's rose and book startlingly powerful emblems for the lifework of Charles Darwin and Emily Dickinson.

At the start of the nineteenth century, botany was associated with radicals and revolutionaries, largely because of Erasmus Darwin and his friends in the Lunar Society of Birmingham, the engineers of the British Enlightenment. In eighteenth and early nineteenth-century Britain, Charles's grandfather was one of the best-known advocates of the antiestablishment aspects of natural philosophy. He was a physician, a botanist, and a poet.

Erasmus Darwin's Lunar Society was much less formal than a London club or the Royal Society in London, more like a twenty-first-century book club than a professional association. Most of the Lunar Men were busy professionals or tradesmen. Every month they met on the Sunday nearest the full moon so they could take advantage of the moonlight as they journeyed home. Almost all of them had to work on Monday morning.

Though the Lunar Men shared intense curiosity about the natural world, they differed from one another in many respects. Joseph Priestley recorded, "We had nothing to do with the religious or political principles of each. We were united by a common love of science, which we thought sufficient to bring together persons of all distinction, Christians, Jews, Mohametans, and Heathens, Monarchists, and Republicans."[9] In a repressive society, such inclusivity was itself a political stance. In Britain, the Test Acts (passed by Parliament in 1673) excluded Dissenters—those who did not agree with all thirty-nine articles of the Anglican catechism—from attending universities, teaching, and holding public office. A century later, most of the Lunar Men advocated repeal of the Test Acts. Priestley was an antimonarchist Republican who supported the American and French Revolutions wholeheartedly. He would eventually flee to America, in danger of arrest for sedition.

Erasmus Darwin was Josiah Wedgwood's personal physician. Their families grew close in the 1770s, as both men wrestled with trying to educate their children without conforming to the Test Acts. Darwin was a freethinker (a rationalist who did not adhere to any religious doctrine), while Wedgwood was a Unitarian. For both men there was more to the education question than religious dissent. They objected to the standard classical curriculum of the day, with its focus on studying Latin, Greek, and mathematics. Classical languages seemed particularly useless for their children since the Test Acts made it unlikely they would hold any sort of public office—university, church, or government.

After Wedgwood's pottery business became global in scale, he built a model industrial village with a giant family home at its center. Perhaps on Darwin's suggestion, he called the complex *Etruria* since the factories on the property produced burnished, black-and-red neoclassical "Etruscan" pottery.[10] There was an idealistic, almost utopian element to Wedgwood's Etruria.

Wedgwood hired tutors and governesses to staff an "Etruscan" school offering the modern languages and natural sciences not being taught at the schools sponsored by the established Anglican Church.

Erasmus Darwin's children attended the school, sometimes making monthslong visits to Etruria Hall. Robert Darwin, Charles Darwin's father, got to know his future wife, Susanna Wedgwood, when both attended the Etruscan school. By this time Charles Darwin's grandfathers were close friends. Erasmus Darwin had introduced Josiah Wedgwood to the Lunar Society, and Wedgwood often made the long journey (about forty miles) to Birmingham for the monthly meetings.

It can be hard for twenty-first-century readers to understand what the Lunar Society was about since it is difficult to imagine a time before the familiar epistemological boundaries that structure our thought. The Lunar Men lived in a world without disciplinary divisions. To add to our present-day confusion, many of the terms they used to describe their thinking have changed meaning. In her account *The Lunar Men*, Jenny Uglow explains that in the eighteenth century, *science* meant general knowledge, while "when people spoke of the 'arts,' they did not mean only the fine arts but also the 'mechanic arts.'"[11] The Lunar Society brought these inquiries together in a distinctively eighteenth-century manner. What its members described as art and philosophy included the pursuits that we in the twenty-first century know as engineering, technology, and science. When they referred to "science," they usually meant the broad swathe of disciplines now grouped together as the liberal arts.

At Lunar Society meetings, conversation centered on experiments of all kinds. Matthew Boulton and James Watt tinkered endlessly with steam engines. Priestley played with electricity and oxygen. James Keir was interested in metallurgy—making alloys and composites. Wedgwood was interested in chemistry and engineering, both of which were crucial to his expanding industrial pottery. The clockmaker John Whitehurst and the physician William Withering focused on geology, though Withering also shared Erasmus Darwin's and Jonathan Stokes's interest in botany and medicine. Some of the other early members, including Richard Lovell Edgeworth, Thomas Day, and Richard Small, were as interested in political philosophy as in natural philosophy. Edgeworth described the group as "men of very different character but all devoted to science and literature."[12] In truth, all of them now seem more like polymaths than specialists.

They were also revolutionaries, curious about social and political ideas and experiments. The group had active ties to the American colonies. Richard Small had been a professor at the College of William and Mary in Virginia, where Thomas Jefferson had been one of his students. Benjamin Franklin attended a very early meeting of the Lunar group at Erasmus Darwin's house in 1758. At the time—before the American Revolution—Franklin was in England as an agent of the Pennsylvania Assembly. Later, when Jefferson wrote the Declaration of Independence with Franklin's help, many of the Lunar Men cheered privately. They were eager for change.

They supported the French Revolution as eagerly as they had supported the Americans. In 1789, shortly after the fall of the Bastille, Josiah Wedgwood wrote to Erasmus Darwin, "I know you will rejoice with me in the glorious revolution which has taken place in France. The politicians tell me that as a manufacturer I shall be ruined if France has her liberty, but I am willing to take my chances in that respect. A gentleman who has just come from his travels has been here a day or two, & he assures me that the same spirit of liberty is developing itself all over Germany, all over Europe."[13] This joyful letter shows that the "spirit of liberty" brought Wedgwood and Darwin together just as much as their voracious intellectual curiosity.

Of all the Lunar men, Erasmus Darwin may have been the most voracious, socially as well as intellectually. In person he was larger than life. The biographer Jenny Uglow describes him as both "enormously gifted" and enormously large. Eventually, "he cut a semi-circle in his dining table to fit his stomach."[14] In the unbuttoned eighteenth century, his girth was just another sign of his great virility and strength. It was widely known that Erasmus Darwin had fathered at least fourteen children, five with his first wife and seven with his second wife, plus two with his mistress. There were whispers of more children, born to married women in the neighborhood.

At the time, large families were not unusual. Much more notable was Dr. Darwin's intellectual interest in sex and sexuality. In 1787, more than twenty years before his grandson was born, Erasmus Darwin translated Carl Linnaeus's work on *The Families of Plants*, a treatise that classified plants according to a structured hierarchy, with unique organisms corresponding to a particular genus and species, named according a new binomial system using descriptive Latin words. Linnaeus's naming system applied to animals as well as plants. (For example, human beings—who classify themselves as members of the genus *Homo* and the species *sapiens*—are *Homo sapiens* in the Linnaean system.)

The first English translation of Linnaeus censored the original, removing all references to sex. Dr. Darwin insisted on making his own translation. According to Janet Browne, Darwin's translation helped make it clear that "to be a Linnaean taxonomist was to believe in the sex life of flowers."[15] A few years later, in 1791, he published a book-length poem, *The Botanic Garden*. In some ways it was an odd choice of title, since the first part of the book was more about human beings than plants.[16]

Erasmus Darwin started work on *The Botanic Garden* while translating Linnaeus. He published one section, *The Loves of the Plants*, as a stand-alone volume in 1789, testing the market before he published the complete edition. It opened with a brief preface:

GENTLE READER!

Lo, here a CAMERA OBSCURA is presented to thy view, in which are lights and shades dancing on a whited canvas, and magnified into apparent life! – if thou art perfectly at leasure for such trivial amusement, walk in, and view the wonders of my INCHANTED GARDEN.

Whereas P. OVIDIUS NASO, a great Necromancer in the famous Court of AUGUSTUS CAESAR, did by art poetic transmute Men, Women, and even Gods and Goddesses, into Trees and Flowers; I have undertaken by similar art to restore some of them to their original animality.[17]

In the first sentence of this exuberant introduction, Darwin sounded like a carnival barker hawking an exhibit. He likened his book to an up-to-date eighteenth-century visual technology (the camera obscura) often used for public amusements. He invited readers to "walk in, and view the wonders of my inchanted garden." The next sentence made Darwin's intellectual ambitions for the book more explicit. He announced that he hoped to equal the Roman poet Ovid, whom he described as a "Great Necromancer." Ovid's *Metamorphosis* had transformed humans and gods into trees and flowers; Darwin promised to reverse the process and restore the flowers "to their original animality." Though he referred to his garden as "inchanted" and styled himself as a "Necromancer," he was not necessarily presenting himself as a magician performing illusionary magic tricks on a stage. Instead, the magic at the heart of *The Botanic Garden* was transmutation, Dr. Darwin's version of Ovid's metamorphoses. *Transmutation* was his preferred term for the biological change over many generations now known as evolution.

Throughout *The Botanic Garden*, Darwin insisted that botany and zoology could provide insights into human life because plants and animals were inseparable, and humans were animals—kin to all other living beings. The concept of transmutation allowed him to see kinship everywhere. He personified plants as easily as he botanized human beings. *The Loves of the Plants* used personification to take Linnaean taxonomy to libidinous heights. Its explicitly erotic verses about stamens and pistils added an element of sexual titillation to botanical pursuits. In Darwin's lexicon, pollen-bearing stamens were *Beaux*, while seed-generating pistils were *Beauties*. *The Loves of the Plants* narrated botanically specific dramas of courtship within each blossom:

> What Beaux and Beauties crowd the gaudy groves,
> And woo and win their vegetable Loves.[18]

The poem described eighty-three separate species of plants and their "vegetable loves."

Since most flowering plants have more stamens than pistils, Darwin's botanical love stories almost always featured a single female "Beauty" surrounded by a large number of male "Beaux." For example, Darwin dramatized the action within a blooming daffodil during an early spring storm. Daffodils are members of the amaryllis family in the genus *Narcissus*. They have one pistil and six stamens.

> When heaven's high vault condensing clouds deform,
>   Fair AMARYLLIS flies the incumbent storm,
> Seeks with unsteady step the shelter'd vale,
>   And turns her blushing beauties from the gale. –
> *Six* rival youths, with soft concern impress'd,
>   Calm all her fears, and charm her cares to rest.[19]

In addition to presenting the all-important number of stamens, this stanza features many specific characteristics. Daffodils tend to bloom before the last storms of winter, and each blossom has a bell-shaped corona that protects the central pistil and stamens from sleet and snow. Darwin's description of the daffodil's pistil as a blushing beauty who seeks refuge from a gale within a "shelter'd vale" was an inspired piece of botanical personification.

At times it was hard to separate the author from his book. Erasmus Darwin had a reputation for promiscuity. He was nearly as well-known for his large social circle and his many children as he was for his explicit sexual writing. An 1803 engraving depicted "Erasmus Darwin, Author of the Loves of the Plants" above a frolicking cupid. In spite of—or perhaps because of—the somewhat prurient rumors about his unacknowledged offspring, *The Botanic Garden* was a great success. It helped to popularize botany across the English-speaking world.

Darwin did not publish the entire poem until 1791, after *The Loves of the Plants* had become very popular. *The Economy of Vegetation* was politically risky. It offered a progressive, even evolutionary account of scientific and political history, arguing that human societies evolved toward greater enlightenment just as plant and animal species advanced toward perfection. Intertwining political history with the history of science, Darwin's work celebrated the American Revolution and the French Revolution while it denounced slavery and colonization. Sections of the poem were widely reprinted in newspapers and

Erasmus Darwin by William Holl Sr., after J. Rawlinson (1803). Wellcome Collection.

journals. His description of European colonization gives the flavor of Darwin's politics:

> When Avarice, shrouded in Religion's robe,
> Sail'd to the West, and slaughter'd half the globe:
> While Superstition, stalking by his side,
> Mock'd the loud groan, and lap'd the bloody tide;
> For sacred truths announced her frenzied dreams,
> And turn'd to night the sun's meridian beams.[20]

Although twenty-first-century readers might be surprised by Erasmus Darwin's explicit condemnation of colonization, radical politics were no more unusual in his social circle than his large family. *The Botanic Garden* was one of the best-selling books published by Joseph Johnson, who also published works by many other radical thinkers (including Joseph Priestley, William Godwin, and Mary Wollstonecraft).

In later centuries it would be almost unthinkable to publish a serious scientific treatise in verse. When Erasmus Darwin wrote *The Botanic Garden*, however, poetry and philosophy were often bedfellows. Darwin's poem exemplifies the late eighteenth-century melding of poetry and natural philosophy that gave rise to British Romanticism. Yet although he was one of the foundational figures of British Romantic science, Erasmus Darwin was not exactly a Romantic poet. The literary critic Robert Richards explains that Romantic science grew from German *Naturphilosophie*, which "focused on the organic core of nature, its archetypal structure, and its relationship to mind, while Romanticism added aesthetic and moral features to this conception of nature."[21] In Britain, the Romantic approach to natural science hued closely to European models, while the Romantic approach to literature took a distinct turn toward less formal, more fluid language. As a naturalist, Erasmus Darwin was closely aligned with European Romantic science. As a poet, he would become the bête noir of British Romantic poetry. He wrote in heroic couplets much closer in style to eighteenth-century poets such as Alexander Pope than to the Romantic poets who followed. *The Botanic Garden* was witty and sometimes risqué. Its language was formal, didactic, and slightly stilted.

A few years later, in 1798, Erasmus Darwin's young acquaintances William Wordsworth and Samuel Taylor Coleridge would publish a book of poetry in a new, less formal style. *Lyrical Ballads* became the literary cornerstone of British Romanticism. One of the poems in *Lyrical Ballads*, Wordsworth's "Tables Turned," offered sharp critique of Erasmus Darwin's materialist approach to studying the natural world, arguing that unmediated experience was more spiritually and psychologically meaningful than methodical study that focused on specific plants and animals. As Wordsworth put it:

> Sweet is the lore which Nature brings;
> Our meddling intellect
> Mis-shapes the beauteous forms of things: –
> We murder to dissect.[22]

For Wordsworth and Coleridge, Erasmus Darwin was the prime example of a "meddling intellect" who destroyed nature by anatomizing it.

Wordsworth and Coleridge lived in the same region as Darwin and wandered through similar landscapes. Wordsworth's description of daffodils in "I Wandered Lonely as a Cloud" offers a powerful contrast to Darwin. The poem starts by describing the poet's experience of coming across a host of daffodils "fluttering, dancing in the breeze." It concludes by remembering that moment. For Wordsworth, the memory is more significant than the flowers.

> For oft, when on my couch I lie
> In vacant or in pensive mood,
> They flash upon that inward eye
> Which is the bliss of solitude;
> And then my heart with pleasure fills,
> And dances with the daffodils.[23]

Here, Wordsworth celebrates the poet's experience of daffodils, first outdoors and then—more importantly—within "that inward eye / which is the bliss of solitude." The flowers are almost completely dematerialized.

After *Lyrical Ballads*, British poets would grow more and more hostile to what Coleridge called "Darwinizing."[24] Coleridge dismissed Erasmus's work as "all surface and no content, all shell and no nut, all bark and no wood."[25] Coleridge hated Darwin's poetry because it was too materialist—too focused on the nitty-gritty details of physical science instead of philosophical ideals. In subsequent decades many poets turned against empirical, observation-based science.

In 1819 John Keats would proclaim the importance of looking inward to emotional and aesthetic experience, writing:

Beauty is truth, truth beauty, – that is all
Ye know on earth, and all ye need to know.[26]

Keats's dictum conflated beauty and truth, reframing the manner in which *The Botanic Garden* had joined poetry and natural philosophy to emphasize imagination over observation.

For the Romantics, as for the earlier generation, poetry and empirical science were profoundly and essentially linked. Truth and beauty were two different dimensions of nature, and philosophy and poetry were ways of exploring and describing them. But the relationship between poetry and science was not as straightforward as it had been in the eighteenth century. Though the statement that "beauty is truth, truth beauty" was not necessarily antiscience, it joined a chorus that had begun to dismiss empiricism as irrelevant. The Romantic poets rejected the materialist, mechanical approach to science personified by Erasmus Darwin. As we will see in a later chapter, Coleridge's objections to "Darwinizing" would spur the separation between science and literature and help prompt the invention of the word "scientist."

The Botanic Garden was just one of Erasmus Darwin's publications. His prose treatises, particularly *Zoonomia* (1794) and *Phytologia* (1800), found large audiences. In his later works, he focused more and more on the transmutation of species. Erasmus Darwin's last poem, *The Temple of Nature* (published posthumously in 1803), offered an evolutionary (in his terms, transmutationist) account of creation that provided a vivid alternative to Milton's *Paradise Lost*. His contention that life had developed gradually over time challenged biblical narratives about the Garden of Eden, but he was radical in so many ways that his stance surprised no one.

Twenty-first-century readers would use the term *evolution* to describe Erasmus Darwin's theory that living things gradually became more complex. In the late eighteenth century, however, the word "evolution" implied a sudden twist or a sharp turn. In *Zoonomia*, Darwin contrasted the gradual process of species development to instantaneous divine creation, proposing, "The world ... might have been gradually produced from very small beginnings ... rather than by a sudden evolution of the whole by the Almighty fiat."[27] For Erasmus Darwin, evolution denoted the opposite of what a twenty-first-century reader would expect. We must navigate a series of dizzying semantic reversals here as we try to follow a story that starts when science included the humanities, philosophy included poetry, and "evolution" was a term used to describe divine creation.

In the dark summer of 1816, Erasmus Darwin's account of the origins of life would give Mary Shelley bad dreams. Her parents, the writers William Godwin and Mary Wollstonecraft, had been members of Erasmus Darwin's large circle of radical thinkers, and her aunt, Everina Wollstonecraft, was governess to Susanna Wedgwood, who would become Charles Darwin's mother. When she began to think about the terrifying possibilities of Erasmus Darwin's materialist natural philosophy, she was inspired to start writing *Frankenstein*. Looking back many years later, she remembered:

> Many and long were the conversations between Lord Byron and Shelley, to which I was a devout but nearly silent listener. During one of these, various philosophical doctrines were discussed, and among others the nature of the principle of life, and whether there was any probability of its ever being discovered and communicated.

> They talked of the experiments of Dr. Darwin, (I speak not of what the Doctor really did, or said that he did, but, as more to my purpose, of what was then spoken of as having been done by him,) who preserved a piece of vermicelli in a glass case, till by some extraordinary means it began to move with voluntary motion. Not thus, after all, would life be given. Perhaps a corpse would be re-animated; galvanism had given token of such things: perhaps the component parts of a creature might be manufactured, brought together, and endued with vital warmth.[28]

Mary Shelley did not base her fictional natural philosopher, Victor Frankenstein, directly on the historical Erasmus Darwin. But though she did not want her readers to imagine Erasmus himself, her novel would explore the horrifying implications of the "experiments of Dr. Darwin." In her account the ability to "bestow animation upon lifeless matter" would turn out to be a terrible curse.

When Charles Darwin was a boy, his parents were a bit shy of this complicated legacy. Erasmus Darwin had been a successful poet and a renowned natural philosopher, to be sure, but in the intervening years William Wordsworth, Samuel Taylor Coleridge, John Keats, Lord Byron, Percy Shelley, and Mary Shelley had built a literary movement in reaction against Erasmus Darwin and his fellow Lunatics—as members of the Lunar Society laughingly called themselves. Weary of such controversies, Susanna Wedgwood Darwin and Robert Darwin aspired to respectability. Their children would not attend anything like the "Etruscan" alternative school where they had studied together. Charles and his siblings were christened in the Church of England, largely in order to prevent their exclusion from educational institutions controlled by the church. Susanna and Robert sent their sons to conventional schools to study Latin, Greek, and mathematics. They educated their daughters at home. Yet though they distanced themselves from Dissenters and radicals, they did not eschew botany. Both were avid gardeners who had acquired much of their botanical knowledge directly from the notorious Erasmus Darwin.

Botany was a source of social connection for their young son. Charles often used the fruits of his family's remarkable gardens to make quiet gestures of friendliness to his neighbors and classmates. Once he took a flower to school with him to show one of the older boys. He had been told that if you look closely enough into the flower's center, you will learn its name, but he couldn't remember exactly how to do it. The schoolboys gazed intently into the blossom, trying to figure out its secret. Years later, William Allport Leighton surmised, "Doubtless his Mother had been giving him a lesson and explaining the Linnaean system of Botany, but his mind had not sufficiently grasped the circumstances. This however excited in my own mind the latent germ of Botany – and the incident has remained in my memory throughout life."[29] Leighton, who would become an accomplished lichenologist, traced his lifelong fascination with botany to that afternoon when he and Darwin leaned over the silent flower together.

For Charles Darwin and his peers, botany was an escape from lessons. It is unlikely that the schoolboys were aware of the political conflicts or the philosophical debates of their grandparents' generation. They did not associate Linnaean classifications with sex, much less with political radicalism or atheism. They had no idea that Mary Wollstonecraft Godwin, a teenager just a few years older than themselves, had eloped to Switzerland with an aristocratic atheist, Percy Bysshe Shelley, and his rakish mentor, Lord Byron. They did not know that Erasmus Darwin's speculations about the origins of life had sparked her imagination or that she had constructed a new kind of monster animated by the terrifying implications of natural philosophy. What they did know was that at school they were instructed in Latin, Greek, and mathematics but never in botany, chemistry, or any other form of natural philosophy. For them, botany was exciting and vaguely antiestablishment simply because it was not taught in school.

Twenty years later, when Emily Dickinson was a student, botany had become a schoolroom subject and lost some of its thrill. One of her early poems complains about the way that classifying plants could diminish them:

> I pull a flower from the woods –
> A monster with a glass
> Computes the stamens in a breath –
> And has her in a "Class"!³⁰

In Amherst, the young Emily Dickinson was surrounded by these kinds of monsters—educators who were determined to teach her science. Her description of a "monster with a glass" slyly counters *Frankenstein* and suggests that botanical investigations had the potential to generate a new kind of monster. Dickinson was aware of the connotations of botany for Romantic poets and novelists in Britain, but the implications were very different in her own community. In her poem the "monster" is the antithesis of Frankenstein's monster. For Dickinson, forced to memorize each flower's number of stamens, the problem with botany was that it sucked the life from living plants. Mary Godwin Shelley had feared that natural philosophers might cross the line and animate dead matter. A few decades afterward, Dickinson feared that scientists might reverse the process and make living beings seem inanimate.

Turning back to Bullard's portrait, we can see hints of the rich conversation about botany that Dickinson was exposed to as a young girl. It is notable that Bullard did not attempt to place the children in a realistic setting. The portrait gestures toward Europe. Flowering vines twine in from the window, out of the background, across the foreground. The vines are decorative rather than botanically specific. Although it was painted in Western Massachusetts in the winter, the landscape is green. At the same time, the palette in Bullard's portrait is much darker than that of the Darwin pastels. By the 1840s, Victorian tastes had turned toward darker, more saturated colors. There is a vaguely Tuscan feel to the architecture and the landscape in the Dickinson portrait. Perhaps Bullard placed the Dickinson children in an Italian Renaissance setting to illustrate the rich educational resources in their family background. At the same time, however, in some respects the preindustrial mood of the portrait accurately reflects early nineteenth-century New England culture.

We must not conflate Victorian England with the United States. In some respects, living in Massachusetts in the nineteenth century was closer to living in the seventeenth century than any of the cosmopolitan Darwins could have imagined. Though it had been 150 years since the witch trials in Salem, early nineteenth-century Massachusetts writers, including Catherine Sedgwick and Nathaniel Hawthorne, had recently brought that history back to life with their celebrated accounts of Puritan New England, where many people believed in witches, demons, and specters, and most believed in some form of magic. Compared to the Sharples pastels, Bullard's painting has magical undertones. Something about the proportion or the perspective is a little off, giving the children uncanny, doll-like qualities. Though the slightly strange mood is probably unintentional, other elements of the painting seem to gesture toward the supernatural. Lavinia holds a sketch of a black cat, an explicit symbol of witchcraft. In Emily's hand is a vivid flower next to an illustrated botany book. The shaggy-leaved rose almost seems to have been conjured from the text. It leaps magically from the pages.[31]

The Dickinsons did not request a spooky painting from Bullard. Their taste in art was cautious; they hoped their children would grow up to be as conventional as possible. As an adult, Dickinson chafed at her father's ambivalence toward her education. She reported, "He buys me many Books – but begs me not to read them – because he fears they joggle the Mind."[32] On one hand the Dickinsons wanted their children to be as educated as possible. On the other they worried about exposing them to destabilizing ideas.

The year before Emily's birth, Almira (Hart) Lincoln Phelps published the first edition of *Familiar Lectures on Botany*. Emily would eventually use it as a textbook. But even when she was an infant, her Amherst community shared Phelps's belief that "The study of Botany seems peculiarly adapted to females: the objects of its investigation are beautiful and delicate; its pursuits leading to exercise in the open air are conducive to health and cheerfulness."[33] Of all the branches of natural history, botany was the most feminine. The Dickinson family hoped their daughters would study the other branches of natural history and natural philosophy as well. When Emily was just a few months old, her grandfather Samuel Fowler Dickinson gave a speech on "Females." He began by distinguishing "the education of daughters from that of sons; because Nature has designed them to occupy places, in family and society, altogether dissimilar." Samuel Fowler Dickinson reserved the study of theology and classical languages for boys. He thought "daughters should be well instructed in the useful sciences; comprising a good English education: including a thorough knowledge of our own language, geography, history, mathematics and natural philosophy." Today, Samuel Fowler Dickinson's female curriculum may seem surprisingly weighted toward science and mathematics. In 1831 Massachusetts, however, nothing was startling in Samuel Fowler Dickinson's belief that such studies would teach young women that "God hath designed nothing in vain."[34]

Congregationalist Christians like the Dickinsons hoped that studying botany would give their children an opportunity to appreciate the beauty—and the natural hierarchy—of God's creation. In the nineteenth century, this divinely ordained natural hierarchy was often referred to as "the great chain of being," the preferred English translation of the Neoplatonic *Scala Naturae* (literally, the "ladder of nature"). Although the *Scala Naturae* is no longer explicitly taught, it continues to infuse our culture today. Most of us could put plants, nonhuman animals, human beings, and angels in their proper places in the ancient hierarchy. The illustration from Diego Valadés's *Rhetorica Cristiana* (1579) shows how the ranks descend from angels to plants. When Linnaeus proposed the hierarchical system of classification that would eventually be codified into the categories in use today (kingdom, phylum, class, order, family, genus, and species), his new system was somewhat analogous to the older ladder of nature (*Scala Naturae*). Both schemes descended from larger, more transcendent categories to smaller, more particular beings. Both offered stable, hierarchical ways of understanding the natural world. Thus, even after Linnaeus, botany did not necessarily challenge established Christian ways of thinking.

Scala Naturae (The Great Chain of Being). Illustration from Diego Valadés's *Rhetorica Cristiana* (1579). Wikimedia Commons.

In the early nineteenth century, many Christians wanted their children—particularly their "beautiful and delicate" daughters—to learn their place on the ladder.³⁵

In the 1840s a unique strain of Christianity was prevalent in Amherst. Self-governing Congregationalist churches descended from the Calvinistic Puritans of the seventeenth century dominated the region. For more than a century, the religious culture of the Connecticut River valley had been shaped by Jonathan Edwards. Amherst College had its roots in Edwards's approach to education, which promoted the union of empirical science and Calvinist theology. Edwards is often remembered as a fearsome Calvinist. A century before the Dickinson children started school, Edwards preached his famous sermon on "Sinners in the Hands of an Angry God" (1741). His terrifyingly vivid description of hellfire fed a revival movement that helped spread Edwards's New England theology across the British colonies.

Despite that sermon's notoriety, Edwards was more interested in observing the natural world than imagining hell. He paid attention to natural philosophy because he believed the nature of the world revealed the nature of God.³⁶ In Amherst, the Enlightenment rationalist attitudes of the Lunar Men did not have the same impact they had in Britain. Interpretations of nature based on Edwards's mid-eighteenth-century thought continued to predominate.

Jonathan Edwards was born in 1703, fifteen years after Isaac Newton published *Principia*, the treatise that translated the mysterious actions of gravity into mathematics, and ten years after the Salem witch trials. Edwards was not particularly interested in demonology or witchcraft. He was more influenced by Newton than by the witch-hunters. Like Newton, Edwards leaned toward natural magic, which focused on mysterious natural phenomena. Giambattista della Porta's *Natural Magick* (1658, first published in Latin as *Magiae Naturalis* in 1558) explained natural magic as empirical observation and experiment-based inquiry into the natural world. It was neither anti-religious nor anti-scientific, but perhaps a little different from the theology and natural philosophy of the day because it focused on the unknown and encouraged participation and experiment alongside description. As natural philosophy shifted away from received wisdom and toward empirical inquiry, the developing scientific method would come to resemble della Porta's *Natural Magick*.

Edwards was an empiricist who believed that the divine was immanent in nature. He tended to focus on the "similarness" and harmony of natural objects (flowers and rainbows) to divine ideals. He believed that denying the magical connections among material things was tantamount to denying the existence of everything intangible. If there was no magic, there was no God.

Like many of his peers, Edwards understood the analogies and correspondences between the natural and the divine in terms of typology, an approach that started by matching events in the Hebrew scriptures to events in the Christian Gospels. Edwards extended Puritan typology from the scriptures to the book of nature, imagining natural phenomena as *Images or Shadows of Divine Things*.[37] In this respect Edwards's thought, like Isaac Newton's, can be described as a mix of science, religion, and magic.

When Edward Hitchcock was appointed president of Amherst College in 1845, he delivered an inaugural series of lectures, published as *Peculiar Phenomena in the Four Seasons*, which was very close to Jonathan Edwards in spirit. Hitchcock moved from "The Resurrections of Spring" to "The Triumphal Arch of Summer" and "The Euthanasia of Autumn" before concluding with "The Coronation of Winter." As a teenager, Emily Dickinson read *Peculiar Phenomena* over and over, since it was one of the few books that did not worry her father. From Hitchcock she absorbed the idea that the seasons of the year were analogous to Christian narratives. Natural phenomena were endowed with spiritual significance. This strain of thinking grew from magical seeds planted by Jonathan Edwards.

Growing up in Amherst in the 1830s and 1840s, Emily Dickinson would learn much more about the contradictory history and complicated cultural significance of botany than Charles Darwin had learned as a child. As a girl in Massachusetts, far from the center of the British Empire, she would study the histories and traditions of both Amherst and Britain. She would be taught to understand the world in ways informed by Isaac Newton and Jonathan Edwards. She would study the Romantic poets and read every thrilling Gothic novel she could get her hands on.

The colorful botany book that Otis Bullard painted in her hands is an obvious symbol of Dickinson's access to botanical education. To Dickinson herself, studying flowers would come to symbolize the contradictory questions that emerged from studying science in the nineteenth century. In 1862, one of Emily Dickinson's only published poems would allude to the paradoxes at the heart of botany:

> Flowers – Well – if anybody
> Can the extasy define –
> Half a transport – half a trouble –
> With which flowers humble men:
> Anybody find the fountain
> From which floods so contra flow –
> I will give him all the Daisies
> Which upon the hillside blow.[38]

For Dickinson, botany was a fountain of contradictions. Her family imagined that studying botany would help shore up Christian belief and remind girls of their place in a social and natural order established by God. However, Dickinson was also aware of botany's alternative implications. Botany was tinged with Erasmus Darwin's atheistic mechanical philosophy, not to mention his sexual libertinism. Dickinson's ecstatic description of "all the Daisies / Which upon the hillside blow" was not as graphic as Erasmus Darwin's flower descriptions, but it was erotic nonetheless. For her, as for other students of botany, flowers were the brightly colored sex organs of plants.

To complicate matters, in nineteenth-century Massachusetts the expert knowledge of plants was associated with witches and Native Americans as well as with libertines and Lunar Men. Botany pushed back against neat social and religious hierarchies from different sides. The supernatural, witchcraft-tinged aspects of botany threatened the church fathers while the native associations contained an element of opposition to European traditions. These aspects tended to amplify belief in a world of nonmaterial spirits along with presumptions that the world of plants was rich with meaning. As an adult, Dickinson was fascinated by these conflicting implications. "Flowers – Well – if anybody" challenged readers to consider the flood of contradictory emotions and ideas prompted by the study of botany.

In Dickinson's mind, the link between poetry and natural science (including botany) was potent and paradoxical. Both Charles Darwin and Emily Dickinson had been born into a world where science, religion, and magic intertwined with poetry, theology, and philosophy. During their lifetimes, however, everything changed. Strands of thought were pulled apart from each other. When they were children, botany was a place where everything seemed to come together in a gloriously contradictory and endlessly fascinating tangle of philosophy, politics, religion, and empirical observation. By the end of their lives, Erasmus Darwin's enchanted garden would be a distant memory. Scholars gained clarity from this reorganization of knowledge, but many also mourned the changes.

As an old man, Charles Darwin would recall his midlife turn away from poetry as a "curious and lamentable loss."[39] Emily Dickinson, the red-haired child who had clutched blossom and book together when Otis Bullard painted her portrait, would not let go so easily.

CHAPTER 2

Darwin the Naturalist

Shropshire, Edinburgh, Cambridge, 1809–1831; Darwin, to Age 22

IT IS a perfect irony: Charles Darwin—the great evolutionist, the bane of creationists from his time to ours—spent his childhood in the Garden of Eden. Or at least he remembered his childhood that way. Years later, in the cramped cabin of HMS *Beagle*, surrounded by indiscriminate piles of dead birds, animals, and plants, all in various stages of dissection and decay, he sought comfort in the pages of Milton's *Paradise Lost*. He wrote to his sister, "I often think of the Garden at home as Paradise; on a fine summers evening, when the birds are singing how I should enjoy to appear, like a Ghost amongst you, whilst working with the flowers."[1] Darwin might not have believed in paradise any more than he believed in ghosts, but his imagination had deep roots in an enchanted landscape.

Bobby Darwin—as Charles was called as a child—was a sweet little boy, agreeable and pleasant. In the garden he flitted about happily, racing to fetch an extra trug, cutting roses just as his mother and sisters had taught him, laying them together with surprising gentleness. He was a hard worker, always willing to carry buckets of water for the flowers. But when it was time to practice reading and writing, he often disappeared across the fields. If the garden was Charles Darwin's Eden, the schoolroom was something close to hell.

Charles looked like a different child when he sat at his books: frustrated, almost sullen. His younger sister, Catherine, chattered and laughed enough for the two of them. Where he was slow, she was quick. Whenever she got the best of her older brother, she sparkled. Charles turned inward, growing distant and quiet.

Aside from his reluctance when it came to lessons, Charles was an affable child. He was very close to the siblings nearest him in age: his brother, Erasmus (named for his grandfather), five years older, and his sister Catherine, two years younger. His three older sisters, Marianne, Caroline, and Susan, were more like aunts than siblings. Caroline Darwin, nine years older than Charles, took primary responsibility for teaching the youngest two. Sometimes Susan, who was two years younger than Caroline, would try to help, but she had even less patience than Caroline. Susan snapped at Charles for the littlest mistakes, interrupting him to correct his grammar when he tried to join the conversation or brusquely supplying the word when Charles had trouble getting it out. The stammer did not worry the family much—both his father and his Darwin grandfather had stuttered as boys, and both had grown into well-spoken, even eloquent, men. Sometimes their father's grandiloquence left little room for anyone else to speak.

Robert Waring Darwin, Charles Darwin's father, was the oldest surviving son of Erasmus Darwin. Robert was a large and imposing man (he recorded his weight at 24 stone—336 pounds, or 152 kilos). Charles remembered his father as "the largest man whom I ever saw."[2] He was a busy physician, constantly on the move between the nearby hospital and house calls to patients at home. His approach to medicine leaned more toward psychological support than surgical intervention; he avoided heroic measures and was known for his keen perception.

Robert Darwin was also a canny investor, always eager to invest in new ventures or to offer mortgage loans to neighboring landowners who needed cash. Guided by his sharp psychological insight, he was a successful venture capitalist and eventually became one of the wealthiest men in England. His family lived comfortably but not ostentatiously. Charles and his siblings did not really know how rich their family was.

Susanna Wedgwood Darwin, Charles Darwin's mother, was Josiah Wedgwood's oldest daughter. Her father's favorite child, Susanna was unusually well educated for the era. She was one of relatively few women who attended boarding school in the eighteenth century, though her attendance at the school in Manchester was relatively brief. She returned home when her father established the "Etruscan" school for his sons and daughters and the children of a few close friends. Young Robert Darwin befriended Susanna there, and their fathers hoped the two would eventually marry. In 1796, when they finally married at around age 30, it seemed the inevitable result of their childhood friendship and the longstanding closeness between their families.

The friendship between Charles Darwin's two grandfathers gave Charles a slew of interconnected relations. On top of Erasmus Darwin's brood of fourteen, Josiah Wedgwood had seven children. Most survived and many married, giving Charles dozens of aunts, uncles, and cousins. Robert and Susanna's marriage was the first of many Darwin-Wedgwood alliances; in time a surprising number of Darwin-Wedgwood cousins would marry one another. Charles was born into a tangled web of family and friends, many of whom had close financial and emotional ties. Maer Hall, where Susanna's brother lived with his wife, Elizabeth, and their six children, was about twenty miles from the Mount—an easy day's travel. Charles loved to visit Maer Hall. His Wedgwood cousins were full of fun, and the grounds around Maer were just as enchanting as his family's gardens.

As a child, Darwin loved to read. He spent many afternoons under the dining room table reading *Robinson Crusoe*.[3] He was very fond of poetry, as long as it was in English. He did not enjoy Latin verses as much. Studying languages was particularly challenging for him. He was a quiet, dreamy boy, often slow to speak. Then again, with so many voluble Darwins (and Wedgwoods) perhaps he never had the chance to say much. Even when their father took the pony cart out to visit his patients and Catherine disappeared with Nanny, even when the older girls were occupied, sitting with their mother in her quiet room or working in the garden, his older brother filled the silences. Charles adored Erasmus—known as Ras—who was five years older and infinitely more

The Mount, Shrewsbury. Wikimedia Commons.

social. When Ras was away at school, Charles played outdoors. He seemed perfectly happy alone, collecting and arranging his pebbles or scouring the garden and orchards for the most perfect fruit.

Charles Darwin started school when he was 7 years old. At first, he attended a day school at the home of George Case, a Unitarian minister and family friend for many years. Although the Darwin children had been christened in the Church of England, most Sundays their mother took them to Case's Unitarian chapel on High Street in Shrewsbury. Case's school was in the Church Manse at 13 Claremont Hill. Darwin's time at the day school was pleasant and relatively uneventful. It was here that he and William Leighton attempted to make a flower speak its name.

Darwin's first school was at the heart of the early nineteenth-century debates about theology, poetry, and philosophy. A few years before, in 1798, it had been the setting for a significant event in intellectual history. In its rooms, George Case convinced Samuel Taylor Coleridge to accept a financial settlement from two of Charles's uncles, Tom Wedgwood and Josiah Wedgwood II. At the time Coleridge was a candidate for a position as a Unitarian minister in Shrewsbury. Josiah Wedgwood II wanted to dissuade his younger brother's friend from becoming a clergyman. At their meeting in the modest building on Claremont Hill, Coleridge agreed to focus on writing instead of preaching.

Coleridge's ideas were somewhat wild. As an undergraduate he had attempted to form a utopian community based on his own philosophy of *pantisocracy*—absolute equality for all. He was a theological radical on the extreme, freethinking edge of Unitarianism. It was not at all clear that Coleridge believed in God, but he needed to find a way to support himself.

Coleridge and Tom Wedgwood were close friends. Tom was freehanded with his Wedgwood fortune, and Coleridge often tagged along with him as he roamed around England. Coleridge likened their relationship to "a Comet tied to a Comet's Tail."[4] Both men were brilliant talkers. Tom was almost as obsessed with trying to capture images and shadows as Coleridge, but while Coleridge stuck to old-fashioned pen and paper, Tom experimented with the techniques that would eventually lead to the development of photography. Both men were prone to illness. In the course of treatment, both became addicted to opiates. Erasmus Darwin, the Wedgwood family doctor, was the physician who originally prescribed opium to Tom Wedgwood. Coleridge may have picked up the habit from his friend. Within a short time, both realized that opiates were harmful, but it was hard to turn back. Later, Coleridge would try to procure a supply of cannabis from Sir Joseph Banks in hopes that cannabis might cure Tom Wedgwood where opium had failed.

Around Shrewsbury, Unitarianism was the comfortable, socially respectable alternative to the Church of England. The Wedgwoods were committed Unitarians. Josiah Wedgwood II feared that Coleridge's outlandish ideas—and behavior—would make Unitarianism seem less innocuous to his neighbors. Josiah II hoped to prevent this, for his own sake and to preserve the respectability of middle-class Unitarians everywhere. He liked his brother's friend, and he knew that Coleridge would find the job difficult, since he was likely to offend people with his radical ideas. Tom agreed that Coleridge should avoid preaching, perhaps because he saw an opportunity to help support his friend's literary ambitions.

With Case acting as mediator, Coleridge and the Wedgwoods met at his house to work out the agreement. The Wedgwoods promised to pay Coleridge an annuity, to patronize him as a poet and a political journalist (and perhaps even a philosopher), on the condition that he refrain from preaching or writing about theology. Freed to concentrate on poetry, Coleridge turned his attention to his writing. He finished *The Rime of the Ancient Mariner* in the first flush of his patronage, and he and William Wordsworth published *Lyrical Ballads* about ten months after he accepted the Wedgwoods' offer.

Today, more than two hundred years later, it seems particularly significant that Charles Darwin started school in the place where his wealthy uncles had persuaded Samuel Taylor Coleridge to choose between poetry and theological speculation. The moment his uncles stepped in to separate theology from literature was a major step toward the specialization and disciplinary division that would shape Charles Darwin's life.

Over time, hostility between Coleridge and the Darwin-Wedgwood clan would develop, eventually becoming public. As early as 1796, Coleridge had written to a friend that Erasmus Darwin's poetry made him sick to his stomach. (The word he used was *nauseate*.)[5] In later years Coleridge attacked the elder Erasmus Darwin on two fronts. On the one hand, he objected to "speculation," to the giant inductive leaps that allowed Erasmus Darwin to theorize about the origins of life based on his observation of familiar household plants and flowers. On the other, Coleridge was an idealist who objected to Erasmus Darwin's insistence on the significance of the material world. In Coleridge's mind, focusing on material bodies risked denying the spiritual core of every being. "Darwinizing" disgusted him.

Coleridge's derogatory use of the Darwin name was partly personal. When Tom Wedgwood died in the throes of addiction in 1805, Coleridge lost the part of his annuity that had come from his friend.

Tom's death was a financial loss as well as a personal loss. Meanwhile, Coleridge lived with an addiction similar to Tom Wedgwood's. He seems to have partly blamed Erasmus Darwin for his poverty and illness. His sallies against Darwinizing had roots in a complex emotional mix of personal resentment and genuine philosophical disagreement.

Charles Darwin was insulated from some of this controversy. He may not have heard Coleridge's "Darwinizing" slur when he was a child. But whether he knew the term or not, he knew that he was a Darwinizer—and he knew that many people in his neighborhood disapproved.

"I was in many ways a naughty boy," he would later recall: "By the time I went to this day-school, my taste for natural history, and more especially for collecting, was well developed. I tried to make out the names of plants, and collected all sorts of things, shells, seals, franks, coins, and minerals. The passion for collecting which leads a man to be a systematic naturalist, a virtuoso, or a miser, was very strong in me."[6] The young Charles Darwin might have equated being a "systematic naturalist" with being a "naughty boy" under the influence of his father, who disliked natural history and wanted his sons to study medicine. Darwin later described Robert Darwin as a man without "any aptitude for poetry or mechanics, nor did he possess, as I think, a scientific mind, though he had a strong taste for flowers and gardening."[7] His father's lack of interest in science puzzled Charles. He wrote, "I cannot analyse why my father's mind did not appear to me fitted for advancing science.... His powers of observation were all turned to the practice of medicine, and still more closely to the human character."[8]

Robert Darwin was not the only person who thought natural history and natural philosophy were a waste of time. In 1817, when the 7-year-old Charles Darwin recited Latin verses at school, no one imagined that scientific approaches to understanding the world would become so important in the next decades. What Darwin was learning, first from his family, then from school, was that studying the natural world was, at best, a hobby, an avocation, a pleasurable escape from classical languages and abstract mathematics.

After a year of study at Case's day school, Charles was ready for boarding school. But that summer his mother, Susanna Wedgwood

Darwin, died. Her death was sudden and painful. Most likely, it would be diagnosed today as peritonitis resulting from a burst appendix. At the time it was a frightening and inexplicable ailment characterized by sharp abdominal pain. She was ill for three days. Her husband, Robert, was a respected doctor, but there was little he or anyone else could do. Though Charles and his younger sister were kept out of the room, they probably heard their mother's cries of pain. When Charles was invited into his mother's bedroom after her death, he focused on a small table off to the side. He later recalled, "My mother died in July 1817, when I was a little over eight years old, and it is odd that I can remember hardly anything about her except her death-bed, her black velvet gown, and her curiously constructed work-table."[9] His experience of her death overshadowed earlier memories.

On some level, however, Charles must have been aware that his mother and father in partnership had designed and created the gardens where he sought refuge and that his mother had managed them until her death. The heavenly atmosphere of the gardens at the Mount was part of her legacy for her children. Susanna Wedgwood Darwin approached her flowers, fruits, and birds as a passionate "fancier," mixing a compelling urge to complete her collection with a deep appreciation of beauty.[10] In *The Ghost in the Garden,* Jude Piesse proposes that Charles Darwin's later enthusiasm for pigeons may have harked back to his mother's carefully tended, meticulously documented flock. She was a passionate pigeon fancier long before her son.

Like his mother, Charles had a fancier's eye. As a boy he lost himself in his own collections, moving from franks and seals to marbles to all the pebbles in the front walk. As a young man, he would go through phases, relentlessly (and methodically) hunting birds, collecting marine animals, or capturing beetles. Later, when he became a parent and settled into his own house, he would try to re-create the atmosphere of the gardens surrounding the Mount.

In 1817, as the Darwin family grieved for Susanna, the older girls stepped ably into their mother's place. With five siblings and a watchful father, Charles was certainly not lonely. He was often surrounded by sisters or playing happy sidekick to his brother. Sometimes during the

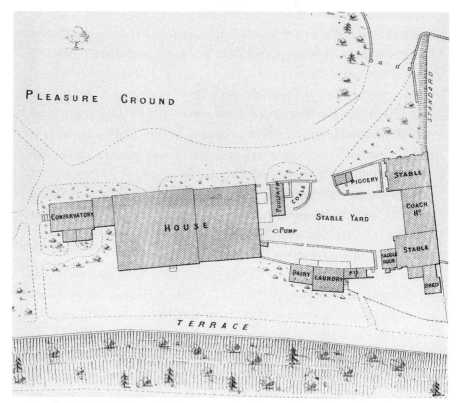

Detail: T. Tisdale, "Plans for Property at the Mount, Shrewsbury" (1867). D3651/B/165/5, Shropshire Archives.

holidays he squeezed into the pony cart with his father, accompanying him as he visited patients. Nonetheless, Charles was most at ease high in the branches of his favorite chestnut tree or splashing rocks into the Severn River at the foot of the garden. He and his sister Catherine devised a private language: they called their secret place *Lorum*.[11] In the magical fields and gardens around the Mount, Darwin had many hiding places. He loved to wander off alone.

Charles Darwin started boarding school a few months after his mother's death. He was 8 years old. He had never been an eager student, but things took a definite turn for the worse when he enrolled at Samuel Butler's Shrewsbury School. The school was housed in a large group of neo-Gothic buildings at the center of town, just a little more than a mile from the Mount. The headmaster, Samuel Butler, was a prize-winning mathematician and a severe disciplinarian. He flogged his students regularly. In 1826 he would protest against his reputation for excessive punishment, writing: "I have never flogged the same boy twice a week more than three times in twenty-six years."[12] Despite the weekly beatings, the school was respected, largely because Butler prepared students for Cambridge University entrance examinations with great success.

Since Cambridge did not require any preparation in natural sciences, Shrewsbury offered no instruction in natural history or natural philosophy. Perhaps this was lucky for the world since the Shrewsbury methods of teaching filled Darwin with loathing for every subject he studied there. In his *Autobiography*, Darwin commented, "Nothing could have been worse for the development of my mind than Dr. Butler's school, as it was strictly classical, nothing else being taught except a little ancient geography and history. The school as a means of education to me was simply a blank."[13]

The mile between school and home was easy enough to traverse. Charles spent many afternoons at the Mount, running across the fields to check in to the dormitories at the last possible minute. In part because he avoided school as much as he could, his schoolmates perceived him as proud. To his family and close friends, however, Charles was affectionate and sometimes very funny. In 1822 he wrote to a "dear friend" with a story about his older sister's efforts to convince the brothers to bathe. It started when he confessed to Caroline that he only washed his feet once a month, "which I confess is nasty, but I cannot help it, for we have nothing to do it with." He and Erasmus had roared with laughter at their sister's disgust.[14] As Charles explained, Caroline "said she did not like sitting by me or Erasmus for we smelt of not washing all over, there we sat arguing away for a good while."[15] At almost 13 years old, Charles was so amused by the exchange that he composed a letter about it and copied it into his memorandum book, signed with a flourish as "Justice, with a nose as big as your fist." The dear friend whom he addressed has never been identified, but editors guess that it may have been Catherine, Charles's younger sister.

In his *Autobiography*, Darwin wrote, "Looking back as well as I can at my character during my school life, the only qualities which at this period promised well for the future, were, that I had strong and diversified tastes, much zeal for whatever interested me, and a keen pleasure in understanding any complex subject or thing."[16] As it happened, none of the subjects that fascinated him were taught—or valued—at Shrewsbury.

If Butler had tried to force Charles to study chemistry, he might well have resisted. As it was, Charles and his older brother, Ras, became obsessed with conducting chemistry experiments. They spent all their extra cash on equipment. Purchasing glass bottles and burners and stopcocks, they set up their very own laboratory in a garden shed. In time the Darwin brothers' backyard chemistry laboratory allowed them to perform experiments worthy of the Lunar Society. Though their father was not particularly enthusiastic, he did not stop them from emulating their scientific grandfathers.

After Ras went to Cambridge University, Charles continued the experiments alone, both at home and—recklessly—at school. His classmates nicknamed him *Gas* because of his predilection for chemistry. None of his "experiments" were particularly original—he just liked to try his hand at transforming substances. The Darwin brothers were particularly fond of making and measuring crystals. They loved the remarkable symmetries, the complex angles, and the beams of refracted light. For them, chemistry was a way of creating beauty, changing simple liquids into startling clouds of gas or extraordinary glittering crystals. The fact that Charles could do it without any assistance from his schoolmasters filled him with stubborn, antischolastic glee.

By the time Charles was 15, it was clear that he was not cut out for studying classical languages or mathematics, the two subjects taught at Shrewsbury. Yet he was a bright boy, strong, quiet, and levelheaded. He seemed just as mature as his brother Ras, who had completed the first part of his degree at Cambridge and planned to enroll in medical school at Edinburgh for the practical scientific coursework that was not offered at Cambridge.

The history of medical degrees at Cambridge University is somewhat puzzling. In 1540 Henry VIII endowed a professorship in medicine known as the *Regius Chair*. However, the university did not offer any instruction in clinical medicine for three hundred years. The university website explains that the Regius professors of medicine "appeared to look upon their positions as a means to an end, enabling them to do their own work without the inconvenience of having to teach students."[17] Up to the 1840s, it was impossible to learn much about medicine at Cambridge, though students could sit for exit examinations and receive degrees in the subject. The older Erasmus Darwin, Charles and Ras's grandfather, had taken this route, matriculating at Cambridge, moving to Edinburgh to study medicine, and then returning to Cambridge to take his degree once he had mastered the subject. Afterward, he refused to send his own sons to Cambridge. He objected to the Church of England, and he was outraged that the university granted degrees without providing instruction. However, in the next generation, Robert Darwin, Charles and Ras's father, saw the value of the prestigious degree his father had denied him. He encouraged Ras to enroll at Cambridge and carefully planned out his oldest son's medical studies in Edinburgh, London, and Paris.

But the family also worried about the plan. One of Robert Darwin's brothers (Charles, after whom he named his second son) had died while studying medicine at Edinburgh. Ras was 21, but he was not physically strong. Their father decided to pull the sturdier Charles out of Shrewsbury and send him to Scotland to keep Erasmus company. Charles had no desire to attempt the Cambridge entrance examinations. There was no reason to stay at Shrewsbury prepping for Cambridge if he could go directly to Edinburgh.

Edinburgh University did not have entrance examinations or any organized admissions procedure. Any young man who was interested in studying could simply move to the city, register with the university, and attend lectures. Professors were not salaried by the university. Instead, they were self-employed performers who supported themselves by selling tickets to their own lectures.

At age 16, Charles Darwin matriculated at Edinburgh University as a medical student. It cost ten shillings—a pittance—to enroll. To maintain good standing, students were required to sign in with the registrar once a month during the term to show they had not left the area. Other than that they were on their own. Charles and Ras rented a beautiful set of rooms on the fourth floor at 11 Lothian Street, light filled and elegant. In a letter to his father, Darwin gloated, "Light bedrooms are very scarce articles in Edinburgh, since most of them are little holes in which there is neither air or light."[18] He was not exaggerating. The city was much bigger than Shrewsbury and far to the north. Compared to rural Shropshire, it was dark, cold, and smoky. For the young Darwin, accustomed to the gentle midlands, Edinburgh was a shock.

The brothers spent almost all their time together. They kept to themselves, two elegantly dressed, noticeably tall young men who spent their days taking long walks on the beach at Leith, borrowing stacks of leather-bound volumes from the library, and reading together by the roaring fire in their rooms. Both brothers continued in the pattern that had started at Shrewsbury. They were passionately engaged in studying outside of school, but they had trouble mustering enthusiasm for schoolwork. The lectures were "intolerably dull," Charles wrote. "To my mind there are no advantages and many disadvantages in lectures compared with reading. Dr. Duncan's lectures on Materia Medica at 8 o'clock on a winter's morning are something fearful to remember."[19] Darwin hated waking in the darkness to attend Duncan's lectures, which he described as "long and stupid."[20]

Other lectures were somewhat better. The brothers enjoyed Professor Thomas Hope's flamboyant chemistry demonstrations, and they were interested in the clinical lectures at the hospital, although Charles was less interested in surgery. In his *Autobiography* he remembered the two occasions when he visited the operating theater in the hospital at Edinburgh. As it happened, he "saw two very bad operations, one on a child, but I rushed away before they were completed. Nor did I ever attend again, for hardly any inducement would have been strong enough to make me do so; this being long before the blessed days of chloroform. The two cases fairly haunted me for many a long year."[21] Disturbed by the violence of the operating theater, curiously unable to remember his mother's illness, Darwin was realizing that hospitals and other people's sickbeds were not the place for him, any more than lecture halls.

The morgue was even worse. Without antibiotics, dissecting corpses could be very dangerous. He had long known that his father's brother Charles, a promising Edinburgh medical student, had died "before he was twenty-one years old from the effect of a wound whilst he was dissecting the brain of a child."[22] What he had not known—and could not have expected—was that the cadavers that Edinburgh's anatomists depended on were often purchased in dubious circumstances. In the 1820s the active trade in human corpses reached appalling heights. As human bodies commanded higher and higher prices, Edinburgh residents resorted to installing iron cages called *mortsafes* around coffins to try to keep graves safe from robbery. Darwin's years in Edinburgh, 1825–1827, were at the height of the corpse trading. Authorities turned a blind eye to the body snatchers until 1828, when it was discovered that William Burke and William Hare had murdered sixteen people to sell their bodies to the anatomists. William Wordsworth's critique of nineteenth-century approaches to learning had proved chillingly prophetic. In Edinburgh, they did, in fact, "murder to dissect."[23]

It was no wonder that Charles Darwin could not face the prospect of studying human anatomy. Instead, he read novels and books of travel voraciously that winter. He also read many works of natural philosophy, including his grandfather's *Zoonomia*, an ambitious prose treatise that explained Erasmus Darwin's ideas about how complex life-forms shared a common ancestor. In *Zoonomia*, Erasmus Darwin unfolded a detailed account that started with microscopic *filaments*—tiny specks—floating in the ocean and then developing over millions of years, progressing from fish, through amphibian and land animal, to eventually develop into the highest form of life—"humankind."[24] Up to that point, Charles had not known much about his grandfather's ideas.

As a child, Charles might have noticed that his family library was full of books whose bookplates were emblazoned with his grandfather's motto, *E conchis omnia*—Everything comes from seashells. He did not know that fifty years before, his grandfather had scandalized his neighbors by painting the motto on the doors of his coach. In response, Thomas Seward, a priest at the nearby Litchfield Cathedral, published a poem mocking him:

> Great Wizard, he! By magic spells,
> Can raise all things from cockle shells![25]

Erasmus had been chastened enough by the public outrage to scrub the motto off his carriage doors. Charles's father had been embarrassed by the incident. In later years Robert rarely mentioned his father's scandalous transmutationist ideas, but he kept the family motto and used it for his bookplate, where it could be kept concealed. When Charles read *Zoonomia*, he was almost as fascinated by the family secret as he was by his grandfather's ideas. For his own family crest, Charles would use the three seashells, though he would adopt a different, more circumspect motto (*Cave et Aude*—Beware and Dare).

Charles knew that his grandfather's publications had embarrassed his father, but he could not help but admire *Zoonomia*. He wondered about his grandfather's theory that life on earth had begun with a single "filament."[26] As he and his brother, Ras, walked along the beach, he tried to imagine grasping that original thread. In his last long poem, *The Temple of Nature*, his grandfather Erasmus had speculated:

> Organic life beneath the shoreless waves
> Was born and nurs'd in ocean's pearly caves;
> First forms minute, unseen by spheric glass,
> Move on the mud, or pierce the watery mass;
> These, as successive generations bloom,
> New powers acquire and larger limbs assume;
> Whence countless groups of vegetation spring,
> And breathing realms of fin and feet and wing.[27]

On the crowded sands of Leith, Charles encountered "countless groups of vegetation" and hundreds of different members of the "breathing realms of fin and feet and wing." It was becoming increasingly clear that he did not want to be a physician. Instead, he started to think about becoming a naturalist. However, at the time natural science was usually more a pastime than a career. If he wanted a profession, he would probably need to become a clergyman too.

In 1825, the year that Darwin enrolled at Edinburgh, his grandfather's critic, Samuel Taylor Coleridge, published *Aids to Reflection*, which criticized naturalists because they had "no light of revelation."[28] Charles's grandfather Erasmus would have agreed about naturalists because he scorned the concept of revelation—the idea that God directly revealed truth to humans. Both Coleridge and the elder Darwin defined naturalists as those focused on the material aspects of the natural world and contrasted them to spiritualists, who saw the world as a revelation of the divine.

For members of Charles's generation, however, the meaning of "naturalist" had changed. Across Britain and the United States, most naturalists were devout Christians and followers of William Paley. Paley's *Natural Theology* (1802) had convinced many members of the Church of England that the close study of nature could uncover truth about the divine. After Paley, the British men who contributed to the new natural history journals tended to be Episcopal priests. Since natural history was really the only thing that interested Darwin, he began to toy with the idea of studying divinity rather than medicine. If the route to becoming a professional naturalist was by becoming a parson, Charles was game. He wrote to Caroline, "I have tried to follow your advice about the Bible, what part of the Bible do you like best? I like the Gospels. Do you know which of them is generally reckoned the best?"[29] His impulse to make a ranked list savored more of his fancier's passion for collecting and making lists than of divine inspiration. At any rate, with or without the "light of revelation," Darwin was beginning to imagine himself as a parson-naturalist.

Despite his lack of interest, Darwin returned to Edinburgh for a second year, at his father's insistence. This time he was alone: Ras went to London to continue along the path of studies laid out by their father. The 17-year-old Charles moved into his own set of rooms and conscientiously set out to make some friends. In November 1826 he joined the Plinian Natural History Society, a casual student group that met every Tuesday. A week after he joined, he was elected to the Society Council.

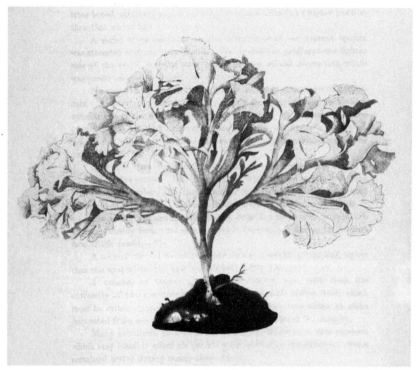

Flustra by J. G. Dalyell. Published in *Rare and Remarkable Animals of Scotland* (London: John Van Voost, 1847).

At their evening meetings in a basement room at the university, the members, most of whom were medical students, would read papers on their discoveries. One member stood out: the 34-year-old Robert Grant, an anatomist who had completed his medical training in Paris. Grant and Darwin struck up a friendship. The two men spent hours on the beach collecting specimens of the fascinating sea creatures that seemed to live right on the blurry line between plants and animals. They were most interested in sponges, corals, and the seaweed-like *Flustra* that lived in the intertidal waters.

Commonly known as *sea-mat*, *Flustra* were sometimes called *zoophytes* or *bryozoans*—compound words that indicated they had characteristics of both animals and plants. They are now classified as *moss animals*. The species that Darwin studied that winter was most likely *Carbasea carbasea*. They can be found clustered in the rocks on the edge of the beach, making webs of hundreds of pale-brown polyps tangled together.[30] Such primitive sea creatures were exactly the sort of beings that Erasmus Darwin had imagined as the original "filaments" of life. Grant and Darwin were interested in them for many reasons. Most significantly, they shared an interest in evolutionary ideas.

In 1826 no one in Edinburgh used the word "evolution" to discuss biological change over generations. Erasmus Darwin's term "transmutation" was still used occasionally in Scotland. In France, however, Jean-Baptiste Lamarck had proposed a slightly different theory known as *transformation*. When these ideas were mentioned publicly, they were condemned as heretical and anti-Christian. The official line at the University of Edinburgh was that species were immutable and fixed. In private, however, students quietly discussed the dangerous idea of species change. Robert Grant had studied in Paris, where Lamarck was a professor at the Jardin des Plantes, the French center for the study of natural history. Darwin listened raptly as Grant explained Lamarck's more up-to-date ideas about how species developed and changed.

After sunset Darwin and Grant would return to the laboratory and study the day's haul, carefully examining the creatures under a microscope. They were very close friends for a while, with the intensity of a brief student friendship. Their relationship hit a crisis with a dispute over which one of them had discovered a particular anatomical quirk of *Flustra*. Late one night, Darwin saw something new. He observed "that the so-called ova of Flustra had the power of independent movement by means of cilia, and were, in fact, larvae."[31] In his notebook he crowed: "That such ova had organs of motion does not appear to have been hitherto observed either by Lamarck Cuvier Lamouroux or any other author."[32] This was just the sort of detail that his mentor Grant had been searching for because it showed that *Flustra* were definitely animals, not plants. However, Grant was not pleased that Darwin had made the first observation of the moving larvae. Darwin was confounded by his mentor's cold reception.

For Darwin the experience was emotionally bewildering. First was the joy of feeling that he had made an original discovery. Next came Grant's puzzling, dismissive reaction. Finally, Grant delivered a paper to the Wernerian Society and seemed to take credit for the discovery. Darwin was hurt and angry. In fact, neither Grant nor Darwin was making an original discovery—the information had already been mentioned in several publications. As a 17-year-old student working with a more established person (not merely older but twice his age), Darwin might have too quickly taken credit for a discovery in his mentor's laboratory. By the same token, Grant might have been ungenerous. Whatever the case, it left a sour taste. Darwin and Grant's friendship flared out as quickly as it had begun.

Edinburgh was a disputatious city, and Darwin found intellectual debate painful. He did not take any pleasure in the rough and tumble of Scottish intellectual battles. When his conflict with Robert Grant erupted, Plinian Society meetings became less appealing. Even so, Darwin tried to change course, enrolling in natural history lectures instead of following the medical curriculum.

He attended lectures by Robert Jameson, the Regius Professor of Natural History. It is likely that Darwin was searching for a new mentor, but he quickly concluded that Jameson would not do. Jameson remained so focused on arguing his side of a debate with a rival geologist that he seemed blind to the fascinating geological features of Edinburgh itself. Jameson's lectures on geology—both boring and contentious—outraged Darwin so much that he turned against geology completely. Later he recalled that Jameson inspired him with "the determination, never as long as I lived to read a book on geology or in any way study the science."[33] We should read this extreme statement with some caution. Darwin was a skillful raconteur, and when he told this story he had become a leading geologist. The grain of truth here is that he did not like Jameson and was frustrated with natural history instruction in Edinburgh.

Darwin always thrived best when he had a mentor. With Ras gone, his friendship with Grant exploded, and Jameson a dead end, Darwin turned to the taxidermist John Edmonstone. He spent much of the winter with Edmonstone at the University Zoological Museum, learning taxidermy. Edmonstone was a Black man who had been born into enslavement in Guyana. He had been trained as a naturalist by the

British explorer Charles Waterton before migrating to Scotland and taking a post at the Zoological Museum. In his *Autobiography*, Darwin recalled that Edmonstone "made his living by stuffing birds, which he did excellently.... I used often to sit with him, for he was a very pleasant and intelligent man."[34] Darwin's family had long boasted of their abolitionist history, but they did not have many direct experiences with people who had been enslaved. Slavery was abolished in England in 1807, before Darwin was born. He had not met many people of color in the midlands. Now Edmonstone became another mentor, teaching Darwin how to preserve animal skins and prepare zoological specimens. These skills would prove useful later in Darwin's career. Without Edmonstone's instruction, Darwin would not have known how to preserve a Galapagos finch.

In his *Autobiography*, Charles recalled, "After having spent two sessions in Edinburgh, my father perceived or he heard from my sisters, that I did not like the thought of being a physician, so he proposed that I should become a clergyman. He was very properly vehement against my turning an idle sporting man, which then seemed my probable destination." It was not uncommon for wealthy young men of Charles Darwin's era to spend most of their time hunting, and Charles was a crack shot. Frustrated, his father berated him, "You care for nothing but shooting, dogs, and rat-catching, and you will be a disgrace to yourself and all your family."[35]

But despite Darwin's obstinate refusal to engage with formal schooling, he was not at all prone to idleness. He wanted to make his mark in the world somehow, and his experiences in Edinburgh gave him some hope that natural history might be a possibility. The problem was that Oxford and Cambridge did not grant degrees in the natural sciences. An ordinary bachelor of arts degree at Cambridge was Darwin's last resort; it would mean brushing up on the Latin and Greek he abhorred, but he hoped he could memorize enough to make it through.

In the 1820s Cambridge University had a strange relationship with the natural sciences. On one hand, the university was famous for natural science. Isaac Newton had been a professor at Cambridge in 1687 when he first published *Principia*, the most influential scientific work produced in Britain up to that time. In subsequent years the university created endowed professorships in many areas that would today be defined as the sciences. Professors of astronomy, botany, chemistry, geology—and a host of other similar subjects—held appointments, conducted research, and published their findings. They offered regular courses of lectures in university halls. However, the university did not offer examinations, much less degrees, in any of these areas. The situation of the natural sciences was exactly the opposite of clinical medicine, which offered degrees but no coursework. In the natural sciences, there were courses but no degrees.

To gain admission to Cambridge, Darwin needed to pass entrance examinations in Greek, Latin, mathematics, and the Bible. In order to complete a bachelor of arts, he would need to pass final examinations in the same subjects. Like other undergraduates, he would meet one-on-one with a tutor to help him prepare for the final examinations. The lecture courses on all the topics that actually interested him were not offered for credit but were enticing nonetheless.

In January 1828, after months of intense tutoring in classical languages and mathematics, Darwin enrolled at Christ's College. He was almost 19. The next three years were easy and pleasant; he would later describe them as "the most joyful in my happy life."[36] The joys were not particularly scholarly. Darwin did not like his course at Cambridge. As he put it, "My time was wasted, as far as the academical studies were concerned."[37] He did manage to keep up with his reading for examinations despite his profound lack of interest in the official curriculum.

He genuinely enjoyed the extracurricular lectures on botany and geology that had attracted him to Cambridge, but the main reason he was happy was because of the camaraderie with his peers. For the first time in his life, he found a group of friends with whom he felt truly at ease. In many ways he was an "idle sporting man" at Cambridge, just as his father had feared. He kept a horse and a succession of dogs and practiced shooting (with caps instead of bullets) in his college rooms. He and his friends were not as dissipated as some of their classmates at Cambridge, nor were they as serious about theology as others. Their set was affable and outdoorsy. They went to the horse races as often as they attended concerts and galleries. They loved to take long walks and rides out into the Fenlands. At one point they set up what they called a glutton's club, with the intention of eating the strangest food they could find. The club faded away after they dined on a scrawny brown owl.

Of all his friends at Cambridge, Darwin's closest was his cousin William Darwin Fox. The two men looked alike and had a similar interest in natural history. Fox was two years ahead of Darwin at the university and a few steps further along the career path toward becoming a parson-naturalist. Charles admired him immensely. During Darwin's first year, his cousin stepped partway into the role of mentor, but the two men were not so far apart. With Fox he did not need to play the worshipful younger brother or the obsequious follower. Together, both took up entomology, and soon they were passionately collecting beetles.

For some unknown reason, entomology was very fashionable at the time. Popularly known as *Beetlemania*, it was the perfect hobby for Charles Darwin. He had always had a fancier's eye and a collector's greed; now, since so many people were also interested in collecting beetles, he could indulge his competitive side. He devised ingenious methods for trapping beetles. He spent his afternoons crawling through swamps and climbing trees as he scrambled after beetles. He hired boys from town to collect extra beetles for him, and he skipped classes and all sorts of social obligations to search for specimens. His friend Albert Way drew a mocking cartoon that depicted Darwin wielding a butterfly net astride a giant beetle, with the caption "Go it, Charlie!"

Once, when he already had a beetle in each hand and managed to capture a third type, he popped it into his mouth. In a letter to Leonard Jenyns, he recalled, "To my unspeakable disgust & pain the little inconsiderate beast squirted his acid down my throat & I lost both Carabi & Panagus!" Years later the incident still made him laugh. It was worth a little pain to be able to add another dead beetle to his collection. Darwin was overjoyed when he found a beetle rare enough to earn him a mention in a London entomological journal. He was good at this. Perhaps he would find a way to get published in more scientific journals.

After Darwin's first year, Fox got his degree. Soon afterward his cousin took orders and accepted a curacy in Nottingham. Eventually, he

Charles Darwin as a "Beetlemaniac." Sketch by Albert Way (ca. 1830). Cambridge University Library.

would become the sort of parson-naturalist that Darwin aspired to be. Before leaving Cambridge, Fox introduced his younger cousin to John Stevens Henslow, an Episcopal priest and professor of botany and mineralogy. Henslow became Darwin's next great mentor. They spent so many afternoons together, walking out into the Fenlands to botanize, that members of the Cambridge community came to recognize the tall, well-dressed Darwin by sight. He was shy, and few people knew him by name. Instead he became known as "the man who walks with Henslow." During their walks, Henslow encouraged Darwin to become a parson-naturalist and persuaded him to try geology again. At Henslow's scientific soirées, he got to know William Whewell, the young professor who would invent the word "scientist" as he campaigned to restructure the university to facilitate a better match between the coursework and the examinations.

If formal education in the natural sciences had been available to Darwin, would he have been interested? How much of his fascination with science was caused by the fact that it was practically forbidden?

In grammar school he had focused on chemistry instead of classics. At Edinburgh he had read natural philosophy and studied sea creatures instead of medicine. At Cambridge he studied beetles with determination and enthusiasm while avoiding his tutors. He shirked the assigned reading as much as possible without failing.

Christ's College, Cambridge, was the perfect setting for this behavior. Unlike Edinburgh, where lectures were central to the curriculum, Cambridge encouraged students to read and study independently and widely. While a student there, Darwin became very fond of a few key books. John Milton, the author of *Paradise Lost* and *Paradise Regained* (among other works), had been a student at Christ's College from 1625 to 1632. During his own student days, Darwin came to love Milton's account of Adam and Eve in the Garden of Eden. Milton offered a different picture of Creation from his grandfather Erasmus's version. As he studied for the priesthood, Darwin immersed himself in the Miltonic view. The critics Gillian Beer and Robert Richards have traced the influence of *Paradise Lost* on Darwin's later writing.[38]

Darwin lived in the rooms where William Paley, the great natural theologian, had lived. Some of Paley's books were part of the official curriculum, and Darwin was required to study them. Not surprisingly, his favorite work of Paley's was *Natural Theology* (1802), which was not required reading. *Natural Theology* had legitimized the study of nature within the Church of England. Without Paley it would not have been possible for Darwin to imagine becoming a parson-naturalist. And then there was John Herschel, a friend of Henslow's, whose *Preliminary Discourse on Natural Philosophy* was published while Darwin was a student at Cambridge. All in all, Darwin felt a part of a wide community of scholars who were passionate about the natural world despite the university's contradictory policies.

In January 1831 Darwin passed the ordinary bachelor of arts examinations. He did surprisingly well, considering how little he had focused on academics: he ranked tenth of 178 candidates. For the next few months, he stayed on in Cambridge. Even though he had passed his examinations, he needed to complete the university residence requirement. He could still attend lectures and had nowhere else in particular to go.

His credentials qualified him for a career in the church, and he presumed that in time he would become a vicar with lots of free time for collecting beetles and interesting rocks. But before that, he wanted to travel.

Darwin had always loved reading about travel. In the years since he had read *Robinson Crusoe* under the dining room table, he had devoured accounts of the Lewis and Clark expedition across North America and Sir John Franklin's *Narrative of a Journey to the Shores of the Polar Sea* (1823), along with William Scoresby's *Account of the ("Polar") Arctic Regions* (1820) and many others.[39] He followed the exploits of the explorer Alexander von Humboldt, the German Romantic, with passionate interest. In 1829 the complete edition of Humboldt's *Personal Narrative* was published in English. Darwin recalled that the book "stirred up in me a burning zeal."[40] During his last year in Cambridge, he frequently visited the university hothouses to "gaze at the palm trees" and imagine following in Humboldt's footsteps.[41]

His first idea was to set out with a group of naturalist friends for the island of Tenerife in the Canary Islands. He seems to have imagined the trip as a cross between a college reading group and a full-fledged expedition. Writing to Henslow in July 1831, Darwin said, "I hope you continue to fan your Canary ardor: I read & reread Humboldt, do you the same, & I am sure nothing will prevent us seeing the Great Dragon tree."[42] Later in the same letter, he passed on greetings from his classmate Thomas Campbell Eyton, remarking that "his mind is in a fine tropical glow."[43]

Predictably, like many fantasies of tropical paradise the Tenerife plan fell through. It was the wrong time of year. There were no ships available for the passage. Henslow had a newborn baby at home. All of Darwin's friends were getting on with their postuniversity responsibilities. Darwin continued to hold out hope for the following year. In the meantime he went on a geological expedition around Wales as assistant to Cambridge professor Adam Sedgwick. He learned to use the measuring tools necessary for geological observations, in hopes that he might one day get to Tenerife.

When Henslow recommended Darwin for a naturalist position with a berth on HMS *Beagle*, Darwin jumped at the chance. He was invited to travel as a supernumerary, an unpaid and somewhat unofficial member of the crew. The captain offered him the chance to sling a hanging cot in the windowless poop cabin, in the small space above the chart table and beneath the skylight that provided illumination for the navigators. When Darwin packed his bags for the voyage, he took along a few scientific instruments (telescope, microscope, compass, clinometer), "a case of good strong pistols and an excellent rifle,"[44] and a large number of books, including a Bible, *Paradise Lost*, William Paley's *Natural Theology*, John Herschel's *Preliminary Discourse on Natural Philosophy*, Alexander von Humboldt's *Personal Narrative*, and dozens of blank notebooks. He knew that fissures were already appearing between naturalists and the others (poets and parsons, theologians and geologists) but never imagined that he would eventually be forced to choose sides. Instead he wanted to emulate Humboldt and bring everything together into a great Romantic synthesis.

To his great disappointment, the *Beagle* was not allowed to dock in Tenerife. There had been an outbreak of cholera in England. In response the Canary Islands imposed quarantine restrictions on English ships. Seasick and disheartened, Darwin rocked in his hammock, reading Humboldt and writing in his journal. "Oh misery, misery," he wrote.[45] The next morning, when the ship was becalmed, he caught a glimpse of the island, "& here I first experienced any enjoyment: the view was glorious. The peak of Teneriffe. – was seen amongst the clouds like another world."[46] Darwin had arrived in the tropics, crossing "that magic line to all Naturalists."[47] The voyage continued as a mix of misery and magic.

Darwin would spend the next five years on the voyage, dividing his time between bursts of observation and collection ashore and long hours of reading and writing aboard ship. In many ways his situation on the *Beagle* was more perfectly tailored to his learning style than any of his formal schooling. His cot was slung in the chart room along with the *Beagle's* four-hundred-volume library. He slept surrounded by books. When he showed his journals to his captain and shipmates, they were full of praise. Darwin began to hope that he could write a book that would allow him to establish himself as a professional naturalist.

CHAPTER 3

Nature's People: Scientific Amherst

Amherst, 1830–1836; Dickinson, to Age 6

EMILY DICKINSON was born on December 10, 1830, in a large brick house near Amherst College. Her life was shaped by this proximity. Founded in 1821, the college was not much older than the poet. Nonetheless, by the 1830s Amherst had become a college town. Her family's efforts to help establish the institution and keep it going affected her profoundly. To be a Dickinson of Amherst was to be identified with one of the newest and most ambitious colleges in the United States.

Just a few weeks after Dickinson's birth, Charles Darwin passed his final examinations at an older, more old-fashioned institution: Cambridge University in England. During Dickinson's first months of life, Darwin was still in Cambridge, wondering what to do next, walking the meadows with John Stevens Henslow, talking incessantly about natural history and botany. Dickinson was a year old when the *Beagle* finally left harbor in Plymouth, and she was almost 6 by the end of Darwin's voyage. From 1831 to 1836, while Darwin was aboard the *Beagle*, Dickinson was being raised in a community—and a family—that was wholly invested in a new model of education that integrated the natural sciences with the classics.

No one in Amherst would know much about the voyage of the *Beagle* until 1839, when Darwin published his account. But although the young Dickinson had no idea of Darwin's walks with Henslow or his subsequent adventures, her childhood community echoed with the same intense conversations about nature: natural history, natural philosophy, and natural theology.

Looking back on childhood, Dickinson wrote:

Several of Nature's People
I know, and they know me –
I feel for them a transport
Of cordiality –[1]

Since these lines occur within a poem about a snake ("A narrow Fellow in the Grass"), readers often interpret "Nature's People" as nonhuman beings—snakes and other animals and plants. It is certainly true that Dickinson knew the natural world around Amherst very well. At the same time, however, it is important to realize that Dickinson grew up in a town that was literally populated by "Nature's People": botanists and chemists, geologists and astronomers. Many of the people she admired most were those who studied nature and the natural world. In time she would be greatly influenced by her own teachers, particularly Caroline Dutch Hunt at Amherst Academy and Mary Lyon, the founder of what is now Mount Holyoke College.

Before she went to school, in the years from 1830 to 1836 Dickinson absorbed the attitudes toward nature and natural science shared by members of the Amherst community. In Dickinson's circle the Massachusetts scholars and writers Edward Hitchcock and Ralph Waldo Emerson were the most prominent of "Nature's People." Dickinson's approach to the natural world was shaped by these thinkers as much as by her experience of the wild corners of Amherst.

As a child, Dickinson was outdoors at every chance. Sometimes she helped in the garden, but more often she wandered the nearby fields and woods. For her the natural world was animated, lively, and meaningful. As she put it, "When much in the Woods as a little Girl, I was told that the Snake would bite me, that I might pick a poisonous flower, or Goblins kidnap me, but I went along and met no one but Angels."[2] In this account the forest is bright with magic. Though she was wary of fairy-tale dangers—goblins, snakes, and poison flowers—the only natural or supernatural beings she remembered were angels, bright messengers of goodness. Flowers and grasses; insects and birds; snakes and worms; boulders and stones; stars and planets; fossilized tracks and ephemeral sunset colors—all "Nature's People" were potential angels.

Dickinson expressed her "transport of cordiality" for "Nature's People" with more warmth than Darwin's recollection of his childhood "taste for natural history." He remembered his "passion for collecting" as naughty, a little more akin to greed than to love.[3] Darwin's family and his schoolmasters had tried, without success, to end his focus on nature and natural science. Dickinson's situation was different. Her family encouraged her to study natural history and natural philosophy, and these subjects were part of her formal studies.

One reason for the differences between Dickinson's and Darwin's school experiences was their families' very different relationships to educational institutions. The Darwin-Wedgwood traditions were somewhat antiestablishment. Though Charles and his older brother had been christened in the Church of England and attended schools and universities sponsored by the established church, they were outsiders at Cambridge, a conservative Anglican institution that was hesitant about tackling new approaches to science and technology. The Dickinsons, on the other hand, had poured their families' energy and money into establishing schools that reinforced their religious beliefs and promoted their scientific and technological interests. They were Amherst insiders, closely connected to a newer, nimbler, much more scientifically inclined institution.

When Darwin recalled his years in the abusive atmosphere of Samuel Butler's Shrewsbury School, he mainly remembered that most of his intellectual interests were excluded from his schooling. He had been a young naturalist at a time when schools had little room for the natural sciences. During Darwin's years at Cambridge,

John Herschel by William James Ward, after Henry William Pickersgill (1835). National Portrait Gallery, London.

however, a significant shift was underway. In 1830 John Herschel published his *Preliminary Discourse on Natural Philosophy*. Within the next decade, Herschel would lead a methodological revolution that would completely transform natural philosophy. This change in attitude toward the natural sciences would prompt the restructuring of the Cambridge University curriculum. The institutional illogic that had bedeviled the Darwin brothers when they tried to study clinical medicine and natural science would end shortly after they left the university.

By the time Dickinson started school in the late 1830s, many of Darwin's undergraduate distractions (collecting pebbles and beetles; experimenting with chemicals; observing plants and animals under a microscope) had been transformed into formal disciplines in many schools in Britain and the United States. Dickinson was able to study geology, botany, chemistry, and astronomy in school because the professors who had encouraged Darwin—including John Stevens Henslow, William Whewell, and John Herschel—had succeeded in making these topics legitimate academic subjects. Ironically, another reason Dickinson studied so much science was because these subjects were not yet completely established—there was still room for girls and women, amateurs and dilettantes.

It was a happy accident that Emily Dickinson was born at a moment, in a place, and to a family that particularly encouraged girls to study natural history and natural philosophy. In the 1830s and 1840s, it was not unusual for Massachusetts girls to study astronomy, botany, chemistry, geology, or zoology. More often, classical languages were reserved for upper-class boys, while girls and working-class boys were steered toward an "English" curriculum that combined natural sciences and mathematics with history, geography, and English grammar. Dickinson's parents wanted her to be "the best little girl in town" (as her father wrote to her when she was 7 years old).[4] Emily would be encouraged to study everything—both classical languages and natural sciences. At the time it was more unusual for Massachusetts girls to study Latin and Greek than natural science.

Dickinson's proximity to Amherst College would afford educational opportunities that were better than Darwin's in some ways. In 1814 her grandfather Samuel Fowler Dickinson was one of the founders and sponsors of Amherst Academy, an ambitious secondary school that attracted male and female students from all over New England. Emily Dickinson's father, Edward, attended Amherst Academy in its earliest years. Eventually, Amherst Academy expanded its mission and developed a men's college on the academy's foundations. Dickinson's father was in his third year at Yale University when the men's college opened. He came home to attend Amherst College for his junior year and then returned to Yale, graduating in 1823. Perhaps one reason Edward did not stay for a second year at Amherst College was because the academy and the college were not separate. Academy students attended lectures at the fledgling college, and the institutions often shared teachers and curricula.

Even after Edward had wrung all he could from the educational institutions of Amherst, he remained deeply loyal to his hometown schools. He had a strong commitment to public education. In 1832 he was elected to the Amherst General School Committee, which supervised Amherst's public primary schools. He served conscientiously. In 1841 he hosted the famed educator Horace Mann at a county school convention. He was even more dedicated to the private secondary and postsecondary institutions of Amherst. Along with other members of the clan, Edward Dickinson would work for and publicly promote Amherst Academy and Amherst College throughout his life.

If Emily Dickinson had been born in England, it is unlikely that she would have attended school. In Britain, wealthy families generally educated their daughters at home. Darwin's sisters were privately educated. Few records show what the Darwin girls studied, but in their milieu, it was common for girls to learn history, geography, arithmetic, and modern European languages. If they had studied botany or any other natural science, the subjects would have been framed as slightly decorative accomplishments, like watercolor painting or playing the pianoforte. Dickinson's education was far different. In Massachusetts, where common schools sponsored by local communities had widespread and robust support, most girls and boys attended town schools. The Dickinson family was committed to the schools of Amherst, and by age 5, Emily was enrolled.

Dickinson spent about five years at the common school, where girls and boys studied reading, writing, and arithmetic together. She was an eager and talented student who loved school. She could not wait to get to Amherst Academy, and she was extraordinarily lucky to have the chance to study there. By the time of her birth, the academy was ranked among the most prestigious secondary schools in Massachusetts. Although they were formally divided into a girl's branch and a boy's branch, boys and girls often studied together. In most years the girl's branch of the academy had fifty or sixty students. The boy's branch had been even larger once, but by the time Emily attended, it had started to shrink because male students could enroll in the college.

Amherst Academy was notable for offering girls the opportunity to attend lectures at Amherst College. When the celebrated author Harriet Martineau visited Amherst College in 1835, she went to

> the lecture-room where Professor Hitchcock was lecturing. In front of the lecturer was a large number of students, and on either hand as many as forty or fifty girls. These girls were from a neighboring school (*Amherst Academy*), and from the houses of the farmers and mechanics of the village. . . . We found that the admission of girls to such lectures as they could understand (this was on geology) was a practice of some years' standing, and that no evil had been found to result from it. It was a gladdening sight, testifying both to the simplicity of manners and the eagerness for education.[5]

In 1837, when the Italian writer Antonio Gallenga visited Amherst College, he was struck by the "long benches crammed with women, old and young."[6] It is not likely that Emily Dickinson spent much time in the college lecture halls in the mid-1830s when she was an elementary school student. When we imagine the poet as a small child, however, we can picture her eagerly preparing for the day when she will be able to study natural science and attend Amherst lectures.

The window of time when Massachusetts girls were encouraged to study science was relatively brief. After the Civil War, as science gained more professional and institutional weight in the United States, girls were increasingly steered away from the topic. A generation later, when Dickinson's niece Mattie was born in 1866, rigorous instruction in the sciences was primarily reserved for boys. Dickinson and her friends were amused by the change in gender association. About 1866, Samuel Bowles, the publisher of the *Springfield Republican*, read Dickinson's poem about "Nature's People," "A narrow Fellow in the Grass." He expressed his amazement that Emily Dickinson knew so much about the natural environment. In response, her sister-in-law jokingly pointed out the line in the middle of the poem where the speaker remembered having been "a boy and barefoot."[7] By then, just a few decades later, the idea that Emily had once been a boy was the best way to explain her interest in natural science.[8]

We must read carefully here. Emily Dickinson's personal history can be hard to pin down, and she did not write an autobiography. In letters and poems, she described childhood in a variety of ways, but we cannot assume that she was always describing her own experience; at times she was probably projecting imagined experiences. In a letter to Thomas Wentworth Higginson, she warned, "When I state myself, as the Representative of the Verse – it does not mean – me – but a supposed person."[9] At times Dickinson wrote from the persona of a schoolboy, while at other times she imagined herself as a drowning man, a bird, or a rose. In all her writing, letters as well as poems, Dickinson often turns to riddles.

The question of identity is central to "Nature and God – I neither knew," a poem that expresses a different perspective from the writings with which we began this chapter. In those instances she wrote that she had "met only Angels" in the woods and that she knew "Nature's People" very well. In this poem the narrator, Dickinson's "supposed person," denies knowing both nature and God, even as she admits that her "identity" is controlled by them:

Nature and God – I neither knew
Yet Both so well knew Me
They startled, like Executors
Of My identity –

Yet Neither told – that I could learn –
My Secret as secure
As Herschel's private interest
Or Mercury's Affair –[10]

At first, this poem from about 1865 seems to directly contradict "A narrow Fellow in the Grass," which was probably written about 1866. Yet although they make different claims regarding the speaker's knowledge of nature, both poems agree that nature, God, and Nature's People know the narrator very well. What they know is not revealed.

The poem is full of unknowns. Readers do not know what "Herschel's private interest" is, or even which *Herschel* the poem intends to invoke. Dickinson's contemporaries would have been familiar with the name

"Herschel," but since there are no contextual clues, no one could have determined which of the three famous Herschels the poem meant to allude to: William, Caroline, or John.

William Herschel was the British Royal Astronomer who discovered the planet Uranus in 1781. (Adding to the possible interpretations, the planet itself was often called "Herschel" in the nineteenth century.) William's sister Caroline Herschel was also a celebrated astronomer, best known for independently discovering eight comets. It is possible that this poem is referring to Caroline Herschel. There was much debate in the mid-nineteenth century about whether Caroline had observed the skies, calculated orbits and trajectories, and built telescopes because she was interested in astronomy or because she was an extraordinarily obedient sister. "Herschel's private interest" could refer to these questions about Caroline.

Finally, the poem's "Herschel" might be John Herschel, William's son and Caroline's nephew. John Herschel was certainly one of "Nature's People," one of the naturalists whose thinking shaped Dickinson's life and identity. In 1830, the year that Emily Dickinson was born, John Herschel completed his *Preliminary Discourse on Natural Philosophy*. The book would change the way the natural sciences were practiced and taught in the Anglo-American world. It would also profoundly affect the lives of Emily Dickinson's neighbor Edward Hitchcock and her favorite Massachusetts author, Ralph Waldo Emerson, just as it had changed the life of the young Charles Darwin.

Herschel's *Preliminary Discourse on Natural Philosophy* was a sort of how-to book, a methodological discourse that carefully explained how to think like a nineteenth-century natural philosopher. It advanced a way of understanding the world. Herschel invited readers to assume a scientific mindset and laid out methodological guidelines for approaching the natural world as a natural philosopher. The first chapter of Herschel's remarkable book described humans as speculative beings, constantly wondering about nature, God, and their own identities. As Herschel explained, when a person (for Herschel, a "man") considers "his sentient intelligent self,"[11] "a world within him is thus opened to his intellectual view, abounding with phenomena and relations of the highest immediate interest. But while he cannot help perceiving that the insight he is enabled to obtain into this sphere of thought and feeling is in reality the source of all his power, the very fountain of his predominance over external nature, he yet feels himself capable of entering only very imperfectly into these recesses of his own bosom, and analysing the operations of his mind, – in this, as in all other things, in short, 'a being darkly wise.'"[12] Herschel capped his discussion of selfhood with an allusion to Alexander Pope's eighteenth-century poem "Essay on Man," but his sensibility was entirely new. The era of the modern scientist was about to begin. When Dickinson mentioned "Herschel's private interest," she might have been referring to the natural philosopher's awareness that passionate curiosity was both a source of power and a profound mystery.

In early 1831 Herschel's *Discourse* was published as the first volume of a cyclopedia, a series of books aimed at a wide popular audience. Charles Darwin read Herschel's book with great enthusiasm during his last months in Cambridge. He wrote to his naturalist cousin William Darwin Fox, instructing him to "read it directly."[13] In his autobiography, Charles Darwin recalled that Herschel's *Discourse* (along with Humboldt's *Personal Narrative*) "stirred up in me a burning zeal to add even the most humble contribution to the noble structure of Natural Science. No one or a dozen other books influenced me nearly so much as these two."[14] While Humboldt's *Personal Narrative* made Darwin dream of sailing toward the peak of Tenerife, Herschel's *Discourse* offered a thrilling invitation to approach the natural world inductively, first observing carefully, then developing theories based on these empirical observations, and later testing the theories in other circumstances, in hope of uncovering simple, universal laws. As Herschel explained it, the inductive method was a thrilling vocation. Herschel described natural philosophers as uniquely sensitive to beauty, people who "walk in the midst of wonders."[15]

While Darwin was aboard the *Beagle*, Herschel's ideas reverberated through Massachusetts. Ralph Waldo Emerson and Edward Hitchcock were inspired. After explaining how much a sage could learn from a soap bubble, Herschel declared, "To the natural philosopher there is

Ralph Waldo Emerson by Henry Bryan Hall. Library Company of Philadelphia.

no natural object unimportant or trifling. From the least of nature's works he may learn the greatest lessons. The fall of an apple to the ground may raise his thoughts to the laws which govern the revolutions of the planets in their orbits; or the situation of a pebble may afford him evidence of the state of the globe he inhabits, myriads of ages ago, before his species became its denizens."[16] Herschel's insistence on the significance of the smallest natural objects made natural philosophy accessible to everyone who had the right attitude. All that was necessary was thinking like a natural philosopher. When Emerson described himself as a "transparent eyeball" in his essay "Nature," he evoked Herschel's shimmering soap bubble. Hitchcock, who was embarking on a geological survey of Massachusetts, drew inspiration from Herschel's claim for the great significance of "the situation of a pebble." Herschel's book was compelling because it linked their interests and endeavors to the great Isaac Newton's.[17]

For the young Emerson, Herschel's book was transformative. He read it at age 28, shortly after the death of his first wife. In his grief he lost the convictions necessary to sustain him as a Unitarian minister and began to imagine another path. Although his Harvard education had not included much formal study of natural science, he began to imagine himself as a natural philosopher rather than a theologian. According to Robert D. Richardson, "No single volume was more important than Herschel's *Discourse* in shaping Emerson's new interest in science and nature."[18] In his *Journal*, Emerson asked, "What is there in 'Paradise Lost' to elevate and astonish like Herschel or Somerville?"[19] On Christmas Day, 1832, Emerson sailed for Italy. By the summer of 1833, he was in Paris. As he wandered the Museum of Natural History, marveling at the carefully curated specimens, a new possibility beckoned. In his *Journal* he recorded, "An occult relation occurs between the very scorpions & man. I am moved by strange sympathies. I say continually, 'I will be a naturalist.'"[20]

From Paris, Emerson traveled to London. His enthusiasm for the "occult relation" between human beings and nature would persist throughout his career, but his experiences in Britain discouraged him from becoming a professional naturalist. When he wrote about the trip in *English Traits*, he reminded readers, "Sir John Herschel said, 'London was the centre of the terrene globe.'"[21] He was thrilled to visit the cosmopolitan center of the British Empire, but perhaps to his surprise, Emerson found himself in Coleridge's London rather than Herschel's. The rift between the more idealistic natural philosophers and the more concrete naturalists of the Royal Society and the British Association for the Advancement of Science had begun to open. Emerson was introduced to the dreamers: Landor, Wordsworth, Coleridge, and Carlyle. There is no record of Emerson meeting Herschel, Whewell, or Somerville.

When Emerson returned to Massachusetts, he became a lecturer and essayist. He often wrote about natural history, but he never became a full-fledged naturalist. He was not really a member of the scientific community. Instead, Emerson imagined himself as a "scholar," a person who was able to synthesize across many fields. Lawrence Buell explains that "Emerson reached maturity just as intellectual labor had begun to specialize."[22] He found disciplinary specialization limiting, and according to Buell, he "used his favorite, self-identifying term, scholar, in a

Edward Hitchcock by H. B. Hall and Sons (ca. 1859). Amherst College Archives and Special Collections.

disruptively antiprofessional sense: to commend independent-minded thinking."[23] Emerson's independent mind was of the particular sort only available to those of independent means. The substantial fortune he inherited upon his wife's death allowed him to maintain his idealism. Because of his inheritance, Emerson did not need to fit himself into any professional constraints.

In contrast, Edward Hitchcock was not independently wealthy. He depended on the institution that employed him, Amherst College. The differences between the two men help highlight the differences between coastal Massachusetts to the east and the Connecticut River valley to the west, between the city of Boston and the village of Amherst, and between Harvard College and Amherst College.

Emerson was a graduate of Harvard College and Harvard Divinity School, the son of a minister who had also graduated from Harvard. Because his father died when Emerson was a child, he endured periods of poverty while growing up, but he was always culturally well connected. Like his brothers, he was able to rely on friends of his father to pay for Harvard and provide him with introductions and references.

He married a wealthy young woman and eventually inherited enough money to be able to stop working.

Hitchcock's story was different. He did not go to Harvard or to any college or university. Instead, he pieced together an education at a few different (short-lived) academies and seminaries aimed at *mechanics*, or working-class men. Although he eventually received honorary degrees from Yale, Harvard, and Middlebury, he was largely self-taught. This was not such a disadvantage for the scientifically inclined; much the same may be said of Charles Darwin despite his wealth and his time at boarding school, Edinburgh, and Cambridge. At Harvard, Emerson had little exposure to science or the developing scientific method. In rural Massachusetts, Hitchcock trained himself to think scientifically.

By 1829 Edward Hitchcock had been appointed professor of chemistry and natural history at the newly established Amherst College. That year, he fell ill. His complaint started with indigestion and progressed toward depression: "Unreasonable fears, despondency of mind, and dismal forebodings of evil." Hitchcock could not afford to embark on a grand tour of Europe to fight his despair. His wife and collaborator, Orra Hitchcock, suggested he spend less time with his books and get more exercise. In response Hitchcock started a survey of the natural history of Massachusetts and began traveling around the state on foot and on horseback. Two books came out of his travels: *Dyspepsy Forestalled and Resisted* (1830), which recommended avoiding meat, alcohol, and caffeine and spending as much time outdoors as possible, and his first *Report on the Geology, Mineralogy, Botany, and Zoology of Massachusetts* (1833). Funded by the state government and illustrated by Orra Hitchcock, *Geology of Massachusetts* would be revised and reissued through the following decade. It would help establish Hitchcock's international reputation.

Hitchcock's book was granular and precise. In the section on "Useful Rocks and Minerals in the State," he included chapters on "Granite and Sienite," "Gneiss," "Greenstone, Hornblende Slate and Porphyry," "Quartz Rock," and so on. The work was a triumph of field research and precise exposition. At times his concrete and specific details made the writing compelling, even beautiful. Describing a fossil fish (an *ichthyolite*), he wrote, "A thin layer of carbonaceous matter usually marks out the spot where the fish lay; except the head, whose outlines are rendered visible only by irregular ridges and furrows. In some cases, however, satin spar forms a thin layer over the carbonaceous matter, and being of a light gray color, it gives to the specimens an aspect extremely like that of a fish just taken from the water."[24] Hitchcock's fossil fish, with its thin layer of satin spar, still glistens from the page.

Hitchcock's first edition of *Geology of Massachusetts* was published in 1833. When Emerson returned to Massachusetts and gave his first lectures on natural history, Hitchcock was the local authority, while Emerson was the amateur. Emerson, who had once longed to be a naturalist, was not immune to the beauties of Hitchcock's geology. Yet Emerson had a different purpose. He probably would have described Hitchcock as a "sensual man" because, despite his asceticism and his religious belief, Hitchcock was more interested in the material than the ideal. Emerson loved both but he leaned toward abstraction. As Emerson explained in *Nature*, "The sensual man conforms thoughts to things; the poet conforms things to his thoughts. The one esteems nature as rooted and fast; the other, as fluid, and impresses his being thereon."[25] Emerson identified with poets like Coleridge. He confidently tried to make nature conform to his thought. Hitchcock, in contrast, was more aligned with Herschel and the inductive scientific method. In this respect Emerson and Hitchcock were at two opposite poles of the debate about natural philosophy, though neither of them was wholly persuaded by the increasingly acrimonious split.

"I fully believe in both, the poetry and the dissection," Emerson announced in "The Naturalist," a lecture presented at the Boston Society of Natural History in May 1834.[26] In time Emerson relinquished his own ambition to become a naturalist, but he still hoped to become both a poet and a philosopher of nature. He would never countenance the split between poetry and scientific method.

In *Nature*, published in 1836, Emerson declared his allegiance to imagination. He argued that this was not an anti-materialist or anti-scientific stance: "The Imagination may be defined to be, the use which

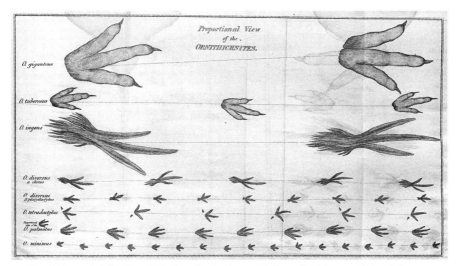

Fossil Footprints by Orra Hitchcock. Amherst College Archives and Special Collections.

Reason makes of the material world." Eight years later, in "The Poet" (1844), he celebrated poetry as "the science of the real" and "true science," arguing that "the poet alone knows astronomy, chemistry, vegetation and animation, for he does not stop at these facts, but employs them as signs."[27]

Although Hitchcock would probably have hesitated before identifying as a "scholar" in the Emersonian sense, he, too, refused to stop at the facts. As he explained in *Geology of Massachusetts*, "I was in doubt for a time whether it would be advisable to add any remarks upon the theory of rock formations, or the explanation of particular phenomena. But such an addition seemed wanting to complete my account of the rocks; and I came to the conclusion to attempt an explanation of all the most important phenomena."[28] He consulted "the most recent, and most able writers upon geological philosophy" and then offered the geological theories he deemed "most satisfactory to my own mind."[29] Emerson and Hitchcock were not so far apart.

Edward Hitchcock made the most exciting geological discovery of his career on the banks of the Connecticut River in 1835. He found hundreds of giant, three-toed footprints embedded in the sandstone. The largest were seventeen inches long. The fossil tracks—locally known as "turkey tracks"—were astonishing because they offered a record of movement from a million years past. Fossilized bones would have been interesting—Hitchcock would search for the rest of his life for bones to match the footprints. But the footprints were more evocative in some ways since they showed how long-extinct creatures had behaved when alive. They filled Hitchcock with so much wonder that he compared the feeling to poetic inspiration. As he explained, when "there came up in my imagination, the bird that formed the enormous *Ornithichnites Giganteous* [fossil tracks], perhaps fifteen or twenty feet high, . . . my long dormant muse was aroused to action."[30]

In January 1836 Hitchcock published a scientific article on the fossil tracks in the *American Journal of Science and Arts*. His wife, Orra, provided the illustrations. The article elicited passionate reactions, including an angry review in the pages of the *Knickerbocker* magazine. Some readers were offended by the idea that such large creatures had existed in Massachusetts. In the aftermath of that critique, Hitchcock submitted "The Sandstone Bird," a relatively long (113-line) poem that was published anonymously (under the sobriquet *Poetaster*) in the *Knickerbocker* in 1836. The poem begins with a frustrated geologist invoking a sorceress to help him see the creature that made the footprints:

> O for a sorceress nigh, to call thee up
> From thy deep sandstone-grave, as erst of old
> She broke the prophet's slumbers! But her arts
> She may not practice in this age of light.
>
> ENTER SORCERESS.
>
> Let the light of Science shine!
> I will show that power is mine.
> Skeptic, cease my art to mock
> Bird of sandstone era, awake!
> From thy deep, dark prison break![31]

The marvelous giant bird rises from the rocky riverbank, only to sneer at the puny geologist and the horrible weather of nineteenth-century Massachusetts. The bird cries:

> These creatures here seem discontented, sad;
> They hate each other and they hate the world:
> O who would live in such a dismal spot?
> I freeze, I starve, I die! With joy I sink
> To my sweet slumbers with the noble dead.[32]

When the sandstone bird describes a nineteenth-century world peopled by creatures that hate each other and the world, Hitchcock may have had in mind the hostile reception to his article about the fossil tracks. It is also possible that he was simply responding to the misery of a Massachusetts winter—the poem was published in December 1836. But although the poem describes a "vex'd geologist," it also conjures forth a lush and verdant geologic past that is enchanted and mysterious. In this context, Hitchcock's "geologic doubts" seem closer to mysticism than any of Emerson's naturalist musings.

As a young girl, Dickinson revered the Massachusetts sages. Hitchcock was a neighbor and a family friend, one of the brightest stars in the Amherst firmament. She would never meet Emerson face-to-face, but by the time she was 11 years old, she had become such a devoted admirer of his writing that she burst into "a passionate fit of crying" when a teacher at Amherst Academy suggested that the European Romantics were better and urged her to read Byron instead of the Massachusetts writers Emerson and Lowell.[33] It is noteworthy that although Dickinson's letters mention her Massachusetts influences repeatedly, neither Emerson nor Hitchcock is named in any of her poems. Instead, in "Nature and God – I neither knew," Dickinson made her cryptic reference to "Herschel's private interest."

We know little about Emily Dickinson's childhood reading. She read voraciously, and the fact that she mentioned Herschel in a poem makes it likely that she read his *Discourse* while she was growing up. There were certainly many copies of the book in Amherst. If so, she may have found Herschel profoundly inspiring, as Emerson and Darwin did. If not, she

The Dickinson Homestead by Carol M. Highsmith. Wikimedia Commons.

absorbed its lessons from the atmosphere around her. Herschel's book described the study of natural science as an "inward spring" of delight.[34] As he put it, "A mind which has once imbibed a taste for scientific inquiry . . . has within itself an inexhaustible source of pure and exciting contemplations."[35] This resonated with Dickinson's own experience. From her earliest years, she looked to the natural world as a source of delightful mysteries. As Dickinson would put it:

> Wonder – is not precisely knowing
> And not precisely knowing not –
> A beautiful but bleak condition
> He has not lived who has not felt –[36]

Perhaps Dickinson acquired her sense of the bleak and thrilling beauty of the unknowable natural world directly from reading Herschel. Perhaps it came more from Hitchcock or another one of "Nature's People" whom she knew personally. Perhaps she learned to feel wonder from bright angels she encountered in her enchanted childhood forest. No matter how it happened, Emily Dickinson's life began "in the midst of wonders."[37]

In many respects Dickinson's childhood may have been even more Eden-like than Darwin's. Like him, she alluded to Milton's *Paradise Lost* when she remembered her childhood home. The Homestead, as her birthplace was known, was very similar to the Mount, Charles Darwin's childhood home. Like the Mount, the Homestead was a well-proportioned brick house surrounded by a large, magical garden. It was grander than the neighboring houses and sat on fourteen acres of land. But here, too, there was trouble in paradise.

Dickinson's mother and father were healthy and active presences in her life, and school was congenial. The Dickinson family troubles echoed *Paradise Lost* in another way, however: like Adam and Eve, the Dickinsons were driven away from their family Eden. Perhaps Emily Dickinson was looking back on her early years when she wrote:

> Eden is that old fashioned House
> We dwell in every day
> Without suspecting our abode
> Until we drive away[38]

Emily's grandfather, Samuel Fowler Dickinson, had built the Dickinson Homestead in 1813, when his law practice was thriving and he was a leading citizen of Amherst. The family's support for Amherst College, however, cost them a great deal. In 1825 Samuel Fowler Dickinson mortgaged the Dickinson Homestead along with his other properties in an effort to prop up the new college, which was struggling with high costs and low enrollments. In 1829, shortly before Emily was born, her father bought half the Homestead to try to keep her grandfather—and Amherst College—afloat. The college survived, but Samuel Fowler Dickinson was forced into bankruptcy. Eventually, Edward Dickinson lost his share of the Homestead too. When Emily Dickinson was born, the Homestead was still awkwardly divided in two. Her family lived in half the house, and the Mack family, the new owners of the Homestead, who also owned the hat factory down the street, occupied the other half.

Even after this financial ruin, Samuel Fowler Dickinson was still personally respected in Amherst, since it was common knowledge that he had "sacrificed his property, his time and his professional opportunities" for the college.[39] Edward Hitchcock, who eventually became president of Amherst College, described Samuel as "one of the most industrious and persevering men that I ever saw."[40] Dickinson's grandparents left Amherst, moving to Ohio, where Samuel was appointed steward of the Lane Theological Seminary. He died in Cincinnati in 1838, when his granddaughter Emily was 7 years old.

Charles Darwin did not learn that he was fabulously wealthy until he reached adulthood. He had no idea that his childhood paradise was supported by his father's mortgages on other properties. In contrast, Emily Dickinson was all too conscious of the mortgage that had cost the family its Homestead. When she was almost 10, her family would move to a nearby house on West Street (now known as North Pleasant Street): a house of their own, all to themselves. The new place was wood instead of brick and not as well situated in the town, but it was still a large, generously proportioned house within a few minutes' walk of everything. The young Emily Dickinson would spend her teenage years looking out toward a graveyard instead of the family orchard.

It would be a mistake to imagine that Emily Dickinson was poor or that her family was socially disgraced. Her grandparents' financial sacrifices had assured their family of a leading position in the town of Amherst, and her parents capitalized on this position. Her father was appointed treasurer of Amherst College in 1835. He held the post until 1873, shortly before his death. His son, Emily's older brother, Austin, succeeded him as treasurer of the college and held the position until he died in 1895. Amherst College trusted the Dickinsons. The family managed the college's finances for most of its first seventy-five years.

When she was a young girl, Emily knew that her family valued school above all else. She felt the same way. As she spelled her first words, added her first sums, and plucked her first flowers, she was eagerly preparing to study everything that Amherst had to offer. She would come to love the British Romantic poets, but she never stopped idolizing Emerson. As Dickinson prepared to enter secondary school at Amherst Academy, she was ready to become a "naturalist" like Emerson. She wanted both, "the poetry and the dissection."[41]

CHAPTER 4

Juggler, Geologist, Dark Horse

Aboard the Beagle, *1832–1836; Darwin, Age 23–27*

CHARLES DARWIN hated sailing. He was always seasick, and he never managed to adjust. For him, time aboard ship was "existence obliterated from the page of life." "I hate every wave of the ocean," he wrote in a letter to his cousin as his voyage drew to a close.[1] During the nearly five-year voyage of the *Beagle,* he would spend more than three years on land, avoiding the cramped poop cabin and the monotonous ocean to explore the astonishingly varied geology, botany, and zoology of different places.

Darwin had never expected to become a geologist. When he left medical school at Edinburgh University, he disliked geology almost as much as he hated medicine. He was determined to avoid both for the rest of his life. Once he got to Cambridge, however, his mentor John Stevens Henslow started encouraging him to attend geology lectures. Darwin liked Adam Sedgwick, the thoughtful professor of geology whom he had met at Henslow's scientific soirées, but even so he was quite slow to take up geology again. After he finished his undergraduate degree, Darwin started learning the rudiments of the subject on his own. In the summer of 1831, he spent a few weeks assisting Sedgwick in a geological survey of Wales. On this brief trip, he learned how to use geological instruments in the field. Perhaps more important, he witnessed Sedgwick's careful, methodical observation—and his refusal to leap to speculative conclusions. When Darwin packed for his voyage on the *Beagle*, he made sure to bring his new clinometer and other geological tools. He certainly did not think of his voyage as a geological expedition, but he knew that a naturalist might be expected to make some geological observations.

At that point Darwin had not read Charles Lyell's *Principles of Geology*, which was published the same year. Adam Sedgwick disliked the book so much that he used his 1831 presidential address to the Geological Society to present a thorough critique of Lyell's approach to paleontology. Henslow thought Lyell's book was worth reading, but his recommendation came with a substantial caveat. Darwin recalled that Henslow "advised me to get and study the first volume of the *Principles*, which had then just been published, but on no account to accept the views therein advocated."[2]

When the captain of the *Beagle*, Robert FitzRoy, gave Darwin a copy of Lyell's book, Darwin kept his reservations to himself. At the same time, FitzRoy also presented him with a sturdy grammar book, a lighthearted reminder that he expected Darwin to write clear accounts of the voyage. Charles Lyell himself had asked FitzRoy to report back on the geological findings from the *Beagle* expedition. FitzRoy entrusted the task to the ship's young natural philosopher, whom he soon nicknamed Philos.

Robert FitzRoy was four years older than Charles Darwin. He was a rising naval officer whose reputation for intelligence and scientific acuity had earned him his command. The main purpose of the voyage was to map the coastline of South America, but FitzRoy had wide interests in natural history and natural philosophy, and he believed the voyage could yield more than naval charts.

In some ways Darwin was a poor messmate. He was much too tall for life belowdecks, and he was terribly prone to seasickness. But he was charming, sturdy, and quiet. When the seas were calm enough, Darwin and FitzRoy ate breakfast together, then spent a few hours working separately before sharing another large meal. As an unpaid supernumerary, Darwin was outside the naval hierarchy. He was the only person aboard who ate his meals with the captain and addressed him informally, speaking to him as a friend rather than a subordinate. But although they were friendly with each other, their familiarity had limits. Each spent most of the day working alone. Darwin noticed FitzRoy's meticulous recordkeeping and his rigorous daily discipline and tried to emulate him. Darwin had found another mentor.[3]

As the voyage progressed, it became clear that Darwin was a multitalented naturalist eagerly interested in every aspect of the natural world—animal, vegetable, and mineral. With FitzRoy as a role model, he worked as hard at writing up his observations and experiences. In addition to his research journals, Darwin wrote long letters to his family—his sisters and his cousin and dear college friend William Darwin Fox—and to his mentor, John Stevens Henslow. But as it turned out, his research journals were addressed as much to Charles Lyell, whom he had never met, as they were to his Cambridge circle. Before the voyage, Darwin had seen geology as the least interesting part of natural history. To his great surprise, he became increasingly convinced that Lyell's version of geology pulled all his interests together. At the same time, he realized that following Lyell would require a significant change of direction.

Darwin planned to become a priest after the voyage. When he started to read Lyell, he understood why Henslow, a devout Anglican priest, had cautioned him about the book. Lyell's ideas challenged Cambridge orthodoxies. Darwin was not quite ready to abandon his Christ's College touchstones. For the voyage he had packed both Milton's *Paradise Lost* and Paley's *Natural Theology*. Milton's version of the Creation story was so compelling that it had replaced the book of Genesis for many English speakers. Paley's theological argument—that the intricacies of natural organisms showed the intelligence of the God who had designed the natural world—was equally influential. Lyell's book offered a completely different way to imagine the earth's origins.

At the time, most geologists tended to imagine that the earth had existed for thousands (not millions) of years and that during these years a few dramatic, catastrophic events had changed the landscape. In fractious Edinburgh, the great debate was between the Plutonians and the Neptunians. The Plutonians believed that massive earthquakes and sudden eruptions of molten lava were the primary geological forces. The Neptunians, on the other hand, thought that the rushing water of violent floods had been the main factor in shaping the landscape. Others argued for a cataclysmic barrage of earthquakes, floods, and volcanic eruptions in quick succession. At the time, most British geologists had doubts about the literal truth of the biblical account of Creation, but even so, many of their competing theories attempted to fit the development of the earth's high peaks and deep fjords within the time span of the five or six thousand years since Adam and Eve. All of Darwin's teachers, from Jameson in Edinburgh (whose lectures he had despised) to Sedgwick and Henslow (whom he wholeheartedly admired), suspected that geological timescales were somewhat longer than scriptural accounts stated. Nonetheless, in the early 1830s they still tended to subscribe to cataclysmic geologies that worked within thousand-year time frames.

In *Principles of Geology*, Lyell argued that the features of the earth had been shaped by ordinary events rather than sudden, violent catastrophes. He described the surface of the earth as relatively smooth and constant on a planetary scale, arguing that gradual increases in elevation in one region were generally balanced by gradual subsiding elsewhere. Lyell explained that it was possible to account for the great variation in landscape features and formations without supposing extraordinary events simply by expanding the scale of time. Rather than thousands of years, Lyell imagined millions and even billions of years. Given enough time, a trickle of water could cut through stone just as forcefully as a violent flood.

Lyell was not the first person to argue that the changes in the earth's structure had happened in the ordinary course of events, without any miracles. James Hutton had made a persuasive case against the cataclysmic view as early as 1795. There were other gradualist, uniformitarian geologists. However, Lyell's book was the boldest and clearest critique of prevailing cataclysmic views. Lyell went further than Hutton or any of the others, not only making a compelling case that the earth had

Charles Lyell by W. H. Mote (ca. 1840). Wellcome Collection.

developed over eons, but also arguing with great force against the idea of geological progress.

Although some geologists were starting to adjust to longer geological time frames, it was hard for them to let go of the progress narrative. They imagined a careful and benevolent deity slowly building the earth, gradually making it more and more perfect (for human beings, presumably). Lyell challenged this slow creationism. He did not believe that the planetary record showed any evidence that the earth had developed or improved throughout successive ages. Instead he thought that the changes were random, due to hundreds of small, chance events governed by natural laws—the laws of physics and chemistry. By implication, Lyell removed God from nature.

Despite his bold opinions, Lyell did not seek controversy. He carefully distinguished between geology and cosmogeny (the study of the origin of the cosmos). His book started with a simple definition of geology that quietly insisted on geology's separation from theological speculation: "GEOLOGY is the science which investigates the successive changes that have taken place in the organic and inorganic kingdoms of nature; it inquires into the causes of these changes, and the influence which they have exerted in modifying the surface and external structure of our planet."[4] By modestly limiting geology to "the surface and external structure" of the earth, Lyell attempted to avoid theological and cosmological debates.

As it turned out, Lyell's book was a clarion call for Darwin. Perhaps he felt a stir of recognition when Lyell announced, "A geologist should be well versed in chemistry, natural philosophy, mineralogy, zoology, comparative anatomy, botany; in short, in every science relating to organic and inorganic nature."[5] Darwin had studied all the subjects on Lyell's list. From his childhood laboratory experiments with Ras to his research on marine life with Grant, his medical training, and the lectures at Cambridge on mineralogy and botany, he had unintentionally assembled the precise set of skills that Lyell called for. It turned out that his seemingly haphazard education provided him with the perfect knowledge base for geology. Furthermore, Lyell's ideas intrigued him. Though Lyell dodged a direct debate with Paley and Milton, he nonetheless painted a picture of the vastness of time and space that was incredibly exciting to the young man confined in a small cabin with his somewhat worn volumes of Paley and Milton.

As Darwin rocked in his suspended cot, riding waves of queasiness, the pages of Lyell's book swam into focus and out again. He read slowly, closing his eyes to let himself imagine the gradual planetary changes that Lyell described. When he went ashore at Saint Jago (São Tiago), Cape Verde Islands, the first place the *Beagle* touched land, he had not yet finished the book. Decades afterward he remembered that the unfamiliar landscape had astonished him by embodying the principles Lyell explained. His *Autobiography* recalled, "I am proud to remember that the first place, namely St. Jago, in the Cape de Verde archipelago, in which I geologized, convinced me of the infinite superiority of Lyell's views over those advocated in any other work known to me."[6] Darwin may have compressed the time line in his memory; geologists P. N. Pearson and C. J. Nicholas point out that the journals and letters written in the moment show more Humboldtian delight than Lyellian method.[7]

"I expected a good deal," he wrote in his diary, "for I had read Humboldts descriptions."[8]

Alexander von Humboldt was a German naturalist, polymath, and world traveler. His ideas about the natural world inspired Goethe and transformed European Romanticism. Humboldt's influence extended far beyond Germany. He published thirty-six books, working toward the goal of representing "the whole material world, – all that is known to us of the phenomena of heavenly space and terrestrial life, from the nebulae of stars to the geographical distribution of mosses on granite rocks."[9] The British Romantics—Wordsworth, Coleridge, and their circle—admired Humboldt's compelling account of the interconnected cosmos. Americans from Thomas Jefferson to Ralph Waldo Emerson adored Humboldt because he valued the natural world and the landscape more than human history. Emerson wrote:

> Humboldt was one of those wonders of the world, like Aristotle, like Julius Caesar, like the Admirable Crichton, who appear from time to time, as if to show us the possibilities of the human mind, the force and the range of the faculties, – a universal man, not only possessed of great particular talents, but they were symmetrical, his parts were well put together. As we know, a man's natural powers are often a sort of committee that slowly, one at a time, give their attention and action; but Humboldt's were all united, one electric chain, so that a university, a whole French Academy, travelled in his shoes.[10]

Of Humboldt's many books, the one that influenced Darwin the most was his *Personal Narrative of Travels to the Equinoctial Regions of the New Continent during the Years 1799–1804*. Darwin was fascinated by Humboldt's wide-ranging account. He wanted to see the places Humboldt had described, and he hoped to catch Humboldt's sense of how everything—the earth, the animals, and the plants—in each place was interconnected.

Saint Jago was the first tropical island Darwin ever visited. The first day thrilled him. He was amazed to see flourishing groves of oranges and coconuts growing outdoors. He tasted his first banana. In his diary he exulted in the new experience of "treading on Volcanic rocks, hearing the notes of unknown birds, & seeing new insects fluttering about still newer flowers." That evening he wrote, "It has been for me a glorious day, like giving to a blind man eyes."[11]

At first Darwin's new eyes were more focused on the tropical plants and animals than the volcanic rock, but since much of the island was bare rock, his thoughts turned to geology. In a letter to Henslow, he explained, "St Jago is singularly barren & produces few plants or insects. – so that my hammer was my usual companion; in its company most delightful hours I spent."[12] In a similar vein, he wrote to his father, "Geologising in a Volcanic country is most delightful, besides the interest attached to itself it leads you into the most beautiful & retired spots."[13]

Over the next three weeks, Darwin's raptures barely diminished. Although his time in Saint Jago was relatively brief, for him it was an epoch. On his departure he summed up his experience with another reference to Humboldt: "Every object was new & full of uncommon interest & as Humboldt remarks the vividness of an impression gives it the effect of duration."[14] The record shows that at first Darwin saw Saint Jago through romantic Humboldtian eyes. His emotions were intense. He wrote to William Darwin Fox, "My mind has been since leaving England in a perfect *hurricane* of delight & astonishment."[15] In his diary he recorded experiencing "a chaos of delight" that left him "fit only to read Humboldt."[16] This storm of feeling was Humboldtian.

After he left Cape Verde and his emotional hurricane began to subside, Darwin carefully reviewed his detailed notes. He was still in a high romantic mood, but at this later stage of processing, which William Wordsworth might have described as "emotion recollected in tranquility,"[17] he finished reading Lyell's book and began to see Saint Jago through the eyes of Charles Lyell rather than Alexander von Humboldt. Where the Romantics tended to emphasize the correlations between themselves and the natural world, Lyell looked for a different kind of correlation. He was interested in the relationship of very small, ordinary things to the large and the grand. Lyell's approach to geology was to try to explain very large geological features by tracing out a series of tiny events that could have shaped them.

Many years later, in his *Autobiography*, Darwin would remark, "I have always thought that the great merit of *The Principles* was that it altered the whole tone of one's mind, and therefore that, when seeing a thing never seen by Lyell, one yet saw it partially through his eyes."[18] When Darwin reached Brazil, he wrote to Henslow, describing his sojourn on Saint Jago: "The geology was preeminently interesting & I believe quite new: there are some facts on a large scale of upraised coast (which is an excellent epoch for all the Volcanic rocks to [be] dated from) that would interest Mr. Lyell."[19]

In South America, Darwin's geological hammer would turn out to be among the most valued of his scientific tools. He would find majestic and exciting fossils and eventually experience a great earthquake and a volcanic eruption. But the most astonishing thing that happened to Charles Darwin was learning to see the world through Lyell's eyes. By 1835 he would write to his cousin Fox, "I am become a zealous disciple of Mr Lyells views, as known in his admirable book. – Geologizing in S. America, I am tempted to carry parts to a greater extent, even than he does. Geology is a capital science to begin, as it requires nothing but a little reading, thinking & hammering."[20]

Reading, thinking, and hammering. Was this enough to build a life on? Starting in Saint Jago, Cape Verde, Darwin had wondered if he could write a book. In his *Autobiography*, his memory glowed with confidence:

> It then first dawned on me that I might perhaps write a book on the geology of the various countries visited, and this made me thrill with delight. That was a memorable hour to me, and how distinctly I can call to mind the low cliff of lava beneath which I rested, with the sun glaring hot, a few strange desert plants growing near, and with living corals in the tidal pools at my feet. Later in the voyage Fitz-Roy asked to read some of my Journal, and declared it would be worth publishing; so here was a second book in prospect![21]

In the moment, however, Darwin was much less certain about his prospects.

On the coast of South America in the wintery months of May and June 1833, Charles Darwin stayed ashore while the captain and crew of the *Beagle* charted the east coast of South America. A poor seaman, he could not help with the charts. Instead he spent the winter on the Banda Oriental, the coastal territory now part of Uruguay. During this time Darwin worked on his journal and wrote long letters to his family. At the end of one letter to his sister back in Shropshire, he scrawled a somewhat embarrassing postscript in red ink. He felt "like the Midshipman in *Persuasion* who never wrote home, excepting when he wanted to beg," but he needed more books (he listed eleven titles), a tape measure, four pairs of strong walking shoes, and an extra lens for his microscope. Toward the end of the list, he asked for some matches: "Also another box of Promethians (I blush like this red ink, when I ask for it)."[22] There was nothing inherently embarrassing about Promethean matches. Perhaps Darwin was simply ashamed of the length of his list. On the other hand, he might have felt some uneasiness about why he wanted them. He planned to use the matches to entertain a credulous audience.

On the grassy coast of Uruguay in the 1830s, the laws of hospitality required householders to welcome every stranger. At times Darwin felt he was taking advantage of the local ways of the Banda Oriental. In *The Voyage of the Beagle*, he recalled, "It is the general custom in this country to ask for a night's lodging at the first convenient house. The astonishment at the compass, and my other feats of jugglery, was to a certain degree advantageous, as with that, and the long stories my guides told of my breaking stones, knowing harmless from venomous snakes, collecting insects, etc., I repaid them for their hospitality."[23]

Of all his acts of "jugglery," igniting Promethean matches was Darwin's most spectacular trick, the one that showed off the newest technology. Patented in 1828, Prometheans were small glass beads filled with acid and wrapped in paper coated with reactive chemicals. To ignite them, one simply broke the glass. When the acid hit the paper, the reaction produced a flame. To make it more dramatic, Darwin would bite the ampule: "It was thought so wonderful that a man should strike fire with his teeth, that it was usual to collect the whole family to see it; I was once offered a dollar for a single one."[24] At the time, a dollar was more than a day's wages. Darwin was paying two dollars a day for his

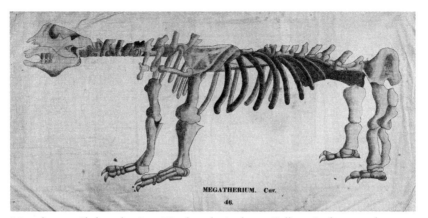

Megatherium Skeleton by Orra Hitchcock. Amherst College Archives and Special Collections.

adventure, with two armed guides and a dozen horses. His spectacular "feats of jugglery" helped him earn his welcome.

Darwin's base was in Maldonado, a "most quiet, forlorn, little town" on the grassy plains of coastal Uruguay. He thought the landscape was very "uninteresting," with "scarcely a house, an enclosed piece of ground, or even a tree, to give it an air of cheerfulness." But he was happy to be on shore. "After being imprisoned for some time in a ship, there is a charm in the unconfined feeling of walking over boundless plains of turf. Moreover, if your view is limited to a small space, many objects possess beauty. Some of the smaller birds are brilliantly coloured; and the bright green sward, browsed short by the cattle, is ornamented by dwarf flowers."[25] Although his published account was polite about the grasslands, Darwin was less positive in his private correspondence. In a letter to his sister Catherine, Darwin confessed that he was tempted to give up and "make a bolt across the Atlantic to good old Shropshire." As he put it, "I am become throughily [sic] tired of these countries: a live Megatherium would hardly support my patience."[26]

The *Megatherium*, like the mastodon, was a giant mammal that had long been extinct. Compared to the elephant-like mastodon, fossil relics of the *Megatherium americanum* (the "giant beast of America"), a twelve-foot-tall sloth-like creature, were relatively hard to find. There was a fossilized *Megatherium* skeleton in Madrid, but no complete *Megatherium* fossils in England or France, or anywhere else, really—except deep in the rock of the Americas. In Argentina the year before, in September 1832, Darwin had chipped the fossilized skull of a *Megatherium* out of a bank of sandstone. Excavating the *Megatherium* took hours; carrying it back to camp had been a feat of endurance. He had shipped the skull and an assortment of bones back to England months before, hoping his collection would interest the natural historians at the British Museum, at Cambridge University, and at the Royal College of Surgeons, where the top British anatomists were assembling a collection of fossilized skeletons. Now, in Uruguay, he was impatient to find out if anyone cared about his contribution.[27]

The 23-year-old Darwin had hired the guides and horses so that he could travel out into the grasslands to hunt. His hunting served two purposes: science and entertainment. In *The Voyage of the Beagle* he recounted: "A nearly perfect collection of the animals, birds, and reptiles, was procured."[28] Collecting also satisfied his longing for sport. His letter to Catherine put it more baldly: "We here got 80 birds & 20 quadrupeds."[29] Perhaps there was some truth to Darwin's father's claim that he "cared for nothing but shooting, dogs, and rat-catching."[30] In South America, however, he was beginning to hope that he could make a career out of such pursuits. His letter continued, "I have worked very hard (at least for me) at Nat History & have collected many animals & observed many geological phenomena: & I think it would be a pity having gone so far, not to go on & do all in my power in this my favourite pursuit; & which I am sure, will remain so for the rest of my life."[31] Uncertain of his professional prospects, bored with the grasslands and the low winter sky, Darwin pinned his hopes on natural history. Perhaps, after all, he would be a geologist.

Midway through his long letter to Catherine, he received an encouraging packet of mail from home. His tone grew more optimistic: "I am quite delighted to find, the hide of the Megatherium has given you all some little interest in my employments." With fresh determination he declared his intention to continue the voyage: "What fine opportunities for geology & for studying the infinite host of living beings: is not this a prospect to keep up the most flagging spirit? – If I was to throw it away; I dont think I should ever rest quiet in my grave; I certainly should be a ghost & haunt the Brit: Museum."[32] In this letter the homesick young man gave a hint of how his ideas were beginning to develop. Increasingly, Darwin was wondering if he could think about "the infinite host of living beings" the way Lyell thought about geology. Biology fascinated him even more than geology. He was beginning to realize that he would be filled with regret if he focused only on the geological past. He liked fossils, but he loved living beings even more.

The invocation of an imaginary live *Megatherium* and a ghost in the British Museum, along with accounts of feats with the compass and the Promethean matches, gives a carnivalesque, magic-show atmosphere to Darwin's letters from the Banda Oriental. Yet although his tone was jocular, more serious questions about his own profession underlie his letters. In June of 1833, it would have been almost as hard for Darwin to distinguish between science and magic as it was for his Uruguayan hosts. Was sparking a Promethean match a technological demonstration or a feat of jugglery? Was Darwin himself an amateur magician, a frivolous sportsman, or a serious natural historian? Was he a "Philosopher," as his shipmates called him? He did not really know.

While Darwin gamely juggled for his supper in South America, scholars in England and the United States were trying to explain the difference between magic and natural philosophy (which was starting to become known as science). The two had not always been so far apart. In 1830 John Herschel's *Discourse* had celebrated the "craving for the marvelous, which might be regarded as a remnant of the age of alchemy and natural magic, but which, under proper regulation, is a most powerful and useful stimulus to experimental enquiry."[33] Along with Humboldt's *Narrative* of his exploits in South America, Herschel's *Discourse* had inspired Darwin's own "craving for the marvelous." Now, depressed, anxious, and homesick in the grasslands of Uruguay, he was beset by uncertainty and self-doubt. Perhaps his hunting trips were mere sport, his scientific instruments good for nothing more than parlor tricks. Maybe he was just a mountebank.

When Charles Darwin bit down on Promethean matches and entertained his South American hosts, he did not intend to deceive them.

Nonetheless, the fear that he was some sort of charlatan was as real for him as it would be for any aspiring scholar in their early twenties. Though it is not likely that Darwin was troubled by the supernatural aspects of Promethean matches, he was certainly wrestling with uncertainty about his own professional future. In his case the question was complicated by his awareness that in many respects he was exactly the sort of well-heeled amateur that the new British Association for the Advancement of Science hoped to supplant.

Just a decade before, Charles Darwin had been a wealthy child, able to indulge his fascination with the "science of wonders." Known as "Gas" in prep school, he had loved chemistry for its beauty as well as the stinks and bangs that he and his teenage brother had sometimes managed to produce in their opulently furnished backyard laboratory. By 1833 the founders of the BAAS had started to express impatience with amateur scientists. In South America, Darwin may not have had a chance to read Charles Babbage's and David Brewster's latest books, but he pored over John Herschel's *Discourse*. Herschel was emphatic about the importance of publication. As Herschel put it, "Nothing has exercised so powerful an influence on the progress of modern science as the publication of monthly and quarterly scientific journals."[34] To contribute to science, it was necessary to publish. From this point of view, amateurish appreciation of the wonders of nature was selfish at best. At worst, it risked pushing natural science back toward the dark side of magic.

Darwin was well informed about the changes in attitudes toward studying the natural world. He was determined to do his best to gain entry to the rapidly professionalizing institutions, but he feared he might be shut out of the study of natural history. Becoming a professional geologist would require tamping down his sense of wonder. Just as important, Darwin would need to publish.

Darwin's first attempt to make his findings public was the shipment of South American specimens carefully packed into crates and barrels and dispatched to his mentor Henslow in Cambridge in 1832. Darwin sent botanical specimens, pressed and dried and mounted onto sheets of paper in a loose-leaf herbarium. He sent glass jars containing whole animals floating in spirits. He sent bones and carefully preserved skins.

He sent rock samples. Perhaps most excitingly, he sent fossils, many of which he thought were fossilized *Megatherium* remains. Shipping was slow and unpredictable. The mail was equally bad. Some letters never arrived; others could take up to a year to be delivered. Would his collections make it to England? Had he found anything significant? He could not wait to find out.

In June 1833, while the homesick Darwin hunted for specimens on the South American plains, the BAAS held a meeting in Cambridge. This meeting would prove to be very significant in the history of science because an unexpected argument between Samuel Taylor Coleridge and William Whewell would prompt Whewell to coin the word "scientist." Their confrontation heralded the great divorce between the arts and sciences. Darwin did not witness the debate.

Since Darwin could not attend the BAAS meeting, Henslow made sure that his student's *Megatherium* fossils were present. Months afterward, Darwin would receive a letter from Henslow:

> The fossil portions of the Megatherium turned out to be extremely interesting as serving to illustrate certain parts of the animal which the specimens formerly received in this country & in France had failed to do. Buckland & Clift exhibited them at the Geological Section (what this means you will learn from the Report I send you) – & I have just received a letter from Clift requesting me to forward the whole to him, that he may pick them out carefully repair them, get them figured, & return them to me with a description of what they are & how far they serve to illustrate the ostuology of the Great Beast.[35]

As it turned out, Darwin's fossils had arrived just in time. His *Megatherium* bones were not the first to arrive in England. Woodbine Parish, the British Chargé d'Affaires in Buenos Aires, had donated an almost complete fossilized *Megatherium* to the Royal College of Surgeons in 1832. Luckily for Darwin, the Parish fossil was missing a few crucial bones that his own collection supplied.³⁶ His mentor had made sure they were part of the exhibit, and now they were on the way to London. Darwin was on the path to scientific publication.

Although Darwin was pleased when he eventually heard that his *Megatherium* specimens had been exhibited at the meeting, his name was not mentioned in the published reports about the conference. The young naval naturalist was not even a footnote for the BAAS in June 1833. Instead, the big story of the meeting was the surprise appearance of his grandfather's great critic: Samuel Taylor Coleridge.

A somewhat frail man in his early sixties, Coleridge had been at the forefront of British thought for more than thirty years. Since his Shrewsbury days, when the pension from the Wedgwood brothers had helped to launch him as a poet, Coleridge had become a celebrated critic and philosopher. Though his body was wracked by years of opium use, Coleridge was still intellectually pugnacious.

Coleridge and his fellow poets and philosophers had grown more and more hostile to what Coleridge called "Darwinizing." In contrast, an emerging network of clergymen, schoolteachers, and amateur naturalists in England and America moved away from the previous generation's Romantic views of nature. These naturalists, who focused on collecting and describing specific natural objects such as beetles, seaweeds, and skeletons, paid attention to Herschel, Babbage, and Brewster instead of idealists like Wordsworth, Coleridge, and Carlyle. The Romantics worried about their diminishing relevance in the face of the new attitudes. In 1829 Thomas Carlyle had railed against the "Mechanical Age," complaining that "the science of the age, in short, is physical, chemical, physiological; in all shapes mechanical."³⁷

Day by day, science was growing increasingly empirical. The new generation of naturalists—geologists, chemists, botanists, and the like—shared a commitment to methods of inductive reasoning grounded in collecting and sorting material things, searching for patterns and laws. At the same time, as Brewster's *Letters on Natural Magic* had revealed, many naturalists were so wary of magic and mystery that they were becoming suspicious of wonder. At times, they were even hesitant about the "beauteous forms of things" celebrated in *Lyrical Ballads*. Coleridge and his allies believed that natural philosophy was in danger of losing its idealism, becoming completely unphilosophical.

On that June day, Coleridge stood to address the crowd at the BAAS. He insisted that people who got their hands dirty digging for fossils, murdering beetles, or dissecting songbirds did not deserve to be called natural philosophers. William Whewell, chair of mineralogy at Trinity College and the host of the BAAS meeting, sidestepped the conflict that Coleridge had hoped for and politely agreed. Whewell conceded that "Philosopher" was "too wide and lofty a term." Instead, he suggested, "by analogy with artist, we may form *scientist*."[38]

While Coleridge was defending idealism against what he saw as crude, mechanistic materialism, Whewell was fighting for the scholarly legitimacy of a pursuit that had mostly been limited to wealthy amateurs. It seemed absurd that Cambridge University forced promising undergraduates like Charles Darwin to focus on classical languages while geology, botany, and the like were treated as casual electives. In his effort to establish a professional identity for scientists, Whewell offered Coleridge and all the "real metaphysicians" an amicable separation, swerving neatly away from philosophy with the coining of "scientist." No one at the meeting liked his coinage very much; most preferred to be known by their specialization (as geologists, botanists, or zoologists, for example).

"Scientist" was a new word. The identity was equally new. When Whewell invented the scientist, ways of understanding the natural world changed.

The next year, Whewell launched the word "scientist" into print. This time he explained the new word in a different way: not as the opposite of artist but as an alternative to "men of science," a phrase that excluded women. He used the new term in his glowing review of Mary Somerville's *On the Connexion of the Physical Sciences* (1834). Whewell admired her five-hundred-page treatise on astronomy, physics, chemistry, geography, meteorology, and electromagnetism so much that he made it part of his curricular reforms at Cambridge. When degrees in the natural sciences were finally offered, Somerville would be a *set book*, on the list of required reading. *On the Connexion of the Physical Sciences* would be

Mary Somerville by Thomas Phillips (1834). Scottish National Portrait Gallery, Edinburgh.

used as an introductory textbook at Cambridge for decades. Until it was overtaken by *On the Origin of Species*, it was the best-selling science book in England. Whewell knew that Somerville's work would allow new scientific approaches to coalesce and develop, but he balked at calling her a "man of science." Recalling his interchange with Coleridge the summer before, he described her as a "scientist."

The term was slow to catch on. In 1840 the word "scientist" would still be relatively rare. In *Philosophy of the Inductive Sciences* (1840), Whewell would try again to make the case for the term: "We need very much a name to describe a cultivator of science in general. I should incline to call him a Scientist. Thus we might say, that as an Artist is a Musician, Painter, or Poet, a Scientist is a Mathematician, Physicist, or Naturalist."[39]

Gradually, "scientist" gained traction. But Charles Darwin did not adopt it. He described other people as scientists occasionally, but he avoided calling himself a "scientist." He might have preferred "man of science" because he did not want to associate with women who pursued science. He might also have been worried about Whewell's insistence on professionalization and publication. Though he was willing to make some of his findings public, he had other ideas that he wanted to hold close.

In his 1833 letter to Catherine (written about the same time that Buckland and Clift examined his *Megatherium* bones and Whewell and Coleridge argued about philosophy), Darwin had celebrated his "fine opportunities for geology & for studying the infinite host of living beings."[40] He was willing to go public with the geology. But his ideas about "living beings" were more exciting, filled with marvel and wonder. He was starting to think about biology the way Charles Lyell thought about geology. What if species change happened in a series of tiny, incremental changes? What if these changes were random, happening by accident rather than by design? For now, he wanted to keep his questions private.

Darwin knew from his own family history that publicly speculating about evolutionary ideas could stir up negative reactions. He also realized that in the 1830s professional men of science were expected to publish their thoughts and dispute them openly. It was possible that Brewster might accuse him of something verging on the occult if he kept his ideas secret. On the other hand, his grandfather Erasmus had been accused of wizardry, of using "magic spells" to "raise all things from cockle shells!" when he went public with his evolutionary speculations.[41] Whether he spoke or kept silent about evolution, he risked exclusion from the tight-knit group of men of science he longed to join. Ever cautious, Darwin decided to become a geologist. He could write and publish his geological findings and hold his ideas about "the infinite host of living beings" close.[42]

When Darwin boarded the *Beagle*, he planned to be ordained as an Anglican priest when he returned home. But while he was on the voyage, his ambitions changed. Rather than becoming a professional cleric and amateur naturalist, he decided he wanted to write and publish books about the natural world. According to Darwin's account, "Love for science gradually preponderated over every other taste."[43] Though he had started out as a passionate hunter and collector of specimens, his version of science became increasingly contemplative and methodical as he studied Herschel and Lyell. He came to enjoy "the pleasure of observing and reasoning," and to his delight he discovered that he could write very well.

Toward the end of the journey, he wrote excitedly to Caroline, "I am in high spirits about my geology. – &even aspire to the hope that, my observations will be considered of some utility by real geologists."[44]

Writing was not easy, but Darwin worked at it with determination. "Whilst we are at sea & the weather is fine, my time passes smoothly, because I am very busy. My occupation consists in rearranging old geological notes; the rearranging generally consists in totally rewriting them. I am just now beginning to discover the difficulty of expressing one's idea on paper. As long as it consists solely of description it is pretty easy; but where reasoning comes into play, to make a proper connection, a clearness & moderate fluency, is to me, as I have said, a difficulty of which I had no idea."[45] He was beginning to imagine himself not only as a geologist but also as a "real geologist" who published his work.

By 1836, as Emily Dickinson approached her sixth birthday, Charles Darwin was homeward bound. He wrote to his sister that his seasickness was getting worse and worse. He was so homesick he had lost all interest in new discoveries. To Henslow he remarked, "If, at present, the wonders of another planet could be displayed before us, I believe we should unanimously exclaim, what a consummate plague." "Oh the degree to which I long to be once again living quietly, with not one single novel object near me," he exclaimed. "No one can imagine it, till he has been whirled round the world, during five long years, in a ten Gun-Brig."[46]

Yet despite his lack of interest in further explorations, Darwin was excited about the prospect of meeting John Herschel in Cape Town, South Africa. To Caroline he wrote, "I have a high Curiosity to see the great Man."[47] After his visit, on June 3, 1836, he reported to his mentor Henslow that he had

> enjoyed a memorable piece of good fortune in meeting Sir J. Herschel. We dined at his house & saw him a few times besides. He was exceedingly good natured.... He is living in a very comfortable country house, surrounded by fir & oak trees, which alone, in so open a country, give a most charming air of seclusion & comfort. He appears to find time for every thing; he shewed us a pretty garden full of Cape Bulbs of his own collecting; & I afterwards understood, that every thing was the work of his own hands. What a very nice person Lady Herschel appears to be, – in short we were quite charmed with every thing in & about the house.[48]

Visiting Herschel's household in South Africa, Darwin found a new model for how a man of science might involve his entire family in his pursuits. Though setting up a scientific household was impossible for parsons or professors, John Herschel was independently wealthy. As he had planned his research in South Africa, he had refused offers of support from the Royal Navy and the Royal Society. He insisted on paying his own way so that he could conduct his research in precisely the way that "may suit my *caprice*."[49] During their four years in South Africa, Richard Holmes reports, "Their packed notebooks show ceaseless family activity: daily meteorological observations, zoological and botanical notes, and hundreds of beautiful plant drawings made, with infinite care, using a *camera lucida*."[50] The Herschel family's interest in science was amply supported by the interest on their extensive investments.

In South Africa the young Charles Darwin caught a glimpse of what the life of an independent scholar with a large fortune and an equally large family could be. Herschel's *Preliminary Discourse* had helped inspire Darwin to travel the world; now Herschel's personal life opened a new possibility. Charles Darwin began to dream of "a very comfortable country house."

After he returned to England, Darwin worked full time on revising and publishing the books he had imagined in Saint Jago. Most of his time was spent sorting through his notes and collections and writing about them. No longer an amateur naturalist, Darwin was becoming a man of science. Publication was a key part of this professionalization.

As he had always done, Darwin also made time to read poetry. He put away the worn-out copy of *Paradise Lost* that he had read a thousand times on his voyage and explored the Romantics. He was interested in alternative ways of thinking about the natural world. The fact that Coleridge had denounced modern science intrigued him more than it alienated him. Romantic poetry of nature had a mysterious power. He recalled, "As I was not able to work all day at science, I read a good deal during these two years on various subjects, including some metaphysical books; but I was not well fitted for such studies. About this time I took much delight in Wordsworth's and Coleridge's poetry; and can boast that I read the *Excursion* twice through."[51]

Darwin had no ambition to write poetry; he would not follow his grandfather in that way. He had no interest in magic shows and little patience with talk of the supernatural. But he could not rid himself of his craving for the marvelous. In public he went to meetings of the geological society in London; he took notes and corrected proofs and worked hard at publishing. But Darwin was a bit of a dark horse. He knew that the emerging group of professional scientists defined themselves in opposition to the Romantic poets. He knew that they disapproved of his grandfather's evolutionary speculation. In private he read the Romantic poets and wondered about the "infinite host of living beings."[52]

CHAPTER 5

Dickinson the Bold

Amherst, 1836–1847; Dickinson, Age 6–16

LOOKING BACK at her childhood, Emily Dickinson vividly remembered roving the woods in search of wildflowers. In a letter to Thomas Wentworth Higginson she wrote, "I had long heard of an Orchis before I found one, when a child, but the first clutch of the stem is as vivid now, as the Bog that bore it."[1] In a more private note to herself, she recalled "the rapture of losing my shoe in the Mud and going Home barefoot, wading for Cardinal flowers and the mothers reproof which was more for my sake than her weary own for she frowned with a smile."[2]

Cardinal flowers are a wild species of Lobelia with brilliant red spikes. In Dickinson's memory collecting them, getting muddy, and going barefoot were causes for rapture. She treasured the memory of her mother's gentle, weary reprimand. It was evidence of a deep understanding between mother and daughter, a time when her mother understood her excitement and approved of her determination. Though her mother had frowned at the lost shoe and all the mud, she had smiled at the flowers.

A similar memory shows up in a Dickinson poem:

I do remember when a Child
With bolder Playmates straying

To where a Brook that seemed a Sea
Withheld us by its roaring
From just a Purple Flower beyond
Until constrained to clutch it
If Doom itself were the result,   Were Doom itself the penalty –
The boldest leaped, and   [The] bravest
   clutched it –³

If this poem is based on real events, we can surmise that Dickinson, who was not afraid of mud or particularly careful with shoes, was most likely among "the boldest" of her playmates. There would have been nothing particularly rebellious about her bravery in this context. When Dickinson was young, in the 1830s and 1840s, girls were urged to botanize—to collect plants for the sake of studying botany. When she risked crossing the brook to clutch a rare flower, she was fulfilling her parents'

Cardinal Flower by Clarissa Munger Badger (1859). From *Wild Flowers, Drawn and Colored from Nature* (New York: Charles Scribner, 1859). Biodiversity Heritage Library, New York Botanical Garden.

expectations. In her early years, Emily Dickinson's bold brightness made everyone around her smile.

By all accounts, Emily Dickinson was a golden girl: loved and loving, extravagantly social, and high-spirited. As she approached the age of 10, she began to stand out. She was brilliant, precocious, vivacious. Her hair was like flame—gloriously red, with gleams of gold and undertones of purple-brown. Her bright wit was even more noticeable than her coloring. Long before she described herself as the "belle of Amherst,"[4] her parents expected their "dear little Emily" to be "one of the best little girls in town,"[5] and she did her best to live up to their expectations. She was extraordinarily gifted, a fire-crowned, silver-tongued prodigy.

In the schoolyard Emily was often surrounded by a crowd of laughing girls. They loved her outrageously inventive stories and jokes. In the classroom, teachers tended to smile and lean forward when she spoke. In the stores the small, well-dressed girl was greeted with gladness on account of her open purse and her quick laugh. Biographer Alfred Habegger believes that an early poem that describes a girl's growing consciousness of her own good fortune is autobiographical:

> It was given to me by the Gods –
> When I was a little Girl –
> . . .
> I heard such words as "Rich" –
> When hurrying to school –
> From lips at Corners of the Streets –
> And wrestled with a smile.
> Rich! 'Twas Myself – was rich –
> To take the name of Gold –
> And Gold to own – in solid Bars –
> The Difference – made me bold –[6]

As a schoolgirl, Dickinson knew she was "Rich." She was a Dickinson, and in Amherst, hers was a "name of Gold." But there was much more to her awareness than a realization that her family had plenty of money. Dickinson was intellectually and temperamentally different from her neighbors. She was surrounded by intelligent people; her golden gift was

Dickinson Family Silhouette (1847 or 1848). Yale University Library.

not ordinary cognitive ability. It was rarer than that. Part of it was courage. She was daring—not only bold in speech and thought but also physically bold. She roamed the swampy places around Amherst in search of wildflowers, with no fear of getting dirty or wet. She loved the mud. Another part of her gift was joy. In one poem she described herself as an "inebriate of air."[7] She could get drunk on words. Perhaps the key to her good luck was that she was allowed—even encouraged—to be a "free spirit."[8] Her family and friends, her school, and her town loved the surprising way she put her ideas together. They genuinely enjoyed her wit and her way with language. Sometimes, she felt that everyone in Amherst admired her.

Of the three Dickinson children, Emily took after their father the most. Like him, she could get interested in almost anything. She was a hungry, eager learner. At times her intelligence and the intensity of her affections overwhelmed others. No one could keep up with her. She loved the dry, understated humor of New England farmers, the extravagant poetry of the King James Bible, the soft Gaelic echoes of the Irish men and women who helped in the house and the yard. Dickinson's childhood friends would remember her droll humor, her wild flights of fancy, and her hilarious juxtapositions. She was bold. Equally important, she was "funny—sparkling with fun."[9]

The Amherst of the 1830s and 1840s was a small town; some might describe it as a large village. Hayfields and pastures radiated out from the houses toward the hazy blue Pelham hills. The Dickinson family had left the Homestead and moved a little further out to the edge of Amherst, as close to the fields and the woods as to the shops. But although their large white house was not at the very center of town, the Dickinsons were at the center of everything. Alfred Habegger describes Emily as "admired and envied" by her circle of friends.[10] Among her peers, Emily was a linchpin, pulling others together with stories, notes and letters, little gifts, and sometimes overpowering expressions of affection. When she and her "circle of five" (Emily Fowler, Mary Warner, Sophia Holland, Mary Louisa Snell, and Jane Hitchcock) launched a handwritten literary journal, *Forest Leaves*, Dickinson contributed a regular column of jokes and humor. She also wrote letters to a school friend who had moved away, Abiah Root. According to biographer Martha Ackmann, the letters to Abiah Root expressed Dickinson's increasing "confidence, independence, and self-awareness." The surviving letters offer an early example of how Dickinson played with language and exulted in her ability to write. As Ackmann explains, they show the young poet's awareness that "possibility circled her young life."[11]

Meanwhile, Edward Dickinson, Emily's father, was also surrounded by inviting possibilities. He was involved in all aspects of town life, from serving on the local school board to helping select the minister for the Congregational Church. The fact that he did not consider himself a Christian believer did not prevent him from leading the congregation. In 1838 Edward Dickinson reluctantly agreed to serve as a state senator, spending many weeks of every legislative session in Boston, far from his young family. Letters between Amherst and Boston give us a good picture of the family dynamics. Edward and his wife, Emily Norcross Dickinson, sent regular, affectionate letters to each other. Edward wrote to the children too, urging them to stay warm and to study hard. He often brought them treats from the city, including all the latest publications.

By the time Charles Darwin published the narrative of his voyage on the *Beagle* in 1839, Emily Dickinson was old enough to read it for herself. Perhaps the Dickinson children got their hands on a copy of Darwin's book. An adventure tale packed with information about natural history, it was just the sort of educational literature their father loved to buy for them. "Teneriffe," "Domingo," and "the Cordillera"—places Darwin described—would all become significant features in Dickinson's imaginary geography, and she would refer to them in her poetry. It is easy to picture Emily and her brother, Austin, sharing a copy of *The Voyage of the Beagle*, perhaps trying to keep it out of their parents' sight during a long Sunday afternoon when they were supposed to keep quiet and stick to devotional reading. Once they had their own piano, they could have hidden it with the sheet music and their cache of forbidden novels, though it is unlikely that such an improving work of nonfiction would have been frowned upon in the Dickinson household.

Despite her allusions to places Darwin described, there is no direct evidence that Emily Dickinson read *The Voyage of the Beagle* as a child. Most of Darwin's landmarks had also been described in Alexander von Humboldt's *Personal Narrative*, and many American writers referred to the same places. Except for Herman Melville, who made a Pacific voyage of his own, Dickinson's contemporaries usually gleaned their geography from books. They all read Humboldt. Many probably read *The Voyage of the Beagle* too, but at the time it was a relatively minor work. Humboldt described Tenerife in exhilarating detail, while Darwin's published book offered only a brief, frustrated glimpse of the receding peak seen from the deck of the *Beagle*, which had been denied entry to the harbor because of a quarantine. Nonetheless, Darwin's adventures may have helped form the young Dickinson's imaginary geography, either directly or secondhand. It is notable that the "Teneriffe" of Dickinson's imagination is also receding into the distance:

> Ah, Teneriffe – Receding Mountain –
> Purples of Ages halt for You –
> Sunset reviews Her Sapphire Regiments –
> Day – drops You His Red Adieu –[12]

Dickinson's description faintly echoes Darwin's description of the moment when he saw the sun suddenly "illuminate the Peak of Teneriffe, whilst the lower parts were veiled in fleecy clouds" as he sailed regretfully away.[13] Whether or not Dickinson read *The Voyage of the Beagle*, we can be certain that the young Dickinson and the young Darwin imagined the earth in similar ways, as part of a Romantic Humboldtian *cosmos*, a planet studded with volcanic peaks that pierced the tropical sky.

Dickinson's frames of reference were similar to Darwin's in ways beyond their Romantic conceptions of geography. She was also extraordinarily well trained in natural history topics such as geology and botany. "Nature study" was encouraged in the United States in the 1840s, and the schools that Dickinson attended were particularly strong in geology (Hitchcock's specialty), botany (Caroline Dutch Hunt), and chemistry (Mary Lyon). In fact, her formal training in the natural sciences was more extensive and thorough than his. She had the chance to study chemistry and astronomy, which Darwin never got. This was partly an accident of gender and location since schools in the United States were quicker to institute formal instruction in the natural sciences than schools in Britain and tended to encourage girls to study science more often than boys. It was also a matter of timing. Twenty-one years (nearly a generation) behind Darwin, Dickinson was much younger when John Herschel and Charles Lyell published their influential books. Because of his age, Darwin studied these books on his own time, as an amateur aspiring to become a real geologist. In contrast, Dickinson approached natural science as a conscientious (and somewhat competitive) student, excelling in her classes at the best schools in her region. In one letter she confessed to gasping with dread about the public examination at the end of the term, but her greatest fear seemed to be the following: "I shall lose my character if I dont recite as precise as the laws of the Medes and Persians."[14] Although she may have been anxious because of high expectations, she also knew that her reputation was extraordinarily good.

Dickinson's experience in school was the opposite of Darwin's. He was not interested in his studies and was never particularly successful.

The principal of his boarding school, Samuel Butler, publicly described him as a "poco curante"—an uncurious person, indifferent to learning. (Decades later, Butler's insult still stung.)[15] Dickinson's attitude was different. She was interested in everything she studied. In her letters she mentioned "Latin, botany, geology, history, ecclesiastical history, 'mental philosophy,' arithmetic, algebra, and geometry."[16] Her biographer Alfred Habegger is sure that she studied other subjects too. Though she did not mention German in any letters, she certainly studied the language. The "English" course at Amherst Academy featured Milton's *Paradise Lost* and many other works of blank verse. In Latin she made it through the *Aeneid*—and she seems to have enjoyed that too, judging from her fond inscription on the flyleaf of her friend Abby Wood's copy of Virgil. She wrote:

> forsan et haec olim meminisse juvabit Aeneid 1–203.
> *Afterwards, you may rejoice at the remembrance of these*
>   *(our school days)*
> When I am far away then think of me – E. Dickinson[17]

Every other week she was required to write a composition. Each day, school began and ended in prayer. Enrollment at Amherst Academy fluctuated from fifty or sixty students (both girls and boys) to over a hundred. Students moved through the curriculum at their own pace.

In 1845 she wrote to Abiah Root, "I never enjoyed myself more than I have this summer. For we have had such a delightful school and such pleasant tea[c]hers." Her teachers enjoyed her just as much. Daniel T. Fiske, Dickinson's teacher at Amherst Academy in 1842–1843, described her as "a very bright, but rather delicate and frail looking girl: an excellent scholar: of exemplary deportment: faithful in all school duties: but somewhat shy and nervous. Her compositions were strikingly original: and in both thought and style seemed beyond her years, and always attracted much attention in the school, and I am afraid, excited not a little envy."[18] Fiske and his colleagues appreciated Dickinson's "strikingly original" writing and celebrated her for it. Because she seemed delicate, they often urged her outdoors, encouraging her to build her strength with vigorous physical activity.

Dickinson loved wandering around in the woods as her teachers urged. She was also happy to be indoors. Home was warm and pleasant. School was "delightful." At 16 she wrote to her former schoolmate:

> We have a delightful school this term under the instruction of our former principals, & Miss R. Woodbridge – daughter of Rev. Dr. W. of Hadley, for preceptress. We all love her very much. Perhaps a slight description of her might be interesting to my dear A. – She is tall & rather slender, but finely proportioned, has a most witching pair of blue eyes – rich brown hair – delicate complexion – cheeks which vie with the opening rose bud – teeth like pearls – dimples which come & go like the ripples in yonder little merry brook – & then she is affectionate and lovely. Forgive my glowing description, for you know I am always in love with my teachers.[19]

Dickinson's description of Miss Woodbridge sounds almost like a parody of a popular novel. She gestures toward a lover's clichés, self-consciously mocking herself even as she expresses genuine admiration. She piles on a series of conventional descriptions, first praising Miss Woodbridge's fine proportions, then her "most witching pair of blue eyes—rich brown hair—delicate complexion." She steps up her praise—and her clichés—by comparing the teacher's cheeks to roses and teeth to pearls and then finally pushes her metaphors into absurdity when she likens Miss Woodbridge's dimples to "the ripples in yonder little merry brook." Her "glowing description" is self-conscious, even self-mocking. If Dickinson's feelings for Miss Woodbridge gave her hints about her own developing sexuality, she found a way to acknowledge being "in love" while laughing off the implications. Although her avowal was cloaked in humor, the fundamental point was clear. Amherst Academy was delightful for Dickinson, and she admired and loved her teachers.

The poet's mother, Emily Norcross Dickinson, had felt the same way when she was a girl. Dickinson's mother was extraordinarily well educated at a time when formal schooling was relatively rare. Emily Norcross had grown up in Monson, Massachusetts, a town similar to Amherst in that it supported an academy. The poet's mother studied at

Edward Dickinson by Otis Allen Bullard (ca. 1840). HNA42, Houghton Library, Harvard University.

Emily Norcross Dickinson by Otis Allen Bullard (ca. 1840). HNA40, Houghton Library, Harvard University.

Monson Academy for a full twelve years. After she finished her studies there, she went on to the Herrick School, a boarding school in New Haven, Connecticut, which offered a full curriculum including "Penmanship, Reading, Grammar, Rhetorick, Logick, Geography, use of the Globes, Arthemetic [sic], Algebra, Geometry, Astronomy, Opticks, and several other branches of Natural Philosophy." In addition to all these subjects, regular writing—"composition"—was "required of all the members."[20] The Herrick School offered Emily Norcross an uncommonly fine opportunity for advanced study, with courses that balanced reading, writing, and arithmetic with a wide array of science topics. At the time, postsecondary education was not all that common for men, but it was even rarer for women. In 1823 Emily Norcross was one of very few young women anywhere in the world who had the chance to study at the postsecondary level.

Emily Norcross and Edward Dickinson just missed each other in New Haven—he graduated from Yale in summer 1823, and she started Herrick that autumn. Their paths were drawing very close. The historian Mary

Bernhard relates that Edward Dickinson's sister Lucretia had begged to attend school in New Haven in 1823, but her family had been unable to afford the steep tuition. The Norcrosses were substantially wealthier than the Dickinsons. Although they did not meet in New Haven, Edward Dickinson eventually struck up a friendship with Emily Norcross's uncle Erasmus, another Yale graduate who practiced law in rural Massachusetts. (Like the Darwin-Wedgwood family, the Dickinson-Norcross family tended to repeat names—there were many Emilys, Austins, and Lavinias. Erasmus is one of the few names that crosses over, appearing in both the British family and the Massachusetts one.)

Emily Norcross and Edward Dickinson first met at a chemistry lecture in Monson. Edward quickly realized that he had found someone he wanted to be his "friend for life."[21] As Mary Bernhard explains, Emily Norcross "had a charm that reflected social experience.... She was sensitive and intelligent, a person with whom he could converse easily and comfortably on a wide variety of subjects—on novels and substantive books, on problems and professional aims."[22] During their two-year courtship, Edward often sent gifts of books and magazines but also urged Emily Norcross not to tire herself out by attending too many lectures and concerts. Later he would express his love and interest in their daughter in a similar pattern.

Before her marriage and her move to Amherst in 1828, Emily Norcross had a busy and full life in Monson. Her family was financially secure and remarkably cultured. She was drawn to Edward Dickinson in part because she liked the similarities between the life he promised in Amherst and her home life in Monson. Although the Norcrosses were concerned about Edward's father Samuel Fowler Dickinson's precarious finances, they shared his priorities. Emily Norcross's father had made substantial investments in Monson Academy, just as Edward's father had done in Amherst. The two families placed a similar value on education. In both generations—the poet's and her parents'—the families were committed to providing their daughters with rigorous formal education. Yet the opportunities for sons and daughters were different. Yale, Amherst, and Harvard did not admit women, and until Mount Holyoke opened in 1837, there was no equivalent institution for women. In 1823 a "higher school" like Herrick was as close to college as a female student could get. At Herrick School, Emily Norcross was probably exposed to more natural philosophy—botany, chemistry, optics, and astronomy—than her future husband had been at Yale. While the Dickinson children turned to their father for help with Latin, their mother was the one who could help them with botany.

As it turned out, Emily Dickinson's education closely replicated her mother's. On top of studying many of the same subjects, she also picked up her mother's affectionate attitude toward her teachers. In one instance they even shared a teacher. Alfred Habegger relates that one of Emily Norcross's favorite teachers at Monson Academy, Caroline Dutch Hunt, later took a job at Amherst Academy when Emily Dickinson was enrolled. Perhaps Emily Norcross Dickinson had played a role in getting her favorite teacher to come to Amherst to teach her daughters. At Monson Academy, Caroline Dutch Hunt had praised Emily Norcross's performance in a note that the poet's mother saved for her entire life. She also saved an unsent note that she had written as a schoolgirl, which seems to have been intended for her teacher:

> Oh! My dear Caroline
>   Remember me
> for
>   forever[23]

When Emily Norcross's daughter wrote to her friend Abiah that she was "always in love with the teachers," she seems to have been following in her mother's footsteps.

In some ways the poet and her mother were completely unalike. Emily Norcross Dickinson hated to write and was sometimes a little obsessive about housework (her children teased that when the house was perfect, she would start to worry about cleaning the rain gutters). In contrast, her daughter hated to clean the house and was exhilarated by writing. However, mother and daughter followed similar educational pathways and thrived in school. Of all the subjects they studied, botany was the one that brought them closest. Flowers fascinated both of them, mother and daughter.

When she was 14, Dickinson sent a geranium leaf to Abiah Root and urged her to press it. In her letter she reported that she was studying "Mental Philosophy, Geology, Latin, and Botany" and commented, "How large they sound, don't they?" At this point Dickinson was proud of her studies and genuinely fascinated by botany. She mentioned that she had been out walking and gotten "some very choice wild flowers" and asked, "Have you made you an herbarium yet? I hope you will if you have not, it would be such a treasure to you; 'most all the girls are making one. If you do, perhaps I can make some additions to it from flowers growing around here."[24]

The contrast between Dickinson's *Herbarium* and Darwin's is particularly interesting. On his great voyage (1831–1836), Darwin picked and pressed plant specimens for his mentor Henslow. The first batch he sent to Cambridge was sloppy, but after Henslow's remonstrance reached him, he began to do a better job. He pressed plants one to a sheet and packed them in wooden chests to be transported to England. The sheets were never bound together because they were intended to be circulated among botanists. They are still stored at Cambridge. The Darwin collections are of great historical and scientific interest, but they are not intended to be beautiful.

Dickinson's *Herbarium* is different. It was never intended for circulation. Though it is now stored in the Rare Books Library at Harvard, for Dickinson the book was a private pleasure. She dried specimens separately and then mounted them into a large bound book, arranging them with as much attention to their metaphorical implications and relationships on the page as to their botanical classifications.[25] She loved to botanize, but she was determined not to allow her herbarium to drain the meaning or the beauty from her flowers. For Dickinson, botany always retained its association with female friendship.

For years Dickinson had roamed the woods, gathering flowers. Now she was studying botany in school, pressing plants into dark, flat shapes. She mounted dried plants in her herbarium with small slips of paper that listed their scientific names and the number of stamens and pistils. The fine script on the label for the cardinal flower, *Lobelia cardinalis 5.1*, indicated that cardinal flowers have five pollen-bearing stamens and one central seed-generating pistil—five male parts around one female part. Dickinson's botany textbook, Almira Hart Lincoln's *Familiar Lectures on Botany*, did not sexualize flowers very much, but it did not avoid explaining that the class of a plant was "signified by prefixing Greek numerals to the word *ANDRIA*, signifying stamen," while the order was "signified by prefixing Greek numerals to the word *GYNIA*, signifying pistil."[26] For a student as attentive to language as Dickinson, the sexual implications of botany were clear.

Erasmus Darwin's *Loves of the Plants*, which used frank sexual imagery to explain botanical classifications, was probably not available to the Massachusetts girl, but she was able to make the connections herself. Erasmus Darwin's poem described the flower now known as shooting star, *Primula meadia*, a one-pistil, five-stamen flower, as a laughing belle surrounded by five suitors:

> *Meadia*'s soft chains five suppliant beaux confess,
> And hand in hand the laughing Belle address;
> Alike to all, she bows with wanton air,
> Rolls her dark eye, and waves her golden hair.[27]

The titillating implications of the many stamens surrounding each pistil were obvious to everyone. Janet Browne comments, "In his poem Darwin listed a procession of female images ranging from virtuous brides and tender mothers to attentive sisters, nymphs, and shepherdesses. Laughing belles and wily charmers were followed by queens and amazons."[28] Flowers with fewer stamens were personified as virtuous women, while those with four or five were labeled belles and charmers.

In her 1845 letter to Abiah, Dickinson made a similar imaginative leap. After encouraging her friend to start her own herbarium, the letter goes on to discuss the different types of girls. Dickinson starts with the virtuous ones: "I expect you have a great many prim, starched up young ladies there, who, I doubt not, are perfect models of propriety and good behavior. If they are, don't let your free spirit be chained by them."[29] According to Martha Ackmann, Emily admired Abiah Root for being "independent and unbridled by convention."[30] Emily wanted Abiah to know that she was a free spirit too. She continued, "I am growing handsome very fast indeed! I expect I shall be the belle of Amherst when I reach my 17th year. I don't doubt that I shall have perfect crowds of admirers at that age. Then how I shall delight to make them await my bidding, and with what delight shall I witness their suspense while I make my final decision."[31]

Thus far, Dickinson's letter had followed Erasmus Darwin's progression from prim young ladies to handsome belles. But now her imagination veered toward other possibilities. She abruptly turned from botany to composition: "But away with my nonsense. I have written one composition this term, and I need not assure you it was exceedingly edifying to myself as well as everybody else. Don't you want to see it?"[32] In Dickinson's mind the prospect of "perfect crowds of admirers" was as nonsensical as the primness of the "starched up young ladies." She was beginning to chafe at some of the expectations imposed on her as a 14-year-old girl.

Dickinson herself had not changed: she was still the same bold, funny person who had once been a golden girl. But as she entered adolescence, her boldness—and even her humor—became less acceptable. She was expected to study botany, but she was not really expected to admit that she understood all of its implications. In a postscript to her letter, she mentioned that "Miss Adams our dear teacher . . . sent me a beautiful bunch of pressed flowers."[33] The fresh, unpressed geranium leaf that she sent to Abiah was similar to the delicate dry blossoms from the teacher, but it was also surprising: fragrant and green.

When Dickinson sent this fresh geranium leaf in the letter praising herbariums and pressed, dried flowers, her message contained an active, interesting juxtaposition. Piquant contradictions appealed to her. A few years before, she had posed for Otis Bullard's portrait holding both a botany book and a living rose. Now a teenager, she was still fascinated by the interactions between flowers and books. She loved ambiguity and double meaning—the way simple things could have wildly different implications depending on how they were angled together. For the rest of her life, she would send botanically mixed messages—a bouquet of flowers with a paper note twined about a stem, a scatter of rose petals folded inside a poem, a woven wreath of leaves accompanied by an unorthodox prayer. Women often expressed their friendship for each other by exchanging plants and flowers, but Dickinson always seemed to twist away from conventional expressions and toward quirky, often startling mixed messages.

Of course, both her parents had been well practiced in the art of ambiguous messages. Dickinson recalled that her mother had "frowned with a smile" at her botanical exploits, while she gently mocked her father: "He buys me many Books—but begs me not to read them—because he fears they joggle the Mind."[34] Despite these observations, it is not likely that the poet's parents were unusually self-contradicting. Instead their daughter Emily was unusually observant. For her, such ambiguities sparkled with fun—and with meaning.

But as she approached adulthood, the ambiguities of womanhood began to trouble her. Despite her joke about becoming the "belle of Amherst," Dickinson was very curious about the possibilities open to single, unmarried adult women. Although she enjoyed "composition," the prospect of becoming an author was not appealing at first. In her deeply conservative Christian town, novels were generally frowned upon, and women novelists were not necessarily respectable. In 1826, in a letter to his fiancé Emily Norcross, Edward Dickinson had mused, "I should be sorry to see another Madame de Staël." He had recently met Catherine Maria Sedgwick, whose novels he admired, and he was filled with "a conscious pride that women of our own country . . . are emulating not only the females, but also the men of England & France & Germany & Italy in works of literature."[35] Nonetheless, meeting Sedgwick reminded Edward Dickinson of his fear of de Staël. His Massachusetts compatriot's literary genius was a source of pride and happiness, but it was also worrisome.

Edward Dickinson and Emily Norcross's daughter shared this ambivalent attitude toward novels and novelists—particularly women novelists. In the first letter she sent to Abiah, she wrote, "Please send me a copy of that Romance you was writing at Amherst. I am in a fever to read it. I expect it will be against my Whig feelings."[36] In this ambiguous postscript, Dickinson both identified with her parents' conservative New England politics and rejected them. (The Whig Party

Mary Lyon (1832). Library Company of Philadelphia.

formed in the 1830s in opposition to Andrew Jackson's more populist Democratic Party.) For a twenty-first-century reader, it may be tempting to imagine that Dickinson's "fever" to read her friend's story was authentic, while her avowal of "Whig feelings" was more perfunctory. But we should not be too quick to dismiss Dickinson's deep ambivalence. There was something about "Romance" that genuinely went against the grain for her. Perhaps her uneasiness added to her excitement as a reader, but there is no evidence that Dickinson ever gave serious thought to writing fiction. She did not aspire to become a novelist.

It is also clear that as much as she adored her teachers, Dickinson had no desire to become a teacher. The work paid poorly and offered few guarantees of stability. Most of her women teachers at Amherst Academy taught for a year or two before marriage. Her older teachers (like Caroline Dutch Hunt, who had taught her mother at Monson) were often widows forced back into teaching by poverty.

One of the few exceptions to this dispiriting pattern was the dynamic Mary Lyon, who had lived with Edward and Orra Hitchcock in Amherst in the 1830s as she raised the money for her own female seminary. If Dickinson thought of Lyon as a teacher, then she had one example of an unmarried woman teacher whose life was accounted a great success. But Lyon was more celebrated for her leadership of the school she had founded than for her teaching, and Dickinson knew for certain that she would never want to manage a large group of people. Lyon was also very certain about her conservative Christian convictions, and Dickinson had trouble with absolutist approaches to questions of religion.

Despite these caveats, Mary Lyon was a remarkable local role model. Her biography, *The Life and Labors of Mary Lyon* (1858), drafted by Edward Hitchcock, placed her in "the first rank in human intellect": "She could grasp and handle abstract truths with the ease and skill of a practised philosopher. She could at the same time illustrate them with a woman's fertility of invention, and enforce them with a woman's earnestness. Her mental eye was originally steady, clear, and far-seeing; and its angle of vision was particularly large.... She spoke with so much ease and pleasure to herself, that it seemed not to disturb, but rather to quicken and array in proper order the train of thought which was going on within."[37]

In some ways Dickinson saw herself in Mary Lyon. She took a similar pleasure in putting her words together, and she was capable of seeing the world from an extraordinarily wide perspective. Most of all, she liked the idea of combining a philosopher's comfort with abstract truths and "a woman's fertility of invention." Before she had started working with the Hitchcocks in Amherst, Mary Lyon had studied chemistry at Troy, the same place that Almira Lincoln, the author of Dickinson's botany textbook, had studied. These women, Lyon the chemist and Lincoln the botanist, were the sorts of role models for which Dickinson was searching. In Britain, Caroline Herschel was celebrated for astronomy. Mary Somerville's books on the physical sciences were required reading for undergraduates who studied the physical sciences at Cambridge.

Perhaps the best early nineteenth-century term for a person like Mary Lyon was "scientist," the awkward word that William Whewell had coined to describe Mary Somerville. In the 1840s, one of the most intriguing possibilities for Emily Dickinson was to imagine herself as a "scientist."

CHAPTER 6

The Leading Scientific Men
London and Amherst, 1836–1845

ON BOARD the *Beagle*, Charles Darwin had missed the brief confrontation between Samuel Taylor Coleridge and William Whewell that led to the coining of the word "scientist." He would rarely use the term, either to identify himself or to describe others. But as he imagined his return to England, he began to hope that he might "take a place among the leading scientific men."[1] In the final year of his voyage, the old boys of Cambridge—and even of Shrewsbury, where he had attended boarding school—worked on his behalf to promote his career.

After almost four years in South America, the *Beagle* sailed westward from Lima, Peru, in September 1835. They spent a little more than a month in the Galapagos Islands and then headed further westward on a route that would eventually take them around the world toward home. Meanwhile, John Stevens Henslow and Adam Sedgwick, Darwin's Cambridge professors, started to actively promote his research.

In November 1835, Henslow read extracts from Darwin's letters to a meeting of the Cambridge Philosophical Society. A few days later, Sedgwick read extracts from the same letters at the Geological Society in London. Charles Lyell, the president of the society, was delighted that Darwin's findings were framed in accordance with Lyell's own views. Although the two had never met, they were fast becoming geological allies. Shortly after the meeting Lyell wrote to Sedgwick, "How I long for the return of Darwin! I hope you do not plan to monopolise him at Cambridge."[2] That winter, Lyell also asked the Admiralty to instruct Darwin to send him "all the latest intelligence of the geology of Patagonia and Chile."[3]

In addition to promoting Darwin at the Geological Society of London, Adam Sedgwick supported him outside London. He wrote a pitch-perfect note to Samuel Butler, Darwin's Shrewsbury School headmaster, informing him that Darwin had done "admirable work in South America and has already sent home a collection above all price." Sedgwick might have been aware that Butler and Darwin had not gotten along very well when Darwin was his student. His note continued, "It was the best thing in the world for him that he went out on the voyage of discovery. There was some risk of his turning out an idle man, but his character will be now fixed, and if God spares his life he will have a great name among the Naturalists of Europe!"[4] Sedgwick visited the Darwins at the Mount that winter and praised Charles to his family in similar terms. In his *Autobiography*, Darwin recalled:

> Towards the close of our voyage I received a letter whilst at Ascension, in which my sisters told me that Sedgwick had called on my father, and said that I should take a place among the leading scientific men. I could not at the time understand how he could have learnt anything of my proceedings, but I heard (I believe afterwards) that Henslow had read some of the letters which I wrote to him before the Philosophical Society of Cambridge, and had printed them for private distribution. My collection of fossil bones, which had been sent to Henslow, also excited considerable attention amongst palæontologists. After reading this letter, I clambered over the mountains of Ascension with a bounding step, and made the volcanic rocks resound under my geological hammer.[5]

When Darwin arrived in England on October 2, 1836, he had changed more than anyone expected. He was charged with energy. He clambered around Shropshire, London, and Cambridge with the same bounding steps—and the same optimism—that had filled him at Ascension. When his father saw him for the first time after the journey, he thought even the shape of his son's skull was altered. England was almost equally changed. Shortly after Darwin returned, the 18-year-old Victoria would become queen. The Victorian age was beginning. This was the great age of the serial novel. Charles Dickens, Anthony Trollope, Elizabeth Gaskell, and many others dramatized the startling social changes that came along with industrialization.

Victorian London was smoky and dangerous, but it was also exciting. Part of the excitement was intellectual. Clattering mechanized printing presses reeled off newspapers, weekly and monthly journals, and all kinds of books. On top of the extraordinary novels of the era, nonfiction publishing also exploded, with hundreds of popular works of political theory, travel, and science, including scientific best sellers like Mary Somerville's *On the Connexion of the Physical Sciences*.[6] The explosion of public popular culture also included a great expansion of museums. The British Museum (which Darwin had imagined haunting) gathered every kind of artifact and specimen from the far reaches of a global empire— art and artifacts alongside fossils and seashells. Scientific societies, including the College of Surgeons and the Royal Institution, offered widely popular exhibitions and lectures to the public. Every winter, Michael Faraday presented Christmas lectures that combined entertainment and science for a wide and enthusiastic audience. Faraday did not lose credibility because of his efforts to make science popular and accessible. The results of his experiments in the laboratories of the Royal Institution were published in the most select scientific journals.

Darwin's entry into scientific scholarly publishing was as easy as his entry into the scientific social networks. Before he returned to England, Henslow simply selected what he liked from the letters Darwin had sent and had them printed up by Cambridge University's own press. Darwin laughingly referred to Henslow as his own personal "first Lord of the Admiralty" because he directed the campaign to promote Darwin's career.[7]

By the time Darwin met Charles Lyell face-to-face on October 29, 1836, both had read each other's work. They not only knew they were going to be professional allies but also hoped to be friends. It was not particularly unusual that Darwin and Lyell imagined themselves as colleagues and friends before they met. The proliferating scientific societies in London and around the world were often social as well as professional, while the boom in publishing and postal systems at the beginning of the Victorian age made international friendships surprisingly frequent. Victorian science was based on collegial networks that centered on scholarly publications. When Lyell asked the British Navy to have Darwin send him information about South America, he was extending a professional network that already reached across Europe and North America.

Darwin and Lyell hit it off immediately. Before the week was out, Lyell was urging Darwin to come to dinner before the other guests. He introduced Darwin to Richard Owen, the comparative anatomist and paleontologist who would help him analyze and catalog his South American fossils. In his *Autobiography*, Darwin recalled, "I saw a great deal of Lyell. One of his chief characteristics was his sympathy with the work of others, and I was as much astonished as delighted at the interest which he showed when, on my return to England, I explained to him my views on coral reefs. This encouraged me greatly, and his advice and example had much influence on me."[8] Darwin also remembered, "I saw more of Lyell than of any other man, both before and after my marriage. His mind was characterised, as it appeared to me, by clearness, caution, sound judgment, and a good deal of originality."[9]

Lyell's support for Darwin was particularly generous because some of Darwin's geological ideas—particularly his theory about the importance of coral reefs to the formation of tropical islands—directly contradicted Lyell's own published theories. Lyell thought they were formed by volcanoes; Darwin convinced him they were made by corals—"myriads of tiny architects."[10] It took Lyell a few months to be convinced, but in May 1837 he wrote to John Herschel, "I must give up my volcanic crater theory for ever, though it costs me a pang at first, for it accounted for so much – the annular form, the central lagoon, the sudden rising of an isolated mountain in a deep sea.... Yet spite of all this, the whole theory is knocked on the head.... Coral islands are the last efforts of drowning continents to lift their heads above water."[11]

Despite his pangs about Darwin's coral theory, Lyell used his Geological Society Presidential Address in February 1837 to promote his new protégé's work. When he sent a copy of the printed speech to Adam Sedgwick, he wrote in his cover letter, "It is rare even in one's own pursuits to meet with congenial souls and Darwin is a glorious addition to my society of geologists and is working hard and making way both in his book and in our discussions."[12] The tone of Lyell's letter is noteworthy. He wrote to Sedgwick as if the two of them were joint mentors of the 28-year-old Darwin. Once Sedgwick had introduced Darwin to Lyell, Lyell facilitated Darwin's introductions to all the other scientists in London.

In his *Autobiography*, Darwin described the interval "from my return to England (October 2, 1836) to my marriage (January 29, 1839)." In his view, "These two years and three months were the most active ones which I ever spent."[13] During those years, he recalled, "I finished my Journal, read several papers before the Geological Society, began preparing the MS. for my 'Geological Observations,' and arranged for the publication of the 'Zoology of the Voyage of the Beagle.' In July I opened my first note-book for facts in relation to the Origin of Species, about which I had long reflected."[14] Lyell encouraged Darwin in all his public activities, though it would have shocked him to learn of the private notebooks in which Darwin was starting to record the evolutionary ideas he had long kept to himself.

Although Lyell was not an evolutionist, he provided an important methodological role model for Darwin's approach to the topic. Darwin recalled, "After my return to England it appeared to me that by following the example of Lyell in Geology, and by collecting all facts which bore in any way on the variation of animals and plants under domestication and nature, some light might perhaps be thrown on the whole subject."[15] Lyell's method was a key building block for Darwin's developing theory.

Since Darwin was so well-connected, he had many influences and sources of information. Lyell was probably the most significant, but many other writers and thinkers, scientific and not, served as role models and guides. Among these, one of the most surprising—and surprisingly important—was the author and social theorist Harriet Martineau.

Harriet Martineau (1872).

Like Lyell, Martineau was among the relatively small group of writers whose works Darwin read and pondered during his voyage. In 1833 his sister Caroline had sent him copies of some of Martineau's early pamphlets. Caroline's letter explained, "I have sent you a few little books which are talked about by every body at present—written by Miss Martineau who I think had been hardly heard of before you left England. She is now a great Lion in London, much patronized by Ld. Brougham who has set her to write stories on the poor Laws—Erasmus knows her & is a very great admirer & every body reads her little books."[16] Martineau's Illustrations of Political Economy was a series of twenty-five brief novellas, published once a month. Although the little books were fictional, they were very realistic about social conditions. As the title of the series proclaimed, their primary purpose was to explain and popularize the economic theories of Thomas Malthus and David Ricardo using statistical analysis to argue that human population would always slightly outpace food supply without carefully controlling the birth rate. When Charles got to know Martineau personally in 1836, he was already familiar with her "little books" and with her sociological and economic ideas. He may also have been impressed by (or even a bit jealous of) her publishing success. How, precisely, had she become a "Lion in London"?

Martineau was as generous to Darwin as any of his other mentors. After one of Erasmus's "brilliant" dinner parties, Darwin reported to his sister Susan, "I had a very interesting conversation with Miss Martineau, – most perfectly authorial, – comparing our methods of writing. – it seems wonderful the rapidity with which she writes correctly."[17] His letter concluded, "I forgot to say that Miss Martineau is going to pay me a visit some day, to look at me as author in my den, so we had quite a flirtation together."[18] In another letter to Susan, he remarked, "She is a wonderful woman: when Lyell called, he found Rogers, Ld. Jeffrys, & Empson calling on her. – what a person she is thus to collect together all the geniuses."[19] Darwin admired her remarkable social network. He was also fascinated by her skillful negotiation with publishers. In the same letter, he mentioned, "She is very busy at present in making arrangements about her new novel. One bookseller has offered 2/3 profits & no risk, but I suppose that is not enough."[20] He hoped to emulate Martineau's network building in general and her publishing savvy in particular.

Although both Charles and Erasmus were very fond of Martineau, at times Charles was uncomfortable with her role as a public intellectual. Being a literary lion was just one of the ways that Martineau challenged Darwin's gender expectations. Her behavior did not adhere to Victorian conventions of femininity any more than his brother's behavior followed masculine norms.

Erasmus never married. He had unusually close friendships with a number of women, including a few married women. At times his family worried that some of these relationships were sexual. At other times they seemed more anxious about Erasmus's seeming lack of conventional sexual impulses. Perhaps they thought he was gay, though they may not have thought of sexuality in such terms (the word *homosexuality* came into use many decades afterward). Because Harriet Martineau was unmarried, the Darwin family was particularly puzzled by Erasmus's bond with her. They might have thought—perhaps even hoped—that the two of them had a romantic interest in one another.

Conversely, they may have perceived Erasmus's friendship with this unconventional woman as a signal of his queerness. Whatever the case, Charles was not comfortable with the relationship between his brother and Martineau.

In his first letter after meeting Martineau, Darwin used a racial slur. This 1836 letter was not only his earliest mention of Martineau but also his earliest recorded use of the slur. Throughout his correspondence he used the N-word infrequently—never as a way of describing a person of color but generally as a way of talking about abasement. Although it makes for uncomfortable reading, it is worth examining the letter in some detail.

> Erasmus is just returned from driving out Miss Martineau. – Our only protection from so admirable a sister-in-law is in her working him too hard. He begins to perceive, (to use his own expression) he shall be not much better than her "nigger." – Imagine poor Erasmus a nigger to so philosophical & energetic a lady. – How pale & woe begone he will look. – She already takes him to task about his idleness – She is going some day to explain to him her notions about marriage – Perfect equality of rights is part of her doctrine. I much doubt whether it will be equality in practice. We must pray for our poor "nigger."[21]

Darwin's discomfort was palpable. He distanced himself from the slur by setting it off with quotation marks and introducing it with a parenthetical phrase that explicitly attributed the offensive term to his brother. The Darwin and Wedgwood families had long been abolitionists, and Charles was passionately opposed to racial essentialism. Nonetheless, when he returned from his voyage and found Erasmus cavorting around town with Harriet Martineau, he resorted to racially derogatory language to discuss their unconventional relationship. By describing his white, wealthy, privileged brother with a word that denoted Blackness and low caste status, Darwin expressed his sense that Erasmus was turning his gender expectations upside down. Though Erasmus was a man, he was "pale and woe-begone"; though Harriet was a woman, she was "philosophical and energetic," qualities that Darwin thought unladylike. Darwin turned to racist language in an attempt to make light of his discomfort with the way that Erasmus and Harriet flouted his gendered expectations.

A month later, in his next letter about Martineau, Darwin said that he was getting more comfortable with her: "I called as in duty bound, on Miss Martineau, and sat there nearly an hour. She was very agreeable and managed to talk on a most wonderful number of subjects, considering the limited time. I was astonished to find how little ugly she is, but as it appears to me, she is overwhelmed with her own projects, her own thoughts and own abilities. Erasmus palliated all this, by maintaining one ought not to look at her as a woman."[22]

Despite his backhanded comment on Martineau's appearance (he found her only a "little ugly"), Darwin enjoyed her company. He was surprised that she was more interested in her own projects, thoughts, and abilities than she was in his, but his brother "palliated" her astonishing refusal to subordinate her interests to her male guest's by explaining that "one ought not to look at her as a woman." Darwin did not find it easy to adjust to Martineau's presumption of equality, but though he was off-balance, he found her wonderful to talk to.

A third letter about his brother and Martineau was more successful at hitting a comic tone. Darwin started by describing a visit to the zoo on a marvelous spring day when the rhinoceros was let outside after a winter of confinement:

> Such a sight has seldom been seen, as to behold the rhinoceros kicking & rearing, (though neither end reached any great height) out of joy. – it galloped up & down its court surprisingly quickly, like a huge cow, & it was marvellous how suddenly it could stop & turn round at the end of each gallop. – The elephant was in the adjoining yard & was greatly amazed at seeing the rhinoceros so frisky: He came close to the palings & after looking very intently, set off trotting himself, with his tail sticking out at one end & his trunk at the other, – squeeling & braying like half a dozen broken trumpets.[23]

After his detailed description of the animals in the zoo, he turned to Harriet and Erasmus, remarking that Miss Martineau had been "as frisky lately ⟨as⟩ the Rhinoceros. – Erasmus has been with her noon, morning, and night: – if her character was not as secure, as a mountain in the polar regions she certainly would loose it."[24] Here again, as he likened Martineau to an absurdly frisky rhinoceros and Erasmus to an equally ridiculous elephant, he pushed them outside the bounds of normal Victorian expectations. Even his comment that Martineau's virtue was "as secure as a mountain in the polar regions" verged on dehumanization. And yet the same letter concluded, "She is a wonderful woman."[25]

All Darwin's letters about Harriet Martineau are shot through with ambivalence. He seems to have admired and liked her—and he may have felt some jealousy at her great publishing success. But he was not accustomed to looking up to women or to seeing them as role models. He did not mention her at all in his *Autobiography*. This omission was particularly egregious since it is probable that Martineau was the person who encouraged him to read Malthus. In 1838, while Darwin was living in London and regularly discussing his work with Martineau, he realized that Malthus held the key to his developing theory of evolution. Based on a statistical analysis of human populations, Malthus had argued that larger numbers of people meant greater competition for resources and that such competition would lead to famine and eventually to the death of the weaker people and the survival of the stronger. Darwin wondered if the struggle for survival might explain how species changed.

Darwin had read Martineau's little books illustrating Malthusian political economy while aboard the *Beagle*; it is likely she reminded him to read *An Essay on the Principle of Population* in one of their conversations about writing. Yet although he repeatedly acknowledged the ways in which Lyell had inspired him, he did not record—and probably did not recognize—Martineau's influence. In his *Autobiography*, Darwin recalled the moment it became apparent that "by following the example of Lyell in Geology, and by collecting all facts which bore in any way on the variation of animals and plants under domestication and nature" he might be able to formulate a theory that explained the transmutation of species. He remembered his conversations with "skilful breeders and gardeners" and his extensive reading, but he did not mention reading any of Martineau's works or discussing Malthusian social theories with her. In his *Autobiography*, he wrote:

> When I see the list of books of all kinds which I read and abstracted, including whole series of Journals and Transactions, I am surprised at my industry. I soon perceived that selection was the keystone of man's success in making useful races of animals and plants. But how selection could be applied to organisms living in a state of nature remained for some time a mystery to me.
>
> In October 1838, that is, fifteen months after I had begun my systematic enquiry, I happened to read for amusement "Malthus on Population," and being well prepared to appreciate the struggle for existence which everywhere goes on from long-continued observation of the habits of animals and plants, it at once struck me that under these circumstances favourable variations would tend to be preserved, and unfavourable ones to be destroyed. The result of this would be the formation of new species. Here then I had at last got a theory by which to work.[26]

It is particularly striking to compare Darwin's description of "following the example of Lyell in Geology" with his statement that he "happened to read" Malthus "for amusement." In his memory Lyell was a serious mentor, while Martineau, who had written the best-selling *Illustrations of Political Economy* he read aboard the *Beagle*, was unworthy of direct mention. He saw her—and her work—as a chance diversion, not an important influence.

In 1838, when Darwin and Martineau were both in London, meeting frequently at Erasmus's parties and comparing their "methods of writing,"[27] Darwin was hard at work on transforming his journal from the voyage of the *Beagle* into a publishable book. At the same time, Martineau was engaged in a similar writing project, revising her journals from her trip to the United States into what would eventually be published as *Retrospect of Western Travel*. At this point Darwin was unpublished, and Martineau was a successful author who had written many best sellers. Darwin probably read Martineau's travel book—a letter he received from his cousin Emma Wedgwood mentioned that she was reading it "with great pleasure."[28]

If Darwin read Martineau's *Retrospect of Western Travel* when it was published in 1838, he may have noticed her brief mention of a visit to the geologist Edward Hitchcock in Amherst, Massachusetts: "The professor showed us the Turkey Tracks, the great curiosity of the place; and distinct and gigantic indeed they were, deeply impressed in the imbedded stone."[29] Martineau's cursory reference to the fossil footprints followed her much more detailed description of the "forty or fifty girls" she saw attending an Amherst College geology lecture.[30] She was interested in women's education but not particularly interested in Hitchcock's fossils. Darwin, on the other hand, paid much more attention to geology than to women's rights. He did not strongly object to Martineau's political ideas, and later in his life, he would offer half-hearted support to the movement for women's suffrage. But the girls on the long wooden benches in the lecture halls would not have intrigued him as they did Martineau. She understood that once they were allowed to study, women might make extraordinary contributions to art and science. Indeed, it would not be long until Dickinson started learning geology. By the time Darwin published his ideas about evolution, Dickinson would be ready.

But in 1838 Charles Darwin had little time to imagine the schoolgirls who attended Amherst College lectures. He was frantically trying to finish his own first book and to make his way as a professional geologist. He was almost 30 years old, and he was looking for a wife. His family teased him about his relationship with Martineau—one aunt even jokingly sent him an engagement present. But he had other ideas.

Since he was writing constantly, it should come as no surprise that Darwin's papers contain an entry about whether to get married. When he wrote out a list of reasons to "Marry" or "Not Marry," he did not express any erotic interest in the prospect. His main worry was that marriage would distract him from his work. On the "Not Marry" side, he listed "Fatness & idleness—anxiety and responsibility—less money for books." On the "Marry" side, he mentioned "female chit-chat," commented that a wife would be "better than a dog, anyhow," and hoped that marrying would help him avoid being "friendless & cold & childless."

Eventually, he proposed to his cousin Emma Wedgwood with all the ardor of a man hiring the housekeeper recommended by his family.

His father had married a Wedgwood and his sister Caroline had married her Wedgwood cousin Josiah III (known as Joe). When Charles proposed to Joe's sister Emma, their families were delighted, in part because another cousin marriage consolidated the Wedgwood-Darwin fortunes even more. The two of them had been cordial friends since childhood, with few signs of romantic interest despite their awareness of their families' desire for the match. When Charles did finally propose, Emma was taken completely by surprise. Immediately after Emma agreed to marry him, Charles was struck with a terrible headache and took to his bed.

A few days later, Emma wrote to her Aunt Jessie Sismondi with the good news. It was a slightly awkward letter, since she had denied any attraction to Charles. First, she explained her reticence: "I was not the least sure of his feelings . . . and the week I spent in London on my return from Paris, I felt sure he did not care about me." Then she announced the surprising news that "On Sunday he spoke to me, which was quite a surprise, as I thought we might go on in the sort of friendship we were in for years, and very likely nothing come of it after all. I was too much bewildered all day to feel my happiness." Emma seems to have been genuinely confused, as unsure of her own feelings as she had been of Charles's. She confessed, "I believe we both looked very dismal (as he had a bad headache) for when Aunt Fanny and Jessie [Wedgwood] went to bed they were wondering what was the matter and almost thought something quite the reverse had happened."[31] Charles was uncertain as well. When his father advised him to conceal his religious doubts from his prospective bride, he did the opposite, telling her immediately. If he wanted to back out, the confession backfired somewhat. If he wanted to establish open and sincere communications, it was a good step. Emma praised his sincerity rather than condemning his views.

In January 1839 Charles and Emma married. That year, Darwin published his first book (*Journal of Researches*, which would eventually be known as *The Voyage of the Beagle*) and was elected a fellow of the Royal Society. At the end of the year, Emma and Charles's first child, William, was born. From the start, Emma and Charles were comfortable with

Emma Wedgwood Darwin, marriage portrait by George Richmond (1840). Wikimedia Commons.

each other. They got along very well. But in the first years of their marriage, Charles saved his passion for his work. In his *Autobiography*, Darwin remembered that time as a scientific social whirl. He wrote, "During the early part of our life in London, I was strong enough to go into general society, and saw a good deal of several scientific men, and other more or less distinguished men. . . . I saw more of Lyell than of any other man, both before and after my marriage."[32] Emma was not so enthusiastic about Lyell. The society of scientists could be heavy going. In an 1839 letter, she described one of the first dinner parties she hosted: "In my opinion, Mr. Lyell is enough to flatten a party, as he never speaks above his breath, so that everybody keeps lowering their tone to his. Mr. Brown,

Charles Darwin, marriage portrait by George Richmond (1840). Wikimedia Commons.

whom Humboldt calls 'the glory of Great Britain,' looks so shy, as if he longed to shrink into himself and disappear entirely; however, notwithstanding these two dead weights, viz. the greatest botanist and the greatest geologist in Europe, we did very well and had no pauses."[33]

Yet despite Emma's lack of interest in his colleagues, Charles Darwin and Emma Wedgwood Darwin were remarkably happy. Emma may have been surprised when Charles decided to embark on a study that compared their first child's psychological development to the growth of a young orangutan at the London Zoo, but she was also delighted that he paid such close attention to his infant son. Within three years they decided that the social obligations of London were simply too onerous and began to search for a house in the country. In 1842 they would move to Down House in Kent. As the years went on, Charles Darwin would become a devoted family man. The epitome of a wealthy Victorian family, the Darwins eventually had ten children, seven of whom lived to adulthood.

Perhaps one of the reasons why Charles Darwin was able to tear himself away from the scientific society of London was because his dear friend Charles Lyell had already departed, following Harriet Martineau's example to make a lecture tour of the United States that would eventually be chronicled in a book. Like Martineau, Lyell visited Amherst and spent time with Edward Hitchcock. If he did enter the lecture halls of Amherst College during his visit, he might have glimpsed Emily Dickinson and her Amherst Academy classmates in attendance.

Lyell's description of Edward Hitchcock's fossil "turkey tracks" was more detailed than Martineau's:

> At Smith's Ferry, near Northampton, about eleven miles north of Springfield, I examined, in company with the Professor, the red sandstone on the banks of the Connecticut River, where the celebrated foot-prints of birds are beautifully exhibited. The rock consists of thin-bedded sandstone (New Red, Trias?) alternating with red coloured shale, some of the flags being distinctly ripple-marked. The dip of the layers, on which the Ornithichnites are imprinted in great abundance, varies from eleven to fifteen degrees. It is evident that in this place many superimposed beds must have been successively trodden upon, as different sets of footsteps are traceable through a thickness of sandstone exceeding ten feet.[34]

Martineau had probably seen a few samples of giant footprints on display at Amherst College. Lyell, on the other hand, had the chance to visit the banks of the Connecticut and see the remarkable stretch of sandstone covered with tracks. He was amazed and impressed, and he would correspond with Hitchcock for many years afterward. Perhaps he encouraged Hitchcock to send a copy of the 1845 edition of *Geology of Massachusetts* to his protégé.

In 1845 Darwin wrote back to Hitchcock in Amherst to thank him for the volume; his letter rests in the archives of Amherst College where many fragments of Dickinson's writing are also stored.

Down Bromley Kent

Nov. 6th.

Dear Sir

Absence from home has prevented me sooner acknowledging your truly generous present of the *Final Report on the Geology of Massachusetts.* – I assure you I feel sensibly the honour & kindness you have done me. I have as yet read only a little, but I see that there will be much that will interest me greatly; I allude more especially to your detailed accounts of the alluvial deposits, ice & water action, &c. &c. Your's is indeed a magnificent work with its numerous & striking illustrations. I am delighted to possess the excellent plates on the footsteps, & I daresay I shall find some further information, though I have carefully read your several papers. In my opinion these footsteps (with which subject your name is certain to go down to long future posterity) make one of the most curious discoveries of the present century & highly important in its several bearings. How sincerely I wish that you may live to discover some of the bones belonging to these gigantic birds: how eminently interesting it would be [to] know, whether their structure branches off towards the Amphibia, as I am led to imagine that you have sometimes suspected. The finding the bones of the Rhynchosaurus in the pure hard sandstone of Grindshill in Shropshire (where there are some Reptile footsteps) may give one hopes.

 I am preparing a little volume on the geology of S. America, which, when published next summer, I will beg you to do me the kindness to accept; though it is a miserable acknowledgment for your grand work.

With my sincere thanks and much respect. Pray believe me, dear Sir

> Yours faithfully &
> obliged. C. Darwin[35]

This letter gives evidence that Edward Hitchcock in Amherst was among "the leading scientific men" who were part of Charles Darwin's circle. It also mentions the question of whether the fossil tracks in Massachusetts were avian, amphibian, or reptilian. Richard Owen, the comparative anatomist who cataloged Darwin's fossils in London, had coined the name *dinosaur* in 1841. Hitchcock acknowledged that dinosaurs had roamed the earth, but he insisted that the Massachusetts tracks belonged to the creatures he imagined as sandstone birds. Perhaps a brewing question about the origin of species underlay the controversy over whether the footprints had been made by birds, salamanders, or lizards. In 1845 neither Edward Hitchcock nor Charles Darwin was quite ready to jump into public debates about evolution. To the contrary, Darwin's letter to Hitchcock was modest. He made no assumptions about the implications of the footprints.

In 1846 Darwin published the last of his books about the voyage, *Geological Observations on South America*. As promised, he sent a copy to Professor Hitchcock at Amherst College. With this publication, Darwin fulfilled the vision of Saint Jago. In fact, he had published his journal and three books of geological observations. But he had been back in England for a decade and had yet to make public any of his ideas about living creatures—plants or animals. Darwin became increasingly convinced that his notebooks and collections contained crucial evidence to support his emerging theory of evolution, but he also knew that the Anglican scientific establishment of the 1840s contained even less room for evolutionary thought than his grandfather's more revolutionary times.

For Dickinson, who avoided certainty both scientific and religious, these questions were exciting. She was profoundly attracted to cosmic mystery. She understood that geology was one of the most fascinating ways to plumb these mysteries. Like many of her contemporaries, Dickinson would be awestruck by Lyell's *Principles of Geology*, which argued that human history on earth was miniscule in the scale of deep time. In her corner of Massachusetts, where the riverbanks were traced with fossil footprints and Edward Hitchcock made sure that every schoolchild had a chance to play with fossils, the vast expanse of geological time was particularly vivid. It fired her imagination.

Both Hitchcock and Darwin would have agreed that science was imaginative work. A comparative anatomist could look closely enough at a footprint or a bone fragment that a great bird or lizard emerged from the tiny trace of the past. A botanist could look deep into the heart of a flower and unravel the tangled relationships between plants and pollinators.

A few years later, Dickinson would write a poem about this kind of imaginative science:

> A science – so the Savans say,
> "Comparative Anatomy" –
> By which a single bone –
> Is made a secret to unfold
> Of some rare tenant of the mold –
> Else perished in the stone –
>
> So to the eye prospective led,
> This meekest flower of the mead
> Upon a winter's day,
> Stands representative in gold
> Of Rose and Lily, manifold,
> And countless Butterfly![36]

"A science – so the Savans say" would question the division between poetry and science that had been widening since Coleridge and Whewell's confrontation leading to the new distinction between "scientist" and "artist." In her poem, Dickinson would celebrate imaginative projections like Hitchcock's great "Sandstone Bird" and Darwin's concept of the manifold variation of species implied by "the meekest flower of the mead."

But in 1845, when Darwin wrote to Edward Hitchcock, neither was aware that a poet as talented as Emily Dickinson was starting to pay attention to their ideas. Despite his friendship with Harriet Martineau, Darwin did not imagine that women might be among the "Savans." Perhaps that was why he preferred to describe his circle as a group of "leading scientific men." The concept of a "scientist" may have been a little too inclusive for Darwin.

CHAPTER 7

Religion of Geology

*South Hadley, Amherst, 1847–1851;
Dickinson, Age 16–20*

ON OCTOBER 1, 1847, prospective students at Mount Holyoke Female Seminary were roused at 6:00 a.m., their first introduction to the strict schedule at the school. Sixteen-year-old Emily Dickinson woke up determined to conquer her homesickness. She had made it through her second night in the dormitory of the large brick building in South Hadley, Massachusetts. Though Mount Holyoke was not quite ten miles away from Amherst, the journey took a few hours by horse-drawn carriage. Afterward, she had needed a full day to recover. Leaving home had been hard, and the long day of travel had exhausted her. The building was intimidatingly large—ninety-four feet long and fifty feet wide, with classrooms and common rooms on the lower floors and dormitory rooms on the third and fourth floors.

Now, on her second morning, it was time for Dickinson to leave her small dormitory room and go down to the first floor to start the entrance examinations. She was nervous. Everyone from home expected her to do well, but their high expectations added to the pressure. Even though she had prepared thoroughly at Amherst Academy, the odds were daunting. Because there was very little standardization among schools, there was no way to gain entrance to Mount Holyoke based on a school record. Instead, hopeful applicants traveled from across the

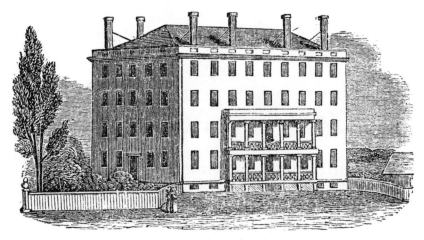

Mount Holyoke (1837).

United States to attempt the three-day-long entrance examinations in South Hadley. That year saw more candidates than usual, so the standards were more severe than ever. Mount Holyoke could accept fewer than half the young women who applied. Looking back, Dickinson wrote, "I never would endure the suspense which I endured during those three days again for all the treasures of the world."[1]

That night, on an island thirty miles off the Massachusetts coast, 29-year-old Maria Mitchell climbed a wooden ladder up to her rooftop observatory atop the Pacific Bank in Nantucket. Mitchell was an accomplished astronomer and mathematician—the first woman to be employed by the U.S. Nautical Almanac as an astronomical observer and calculator. When she scanned the sky that October night, she saw something new: a bright blur that had not been there before. After diligently observing the skies for many years, Mitchell had discovered a comet. Comet Mitchell, as it would come to be called, was a *hyperbolic comet*—its orbit traced out an open-ended hyperbola rather than a closed ellipse. It would never return to earth. No American astronomer (male or female) had made an original discovery on this scale before. The comet made her internationally famous. Mitchell would become the most celebrated American scientist of the decade.

Maria Mitchell by Herminia Dassel (1851). Maria Mitchell Association.

On the night that Mitchell discovered the comet, Dickinson settled to sleep in the unfamiliar dormitory bed, wondering if she would gain admission to Mount Holyoke. It was difficult to picture herself as a college student since none of her women teachers had ever been college students themselves. Growing up in Amherst, Dickinson had known hundreds of college boys. Her brother, Austin, was studying at Amherst College that year. Although Maria Mitchell would eventually become one of the founders of Vassar College, she herself had little formal education and neither did Mary Lyon, who founded Mount Holyoke. The young women who attended Mount Holyoke and the women's colleges founded in the subsequent decades were in the first generation of women given the rare, new opportunity to study at the college level.

Dickinson found it hard to imagine what might come after attending Mount Holyoke. Only a few professions were open to educated women. There was teaching, of course, either in local schools or far-flung missionary schools. Most other professions—law, clergy, military—were out-of-bounds for women. The great exception in the 1840s was the natural sciences. These fields had not yet been completely professionalized. They were still open to all. During the months when Dickinson was at Mount Holyoke, Maria Mitchell's comet would burst into popular consciousness, and Maria Mitchell would become a role model for studious American girls and women. A Mount Holyoke student at the time described the large brick building as a "castle of science."[2] Mount Holyoke students hoped there might be an open niche for them somewhere in the natural sciences. Botany, chemistry, and astronomy beckoned.

At 16 years old, Emily Dickinson was younger than most of the women examined in October 1847, but as her sister proudly recalled, "Emily was never floored."[3] She not only gained admission but also placed in an advanced class that would spend a few weeks reviewing the junior curriculum and then pass to the middle level. In the first few decades, the school offered instruction at junior, middle, and senior levels, though it was rare for a student to attend for three years. Most commonly, young women attended for just one year. She would live in the large, factory-like brick building with 235 students and 12 teachers.

Mount Holyoke was not luxurious. In order to keep costs down, each student was assigned a small daily housekeeping task. Dickinson helped out in the dining room, laying out the knives before each meal and collecting them afterward. Her bedroom was cramped and often very cold. While Dickinson did not complain, she was glad she had not tried to bring any of her plants. They might have frozen. A more significant drawback was the overlap between Mount Holyoke and the schools in Amherst. As it turned out, Dickinson had completed significant portions of the curriculum at Amherst Academy. For example, Dickinson had studied botany and completed her own herbarium before she went to South Hadley. Perhaps this was fortunate; so many students at Mount Holyoke needed to collect plants for herbariums that the area around campus was somewhat denuded. That year, Mary Lyon would enact a rule that flowers could not be picked within three miles of campus.

Though Dickinson was exempted from studying botany (as well as natural philosophy and Latin), she did not lose interest in the topic. From Mount Holyoke, she wrote to Abiah Root:

> How it calms my mind when wearied with study to walk out in the green fields and beside the pleasant streams in which South Hadley is rich! There are not many wild flowers near, for the girls have driven them to a distance, and we are obliged to walk quite a distance to find them, but they repay us by their sweet smiles and fragrance. . . . In our rambles we found many and beautiful children of spring, which I will mention and see if you have found them, – the trailing arbutus, adder's tongue, yellow violets, liver-leaf, blood-root, and many other smaller flowers.[4]

Dickinson's description of going for walks and observing flowers ends with a list of plants that tilts toward botanical specificity. At the same time, this passage also invokes a Wordsworthian, "Tables Turned" style of Romanticism. The blend of sentiment and science was relatively typical of botanical thought at the time. Perhaps most intriguingly, the "sweet smiles" of the wildflowers hint that Dickinson imagines plants as sentient beings that might be capable of communicating. In her discussion of "Dickinson and the Politics of Plant Sensibility," Mary Kuhn explores Dickinson's sense of the "radical possibility" that plants have the "capacity to feel."[5] Kuhn explains that for Dickinson, as for other young botanists in the nineteenth century, plants were paradoxically both "collectible objects" and "vital subjects."[6]

Nineteenth-century natural philosophers were fascinated by questions about life. What was it that made material things alive? Where was the dividing line between living matter and dead matter? Philosophical questions about the "vital" rose from many scientific subjects, including botany and all the branches of biology. After Dickinson passed the junior exams, her coursework focused on chemistry and physiology, astronomy and rhetoric. She found the topics fascinating. In a letter to her brother, she described herself as "all engrossed in the history of

Sulphuric Acid!!!!!" (the five exclamation points are hers).[7] Botany, chemistry, and physiology were directly connected to discussions of vitalism. Questions of life and death, matter and spirit permeated her astronomy studies at Mount Holyoke, just as they had permeated her study of geology at Amherst Academy.

Students at Mount Holyoke were expected to tackle the big questions. Founded in 1837, Mount Holyoke was called a "Female Seminary" at first. It would not be renamed Mount Holyoke College until 1893. At the time, "seminary" was often used for women's institutions of higher learning, though the term could also be applied to men's schools. The name often implied that there would be an emphasis on seminar-style education, where the teachers studied and learned alongside the students. This was particularly appropriate for women's schools since very few teachers had formal training. When Mary Lyon was raising the money for the school, she lived in Amherst with Edward and Orra Hitchcock, who were staunch supporters of the endeavor. Edward Hitchcock had hoped she would call it a "Pangynaskean" (his own coinage from the Greek, perhaps meaning "school for all women" or "school for making whole women"). His suggestion was widely mocked. Wisely, Mary Lyon stuck with "Female Seminary."

In New England in the nineteenth century, the term "seminary" did not necessarily have religious connotations. Nonetheless, like Amherst College, Mount Holyoke was conceived with a clearly defined religious mission. Mary Lyon was a passionate advocate for women's education and a skilled professor of chemistry who had designed a rigorous, science-heavy liberal arts curriculum. She was also a deeply committed evangelical Christian who believed that educating women would help them to become better Christian proselytizers. Many of the students who enrolled at Mount Holyoke shared Lyon's beliefs. They saw their training as preparation for becoming overseas missionaries. Some of the most generous donors to the school hoped Mount Holyoke Female Seminary would be an engine of evangelical revivalism.

In addition to qualifying academic examinations, students at Mount Holyoke also faced religious examinations. During the first week of school, Mary Lyon held a meeting to determine each student's spiritual

Emily Dickinson daguerreotype (ca. 1847). Amherst College Archives and Special Collections.

state. One by one, the students were asked to stand and submit to questioning to determine their "class" of belief. Then, in a process that mirrored the academic sorting, they were divided into three classes. First were the "professors," who had already made public professions of faith and become members of the orthodox Congregationalist Church. Next, in the second class, came students with "a hope," who had started preparing to make this public commitment. Finally, the third class was made up of those who had "no hope" of joining the church. They were known as the "impenitents." The Mount Holyoke *Journal* for 1847–1848 recorded that "more than one" student wept as her name was added to the list of seventy students who had "no hope."⁸ One member of this group, Emily Dickinson, was defiantly "impenitent." Perhaps being relegated to the spiritual third class made her sad, but the proceedings also sparked resistance.

The young Emily Dickinson did not share Mary Lyon's religious feelings. She was committed to the academic vision of Mount Holyoke, but she was skeptical of the revivalist impulse. Indeed, though Dickinson admired Lyon, she did not approach things with the reverence Lyon expected. In a "strictly confidential" letter to her brother, she chafed against "correct ideas of female propriety & sedate deportment" and mocked her teachers' constant search for anything that "savored of rebellion or an unsubdued will."⁹

Dickinson's first term at Mount Holyoke flew by, and in late November she returned home to Amherst for Thanksgiving. Although it had been more than two hundred years since the Massachusetts Bay Colony had been founded as a Puritan theocracy, some traces of Puritanism continued. For the Dickinson family, as for most of their neighbors, Thanksgiving was a much more significant celebration than Christmas, which had been forbidden by the Puritans for many years because they believed the holiday was more pagan than Christian. Emily's brother Austin later recalled, "I was brought up to believe" that Christmas day "was a joint device of the Devil and the Romish church for the overthrow of the true religion and accordingly to be frowned upon by all good people."¹⁰

In keeping with the general Massachusetts preference for Thanksgiving, Mount Holyoke did not offer students a Christmas holiday break. Nonetheless, the mood was different at Mount Holyoke on December 24 and 25, 1847. Instead of festivity, the school was primed for religious revival. Mary Lyon declared December 24 a day of fasting and penitence. That evening she made an impassioned plea with her students to declare their commitment to Christianity. Lyon was very persuasive. Usually only a handful of students held out against her urging; in one year she actually managed to convert every single one of the no-hopers.¹¹ Around Christmas 1847, Hannah Porter, one of the school's most dedicated supporters, came to Mount Holyoke for a twelve-day visit, timed to coincide with the planned revival. Porter was invested in Mount Holyoke's interlocking religious and educational missions. At the start, she had persuaded her husband to fund the construction of the main campus building. According to Alfred Habegger, during her visit in 1847 Porter "did so much praying, preaching, and exhorting" that by December 27 she had lost her voice.[12] For the next week, she met with students individually, whispering hoarse questions about their spiritual lives. As a member of the First Female Praying Circle of Monson, Massachusetts, with ties to Dickinson's mother and her family, Porter was particularly interested in converting Dickinson. After Porter went home to Monson, she corresponded with Dickinson's roommate (her cousin, Emily Lavinia Norcross), her teacher Mary C. Whitman, and her classmate Sarah Jane Anderson about Dickinson's spiritual state. On a cold night in January, when the temperature fell to forty below zero, Cousin Emily wrote to Hannah Porter with the disappointing news that "Emily Dickinson appears no different. . . . She says she has no particular objection to becoming a Christian . . . but still she feels no more interest."[13]

Dickinson's declaration of "no interest" was a carefully considered response to the pressure. In 1844, when she was 13, she had been swept up in an Amherst revival, but her feeling of conviction had faded within a few weeks. Looking back, she recalled that "I seemed to lose my interest in heavenly things by degrees."[14] She was surprised and perhaps even a little frightened at her own indifference. After her friend Abiah committed to evangelical Christianity in 1846, Dickinson wrote, "I shed many a tear & gave many a serious thought to your letter & wished that I had found the peace which has been given to you." Caught up in the intense emotions—and metaphors—of the revival, she added, "I feel that I am sailing upon the brink of an awful precipice, from which I cannot escape & over which I fear my tiny boat will soon glide if I do not receive help from above."[15] Although Abiah probably understood Dickinson to mean that she was on the edge of sailing over a cliff toward damnation, the reference is curiously unmoored. It is not clear which Dickinson dreaded more: belief or unbelief. She may have feared taking the plunge toward evangelical Christianity as much as—or even more than—she feared hell.

In another letter to Abiah a few months later she wrote:

> I am not unconcerned Dear A. upon the all important subject, to which you have so frequently & so affectionately called my attention in your letters. But I feel that I have not yet made my peace with God. I am still a s[tran]ger – to the delightful emotions which fill your heart. I have perfect confidence in God & his promises & yet I know not why, I feel that the world holds a predominant place in my affections. I do not feel that I could give up all for Christ, were I called to die. Pray for me Dear A. that I may yet enter into the kingdom, that there may be room left for me in the shining courts above.[16]

Although she had not yet studied rhetoric at Mount Holyoke when she wrote this 1846 letter, Dickinson's response to Abiah was rhetorically brilliant. By conceding to all of Abiah's religious exhortations and explaining that she was waiting for God to call her (and to fill her with conviction), she effectively ended the discussion. The following year, when Mount Holyoke staged its own revival, Dickinson was well prepared to withstand the pressure of insistent proselytization. At times, she may have wished she could just give in. After the revival subsided at Mount Holyoke, Dickinson wrote to Abiah about how much she regretted her own intransigence:

> I tremble when I think how soon the weeks and days of this term will all have been spent, and my fate will be sealed, perhaps. I have neglected *the one thing needful* when all were obtaining it, and I may never, never again pass through such a season as was granted us last winter. Abiah, you may be surprised to hear me speak as I do, knowing that I express no interest in the all-important subject, but I am not happy, and I regret that last term, when that golden opportunity was mine, that I did not give up and become a Christian. It is not now too late, so my friends tell me, so my offended conscience whispers, but it is hard for me to give up the world. I had quite a long talk with Abby while at home and I doubt not she will soon cast her burden on Christ. She is sober, and keenly sensitive on the subject, and she says she only desires to be good. How I wish I could say that with sincerity, but I fear I never can.[17]

This letter is particularly ambiguous. Dickinson did not argue against any of the tenets of Christianity or even object to the pressure brought to bear upon her to declare her faith. She simply explained that as much as she regretted it, she could not "give up the world."

Dickinson's sorrow may have been genuine. It is quite possible that there were times when she longed to assent to a single, simple, clear-cut creed. On the other hand, her 1848 letter offered an even stronger rhetorical defense than the one she had tried in 1846. By validating the Christians' beliefs—even agreeing with their wish for her to join the church—Dickinson made it nearly impossible for them to argue with her. Throughout the winter she resisted Principal Mary Lyon's insistent efforts to get her to declare her faith unequivocally; Dickinson stayed firmly on the list of students "without hope." Dickinson's cousin Clara Newman Turner recalled that Dickinson had told her about a meeting where "all those who wanted to be Christians" were asked to stand up, and Dickinson was the only person who remained seated. Turner wrote, "In relating the incident to me, she said, 'They thought it queer I didn't rise'—adding with a twinkle in her eye, 'I thought a lie would be queerer.'"[18]

Yet despite her clever self-defense, the pressure must have felt intense that winter. In March one of her Amherst friends visited and reported to her parents that Emily was looking fatigued. More ominously, she had developed a cough. At her family's insistence, Dickinson left for spring break a few weeks early and spent the month of April recovering her health and studying at home. Shortly after she returned to Mount Holyoke in May, she announced in a letter to Abiah that she did not plan to attend the seminary the following year. Attributing the decision to her father, Dickinson did not express great eagerness for a second year at Mount Holyoke, but she did not say much against it either.

One of the primary reasons Dickinson did not complete the course may have been because it was not particularly challenging for her. Because Amherst Academy was so closely intertwined with Amherst College, Dickinson had had the chance to attend college-level lectures for many years. The overlap made the coursework at Mount Holyoke less stimulating than Dickinson had probably expected. Nonetheless, it is notable that Emily Dickinson enjoyed her schooling much more than Charles Darwin ever had. She may not have liked the pressure to convert, but she was perfectly capable of evading the proselytizers, and—though she hesitated to admit it to Abiah—there were times when rebellion filled her with glee.

In February 1848, as Dickinson was resisting intense pressure to join the church, Henry David Thoreau delivered two lectures on "Resistance to Civil Government" at the Lyceum in Concord, Massachusetts. Although Dickinson was an admirer of Emerson, she probably did not know much about his protégé, the man who had lived in a cabin on the Emersons' woodlot near Walden Pond for two years. In the fall of 1847, a few weeks before Dickinson moved to South Hadley, the 30-year-old Thoreau relinquished his cabin to stay with Ralph Waldo Emerson's family while his mentor made a lecture tour of Europe.

While Dickinson studied at Mount Holyoke and Thoreau took care of the Emersons' house in quiet Concord, Ralph Waldo Emerson experienced the tumultuous European politics of 1847–1848 firsthand. Although he was based in England, Emerson traveled to Paris for a few days to witness the revolutionary fervor that spread across Europe. The Transcendentalist writer Margaret Fuller was embedded in Rome, participating wholeheartedly in the Risorgimento—the idealistic revolutionary uprising that led to the unification of Italy. In England the streets were relatively peaceful, though revolutionary ideas were widespread. That year, Charlotte Brontë published *Jane Eyre*; Karl Marx published *The Communist Manifesto*.

This was an eventful period in North America as well. In the Mexican-American War, the United States and Mexico battled for control of the Texas territories. Slavery was legal in Texas, and its admission to the Union would tilt the balance of power in the U.S. Congress toward slavery. In New York, a cholera epidemic raged. In California, a carpenter struck gold. At Seneca Falls, a group of feminists signed the *Declaration of Sentiments*. Perhaps because she came of age in this era of revolutions, Dickinson would always take pleasure in things that "savored of rebellion."

However, at the time both Dickinson and Thoreau felt strangely disconnected from these events. Dickinson wrote to her brother from Mount Holyoke asking for news:

> Wont you please to tell me when you answer my letter who the candidate for President is? I have been trying to find out ever since I came here & have not yet succeeded. I dont know anything more about the affairs in the world, than if I was in a trance, & you must imagine with all your "Sophomoric discernment," that it is but little & very faint. Has the Mexican war terminated yet & how? Are we beat? Do you know of any nation about to besiege South Hadley? If so, do inform me of it, for I would be glad of a chance to escape, if we are to be stormed. I suppose Miss Lyon. would furnish us all with daggers & order us to fight for our lives, in case such perils should befall us.[19]

While Dickinson joked, Thoreau took a more serious tone. In his essay on "Resistance to Civil Government," he remarked: "Practically speaking, the opponents to a reform in Massachusetts are not a hundred thousand politicians at the South, but a hundred thousand merchants and farmers here, who are more interested in commerce and agriculture than they are in humanity, and are not prepared to do justice to the slave and to Mexico, cost what it may. I quarrel not with far-off foes, but with those who, near at home, co-operate with, and do the bidding of those far away, and without whom the latter would be harmless."[20] Their reactions to living in politically disengaged Massachusetts communities were very different. Thoreau's outrage at the unthinking complicity of his neighbors was not nearly as lighthearted at Dickinson's mockery of the way her school community ignored events in Mexico. Nonetheless the parallel between Thoreau's and Dickinson's situations that winter is striking. Both were highly aware that their lives in Massachusetts were disquietingly insulated from the political upheaval of their time.

In addition to his Lyceum lectures, Thoreau spent that winter revising the manuscript of *A Week on the Concord and Merrimack Rivers*, a digressive and at times profoundly philosophical account of a boat trip with his brother John, who had died in 1842.[21] A melancholy book

Henry David Thoreau. National Portrait Gallery, Washington, DC.

suffused with mourning for his brother and for the environmental changes they had witnessed on the industrializing riverbanks, Thoreau's *Week* did not find an audience. Unable to sell it to a publisher, he paid to have it printed in 1849. By 1853, when fewer than three hundred copies had sold, the remainders were returned to Thoreau, who remarked, "I have now a library of nearly nine hundred volumes, over seven hundred of which I wrote myself."[22] When *Walden* was published in 1854, it was much more widely read. After *Walden*, Dickinson probably knew of Thoreau, though it is not certain that Dickinson read any of his writing before 1858, when "Chesuncook" appeared in the *Atlantic Monthly*. It is unlikely that Thoreau ever read any of Dickinson's work.

Nonetheless, unbeknownst to either, a deep resonance between Dickinson's and Thoreau's thought was beginning to develop. They both adored Emerson's writings. Perhaps their interest in nineteenth-century science was prompted in part by Emerson. However, Emerson himself did not end up studying natural science in much depth. Both Dickinson and Thoreau went further. While Dickinson was studying science at

Mount Holyoke, Thoreau, who had studied a traditional classical curriculum as an undergraduate at Harvard, was educating himself in the sciences. Although he described himself as a man of letters,[23] in the late 1840s he grew more and more interested in natural science. He studied Charles Lyell's *Principles of Geology* and Alexander von Humboldt's *Cosmos* carefully. He paid equal attention to Charles Darwin's first book (*Journal of Researches*, a.k.a. *The Voyage of the Beagle*). In his own *Journal*, Thoreau quoted Darwin's precise observation of a guinea fowl and then expressed his sense of the enchantment of the natural world: "It is a certain faery land where we live. You may walk out in any direction over the earth's surface, lifting your horizon, and everywhere your path, climbing the convexity of the globe, leads you between heaven and earth, not away from the light of the sun and stars and the habitations of men. I wonder that I ever get five miles on my way, the walk is so crowded with events and phenomena."[24]

Prompted by reading Darwin's *Journal*, imagining himself looking at the world through the eyes of a guinea fowl, Thoreau described the earth as a "certain faery land," where he lived at the crux of the vital and the material—always walking on a path "between heaven and earth." Wonder almost stopped him in his tracks.

John Herschel had described students of natural science as those who "walked in the midst of wonders."[25] Thoreau went further than Herschel, experiencing the surface of the earth as crowded with such staggering magic that it was a wonder he could walk at all. The philosopher Jane Bennett explains that Thoreau was "swept up, inextricably enmeshed within a vast web of life extending beyond his powers of cognition and imagination."[26] As Thoreau's biographer Laura Dassow Walls relates, Thoreau was inspired by Humboldt and Darwin to become a *green writer*, a poet-naturalist whose work emerged from his experience of a "green world."[27] Thoreau's developing green attitude did not necessarily contradict orthodox Christianity, but it did not mesh very well with it either. In Thoreau's view, Christian churchgoers risked losing sight of what he saw as truly sacred: the enchanted green world, that "certain faery land" that could only be found on the convex surface of the earth.

In general the mid-nineteenth-century approach to science championed by Herschel, Lyell, and Darwin gave both Thoreau and Dickinson ways of understanding the world that allowed them to push back against social expectations and to rely instead on their own experience and observations. For many, geology was the most compelling branch of science in the late 1840s. As Thoreau put it, "As in geology, so in social institutions, we may discover the causes of all past change in the present."[28] In "Resistance to Civil Government," Thoreau based his political reasoning on his observation that his neighbors' unthinking support of local civil institutions made the distant war possible. The local was tied to the global.

Dickinson's increasingly lonely refusal to join the Congregational Church was similarly connected to her sense of the vast scale of geological and astronomical space and time. In the years since 1802, when William Paley published *Natural Theology*, the boundaries of science had expanded. Questions of scale undermined Paley's contention that the earth could only have been designed by a purposeful deity. There was the problem of the vast scale of geological time: If the planet was billions of years old but humans had only existed for a few million, it was hard to argue that God's purpose in creating the earth was to create human beings. Then there was the problem of space: If the universe was as vast as astronomers thought, it was hard to believe that the earth was special. Why would God care about one dust speck in the unimaginably large universe? It was almost impossible to answer such questions with absolute certainty.

Both Dickinson and Thoreau were becoming convinced that the nonhuman world pointed toward alternatives to human social and religious conventions. They believed that guinea fowl, trailing arbutus flowers, hyperbolic comets, and masses of sandstone inlaid with footprints—all the members of the nonhuman world—were vitally linked to the human world. In the cosmos that they imagined, there were few boundaries. All sorts of beings were connected to each other, often in mysterious ways. These ways of thinking harked back to older concepts of natural magic, even as they grew from nineteenth-century natural science.

From a twenty-first-century vantage point, the nineteenth century's shifting associations between science, magic, and religion can be particularly hard to untangle when they intersect with social and political questions about race, class, and gender. In theory, nineteenth-century scientific ideas pushed more and more against orderly hierarchies as the century progressed. Over time the world that science described grew increasingly chaotic, confusing, and strange. But for scientific practice, the movement went the other way. At the beginning of the nineteenth century, scientific education tended to be more inclusive—and much less organized—than other branches of knowledge. Women, people of color, and working-class people were often excluded from formal classical studies and encouraged to study the natural world in less formal institutional settings. One significant reason why outsiders were encouraged to learn about science was the widespread belief that studying the natural world (which was imagined as inherently hierarchical) would make people accept the social and religious order that structured their lives. By the end of the nineteenth century, theory and practice would reverse. Scientific ideas would come to be perceived as challenging, chaotic, and dangerous while scientific education would become rigidly hierarchical and exclusionary.

When Dickinson and Thoreau were young, studying natural science was often associated with pious young women. Neither orthodox Christians nor impenitent rebels against orthodoxy thought of natural science as opposed to Christianity any more than they thought of scientific ideas as radical. However, by 1847–1848, when Thoreau was revising his essay on resistance to civil government and Dickinson was writing letters to Abiah to explain her resistance to religious conformity, the links between natural science and religious conformity were beginning to fray. In June 1848 some thinkers were starting to see that science might eventually come to be associated with disenchantment and secularization. As science and religion began to grow apart, science came to focus exclusively on nature—the material world, while religion claimed the supernatural—the spiritual world. There was little room for crossover—for enchantment—in this emerging dichotomy.

Both Dickinson and Thoreau were fascinated by the moral and spiritual implications of the material world and alert to the transformative possibilities of the sciences. They studied Emerson carefully, for he had been among the first to intuit that the natural sciences might open completely new spiritual paths. Yet although his essays and lectures were full of grand exhortations, Emerson had never really fulfilled his Paris ambition to become a "naturalist." His work was not deeply grounded in the sciences or in close observation of the natural world. In the late 1840s, Dickinson and Thoreau conscientiously studied scientific works. They were both extremely focused on their own experiences of the physical realities of the nonhuman world around them. Both were drawn to the alternative possibilities emerging from these newly vital subjects, but they kept their ideas somewhat private. In different ways they resisted social pressures and tried to think—and to observe—for themselves. They were in the vanguard. Few of their acquaintances would have connected political and religious autonomy to the sciences.

At Mount Holyoke Female Seminary, almost everyone believed that studying science reinforced Christian obedience. Five weeks before the close of her final term at Mount Holyoke, Emily Dickinson and a large group of her classmates and teachers traveled to Amherst for the grand opening of the Octagon, Amherst College's first science museum, known at the time as a natural history Cabinet.

Leaving South Hadley during the semester was generally frowned upon, but the Mount Holyoke contingent was there to support Edward Hitchcock, who had been a generous mentor to Mary Lyon and a strong supporter of the seminary. Hitchcock had become president of the college two years before. He designed the octagonal building himself, with ample space to display his growing natural history collection. It was quite the spectacle. Rows of giant sandstone slabs indented with fossil footprints were the highlight of the central gallery. In addition to the museum space, the Octagon was equipped with a tower that housed an astronomical observatory with a large equatorial telescope. Alongside the fossil footprints were a few fossilized bones and a large collection of natural history specimens, ranging from Assyrian reliefs to stuffed birds.

Amherst College Octagon, 1865. A Hitch in Time: A Digital Humanities Project @ Amherst College.

The women of Mount Holyoke were in their element. They felt completely at home with every aspect of natural history.

Dickinson had spent her teenage years on Pleasant Street, just a few blocks from Amherst College. She had attended dozens, if not hundreds, of college events and lectures. Professor Hitchcock's children were good friends of the Dickinson children, and his books were treasured by the Dickinson family. She had known about his collection of "turkey tracks" since she was a young girl. Even so, when the Octagon opened, Amherst changed. For the rest of Dickinson's life, one of the most celebrated natural history museums in the United States was just a few blocks from her house.[29] She and her Mount Holyoke classmates and Amherst neighbors could peer back into the distant, fossilized past or out into the vast reaches of space.

Since the program of studies at Mount Holyoke emphasized the sciences, the teachers were as interested in the opening of the Octagon as the students. As it turned out, so many Mount Holyoke women attended the celebration that it was necessary to cancel classes in South

Hadley. Although the Mount Holyoke *Journal* noted the disruption of the schedule, no one in the community saw the natural sciences as ideologically disruptive in the least. Like Edward Hitchcock's Amherst, Mary Lyon's Mount Holyoke was built on the assumption that the natural sciences were inherently safe subjects.

As such, the sciences were particularly suitable for women. In the twenty-first century, it is common to imagine science as a predominantly masculine pursuit. Just as frequently, science is framed in opposition to religion. In 1840s Massachusetts, however, the shared assumptions were very different from those of today. In the eastern part of the state, Harvard University was relatively slow to integrate formal instruction in the sciences into the standard curriculum. Though Harvard did launch a school of science in 1847, led by the scientist Louis Agassiz, up until then students at Harvard had concentrated on classical languages. At the time Harvard was exclusively a men's college. Women might occasionally gain access to the Harvard libraries, but they were forbidden from attending lectures.[30] In contrast, Amherst College encouraged girls and women to attend lectures. Amherst actively supported the opening of Mount Holyoke in 1837. Harvard was more reluctant to support Radcliffe, which would not open its doors until 1879, more than forty years later.

In 1848 Massachusetts, studying natural science was generally associated with both religious conservatism and women's education. In matters of religion, Harvard was much less conservative than Amherst. In matters of education, Harvard was more traditional. Thus, Harvard was less open to women's education and less invested in the natural sciences than Amherst, even though Harvard was a hotbed of Unitarianism, which was a radical challenge to Congregationalist Christian orthodoxy. In contrast, in the western part of the state, Amherst College, Amherst Academy, and Mount Holyoke Female Seminary emphasized natural science and adherence to conservative, revivalist Christianity. The Western Massachusetts institutions were more welcoming to women.

When Lucretia Mott, one of the organizers of the Seneca Falls Convention for women's rights, addressed audiences on the subject of women's political rights, she started from the shared presumption that it was socially acceptable for women to study science and to excel in the pursuit and then argued that it should be equally acceptable for women to enter political life: "Do we shrink from reading the announcement that Mrs. Somerville is made an honorary member of a scientific association? That Miss Herschel has made some discoveries, and is prepared to take her equal part in science? Or that Miss Mitchell of Nantucket has lately discovered a planet, long looked for?" Mott's questions were purely rhetorical. She knew that her listeners thought it was perfectly acceptable for women to be scientists. What worried them was the possibility of women entering politics. Mott commented, "Woman shrinks, in the present state of society, from taking any interest in politics.... Who knows, but that if woman acted her part in governmental affairs, there might be an entire change in the turmoil of political life."[31] Mott's entire argument was based on the premise that no one in her audience would be at all surprised by Maria Mitchell's—or any woman's—success in scientific endeavors. At the Seneca Falls Convention in July 1848, it felt radical to advocate for women's involvement in the political sphere. It did not seem radical or dangerous to celebrate the scientific achievements of women like Maria Mitchell, Caroline Herschel, and Mary Somerville.

The Mount Holyoke women who had thronged the Octagon at Amherst College that summer did not necessarily support the political activism that the Seneca Falls *Declaration of Sentiments* propounded. Most of them probably believed that the proper way to respond to the dizzying array of natural specimens displayed in the Natural History Cabinet was to worship the (Congregationalist, Christian) God who had created them. In England, where education was dominated by the Anglican established church, natural history was sometimes perceived as troublingly democratic because of the way it tended to open the doors to working-class students, women, and religious dissenters. The First Church in Amherst was Congregational, not Episcopalian, and the region was extraordinarily open to social mobility. New England Congregationalists did not perceive natural history as a theological danger any more than a social danger. The politics of natural history were so perfectly aligned with the social order in Western Massachusetts that it would have been hard for Dickinson or her classmates or teachers to imagine there could be anything even vaguely political about women studying science.

Dickinson's imagination was extraordinarily vivid and powerful. She could imagine possibilities and see parallels beyond the scope of most people. Even so, there is little indication that she was drawn to Seneca-Falls-style politics. She was amused by the way that Mount Holyoke cut itself off from political news, but her moments of resistance and rebellion tended to arise in response to religious pressures rather than political debates. Although she had sometimes expressed frustration when the academic work at Mount Holyoke was not challenging, her attitude toward school stayed positive. She loved her studies.

When Dickinson left Mount Holyoke for good on August 3, 1848, she had completed more formal instruction in science than most of the figures mentioned in these pages. Neither Mary Lyon nor Maria Mitchell had had the chance to attend college any more than Edward Hitchcock. Like most New England chemists, astronomers, and geologists, all three were self-taught. Men who attended universities in New England, like Darwin and his contemporaries in Britain, studied science informally, as an extracurricular pursuit. Yet it is unlikely that Dickinson—or anyone else at Amherst College and Mount Holyoke Female Seminary—saw science or scientific education as antiestablishment.

Dickinson never resisted learning about sulfuric acid, hyperbolic comets, or fossil footprints any more than she had resisted learning about cardinal flowers or mosses. She saved her resistance for the church. Back in Amherst the revivals continued and the pressure increased. Emily Dickinson's father and her future sister-in-law Susan Gilbert both joined the First Church of Amherst on August 11, 1850. Her dear friend Abby Wood joined at the same time (seventy people professed their faith and joined that day). A few months later, on November 3, 1850, twenty-two more people joined the church, including Emily's younger sister Lavinia. Their brother, Austin, held out until 1856. Emily was the only member of her family who never joined the First Church of Amherst.

Dickinson's refusal to join the Congregational Church distanced her from many of her childhood friends. Abby Wood wrote to Abiah Root, "What shall I say of our darling Emily? How can I tell you that she ridicules and opposes us, and shuts her own heart against the truth. But her very actions show that the Spirit of God is striving in her bosom, and she is perfectly wretched. I went there the other day & she treated me as if she were insane."[32] Though Dickinson knew that many in her circle disapproved of her to the point of describing her as a madwoman, she could not give in. She wrote to Abiah on her own behalf, declaring "I am one of the lingering bad ones, and so do I slink away, and pause, and ponder, and ponder, and pause."[33] The more she pondered, the more she resisted. A few months later, she wrote again, trying to explain. "The shore is safer, Abiah, but I love to buffet the sea – I can count the bitter wrecks here in these pleasant waters, and hear the murmuring winds, but oh, I love the danger! . . . Christ Jesus will love you more. I'm afraid he don't love me *any*!"[34] Dickinson's irreverence may have troubled Abiah as much as it troubled Abby Wood. In part because of these religious differences, they had begun to grow apart.

In contrast, there does not seem to have been much conflict over religion within Dickinson's immediate family. Her accounts are generally more playful than intransigent. A few years later, in 1862, she described her family: "They are religious, except me, and address an Eclipse, every morning, whom they call their 'Father.'"[35] Her description of God as "an Eclipse" is astronomically precise—the image implies that God is invisible, obscured by the shadow of the earth. It is also an exact description of Dickinson theological thought. She did not necessarily question God's existence. Instead, she asserted God's unknowability.

Although Dickinson used scientific metaphors to express her uncertainty about religion, she did not necessarily imagine science and religion in opposition. Her science teachers tended to be devout Congregationalists. Edward Hitchcock had been a clergyman before he became a geologist, and he had joined the First Church of Amherst at the earliest opportunity. He was passionately interested in finding honest and accurate ways to think about God and the earth and was convinced that scientific truth and religious truth were in harmony with each other.

In 1851 Hitchcock published *Religion of Geology*, a book based on lectures he had delivered in the previous decade. If Dickinson read the book when it came out, its contents would have been quite familiar since she had heard so many of Hitchcock's lectures over the years. He must have been an exciting lecturer. Centuries later, *Religion of Geology* is still capable of inspiring wonder and amazement. Hitchcock argued that the largeness of the universe proved God's existence in a new way. "Surely the mind is as much confounded and lost, when it attempts to conceive of the number of the worlds in the universe, as when it contemplates their distances and magnitudes. In respect to number and distance, at least, we find no resting-place but in infinity."[36] This notion—that the infinite cosmos revealed God's infinity—was an elegant update to Paley's arguments. Hitchcock's discussion of the vast expanses of time and space acknowledged that astronomers' rapidly expanding sense of the scale of the universe could be overwhelming and then urged readers to try to make themselves at home with the infinite. These ideas probably made many students and readers uncomfortable. A few, including Emily Dickinson, reveled in the profound uncertainty at the heart of the idea that there was "no resting place but in infinity."

In a chapter titled "The Telegraphic System of the Universe," Hitchcock turned toward electromagnetism. Here, he made a thrilling declaration: *"Our words our actions and even our thoughts make an indelible impression upon the universe."*[37] Hitchcock mused:

> If, as the philosophers now generally admit, there is a subtle and extremely elastic medium pervading all space, why must they not extend to other worlds, yea to the whole universe? Without an accurate acquaintance with the facts, indeed, it will seem a mere extravagant imagination to say that our most trivial word or action sends a thrill throughout the whole material universe; but I see not why sober and legitimate science does not conduct us to this conclusion. Nay, still further, it teaches us that vibrations and changes which our words and actions produce upon the universe shall never cease their action and reaction till materialism be no more.[38]

The president of Amherst College personified "sober and legitimate science" just as much as he embodied Congregationalist Christianity. Nonetheless, Hitchcock's conception of nature was fundamentally interactive. He believed that the material world and the human mind interacted with each other in mysterious, often magical ways.

Earlier in his career, Hitchcock had tried his hand at a poem that called on an imaginary sorceress to conjure forth the extinct creatures who had left their footprints in the Massachusetts sandstone. In *Religion of Geology,* Hitchcock invited readers to imagine an ancient beast (bird or lizard) whose emotion—sudden fear or animal delight—caused electrical impulses to course through its nervous system, making it turn to one side or leap upward. In this conceit the animal's thought becomes an electric signal that becomes a movement, a footprint, and eventually a fossil—a material record of an electrically communicated mood—a telegraphic message transcribed into stone and transmitted across the ages. As Hitchcock saw it, "The solid earth, too, is alike tenacious of every impression we make upon it; not a footprint of man or beast is marked upon its surface, that does not permanently change the whole globe. Every one of its countless atoms will retain and exhibit an infinitesimal, but a real, effect through all coming time."[39] Dickinson fully absorbed Hitchcock's magical attitude toward the natural world. She was convinced that a single, seemingly trivial word could send a "thrill throughout the whole material universe," just as Hitchcock had argued.

At Mount Holyoke, Dickinson had learned that she was as intellectually capable as any of her classmates. Perhaps more important, she had learned that she was capable of withstanding intense social pressure. She did not want to settle into certainty. Instead she craved the thrill of the unknown. Joining the church might be safer, but Dickinson did not care for safety. Her friendship with Abiah Root might not survive, but Dickinson confided her true feelings anyway: "I love the danger!"[40]

As Maria Mitchell's hyperbolic comet raced out into the infinite reaches of space, Emily Dickinson sailed straight toward the philosophical storms looming on the nineteenth-century horizon. Henceforth, she would need more daring companions.

CHAPTER 8

A Slow-Sailing Ship

*Downe, Great Malvern, 1842–1851;
Darwin, Age 33–42*

CHARLES DARWIN's thoughts were just as bold as Emily Dickinson's, but he took little pleasure in defying social expectations. As a young man, Darwin had tended toward religious conformity. Although he would later report that he had given up Christianity in the late 1840s, he was reluctant to talk about his religious ideas. Darwin was a conventional Victorian who was not willing to publicly challenge the established church. Adam Gopnik has described him as "a prisoner of respectabilities."[1] When atheists visited him to talk about theology, he made sure to invite the vicar as well. On Sundays he walked to church with his family, dropped them off, and strolled quietly across the fields. Emily Dickinson's poem about skipping church describes Charles Darwin's attitude in the early 1840s:

> Some – keep the Sabbath – going to church –
> I – keep it – staying at Home –
> With a Bobolink – for a Chorister –
> And an Orchard – for a Dome –
>
> Some – keep the Sabbath, in Surplice –
> I – just wear my wings –
> And instead of tolling the bell, for church –
> Our little Sexton – sings –

"God" – preaches – a *noted* Clergyman –
And the sermon is never long,
So – instead of getting to Heaven – at last –
I'm – going – all along![2]

Like the speaker in Dickinson's poem, Charles Darwin preferred birdsong to hymns. He was more inspired by tree branches arching into the open sky than church timbers vaulting into the dusty darkness. This preference was more socially acceptable in Darwin's milieu than in Dickinson's. Because he had grown up with a mix of freethinking atheists, Unitarians, and Anglicans, Charles Darwin did not view his religious doubts as rebellious in the way Emily Dickinson did. Dickinson's refusal to attend the First Congregational Church of Amherst was shocking to many in her community, where beliefs were quite uniform. Charles Darwin's long Sunday walks were much less noticeable in his village just outside of London.

Darwin had not faced religious pressure at Cambridge University the way Dickinson had at Mount Holyoke, simply because he had not resisted the religious requirements. As an undergraduate he had deeply admired William Paley's *Natural Theology* and had been perfectly willing to prepare to become an Anglican clergyman. In his thirties he grew less and less convinced by Christianity, but it was not until after his marriage that he experienced any sort of proselytization. Once he was married, Darwin faced constant, quiet efforts at persuasion from his wife. She feared that death would separate them forever unless he actively chose to embrace Christian beliefs as she did. He understood how she felt and empathized with her. One reason he worried about making his religious doubts public was that he hated the idea of disturbing Emma.

With Emma Wedgwood, Charles Darwin had met his match. She was brilliant, she was lovely, and she always stood up for herself. In many ways Emma and Charles were opposites. He was very neat; she was perfectly comfortable with household chaos. He was shy; she loved parties. He was fastidious about his dress; she was not interested in clothes at all. He tended to be a bit of a conformist; she was blithely unconventional. His health grew increasingly fragile; she was sturdy and strong.

On the other hand, their similarities were undeniable. Their family backgrounds were almost identical. In fact, as first cousins they had a common grandfather, the potter, chemist, and inventor Josiah Wedgwood I. Their similarities went far beyond the family ties. Both of them were gifted with charm and great emotional intelligence as well as cognitive skill. Like Charles, Emma was an avid reader. In fact, she had been a bit more precocious than he. According to family lore, Emma read Milton's *Paradise Lost* at the age of 5.[3] Charles was only a few months younger than Emma, but it took him years to catch up to her as a reader. Nonetheless, by the time he embarked on the *Beagle*, he loved *Paradise Lost* as much as Emma did. When they married, reading the latest novels together became a large part of their daily routine. They read Dickens. *The Pickwick Papers* had been published before their marriage, but all of Dickens's other books would come out during their married life. They read Elizabeth Gaskell and Anthony Trollope. They read their friend Mary Ann Evans's novels (published under the name George Eliot). In the course of their marriage, the Darwins would read hundreds, if not thousands, of novels together. They spent many happy hours reading to each other beside the fire.

Both Emma and Charles loved the outdoors. Although Charles had enjoyed hunting as a young man, he grew less bloodthirsty as he matured. He preferred long walks across the countryside or contemplative pacing in the gardens near home. In their first years in London, Charles and Emma had tended a long, narrow garden behind their house on Upper Gower Street. In 1842, after they moved to Down House in Kent, they spent much more time outdoors. They purchased extra land so they would have eighteen acres to cultivate. They planned the gardens carefully. They planted dozens of trees and designed a new "Sand Walk," a tree-lined, sanded walkway that recalled the landscape at Maer, Emma's childhood home. They loved animals as much as plants. In London, Emma had often accompanied Charles to the zoo to visit the orangutan Jenny, who fascinated him deeply. Later, Down House had

Down House, Kent. Wellcome Collection.

generous stabling for horses and ponies, and the house itself was full of dogs and cats as well as children.

The first two children, William and Anne, were born in London. The Darwins' third child, Mary, who was born shortly after they moved to Kent in 1842, died in infancy. Henrietta was born in 1843, George in 1845, Elizabeth in 1847, Francis in 1848, Leonard in 1850, Horace in 1851, and Charles in 1856. In addition to their own offspring, the Darwins often hosted their nephews and nieces, sometimes for months.

Charles always tried to keep his study relatively quiet. The space was well organized, though it was packed with specimens: sheets of pressed plants, jars of animals preserved in spirits of wine, feathers, hides, and neatly stacked bones—some that had been fossilized for thousands of years, others that were disturbingly fresh. A steady stream of oddly shaped packages arrived from collectors all over the world, adding to the vast haul from the *Beagle* that Charles was still cataloging when they moved to Kent. In the 1840s he did not see himself as an expert in any particular branch of biology or paleontology. As he sorted specimens from his voyage, he packaged them neatly and forwarded them to a growing network of scientific experts.

Meanwhile, the rest of the house reflected Emma's carefree attitude. As a child she had been called "Little Miss Slip-Slop."[4] Now, as a parent, she encouraged her children to be active and playful and curious—she did not worry too much about rules, and she didn't care at all about putting things away. Theirs was a busy, happy, often wildly chaotic household.

Charles and Emma adored each other. They were besotted with their children too. In fact, their attentive parenting went beyond customary parental attachments of the time. They were much more interested in their children's ideas, emotions, and individual personalities—and far more involved in their day-to-day lives—than typical Victorian parents. Charles took detailed notes on his children's facial expressions and developing emotional sensibilities. He wondered if empathy and kindness were innate or learned. As he observed his children, he concluded that their kindness was wholly innate.

If kindness was inheritable, the Darwin children were certain to be well-endowed with it. Their parents, Charles and Emma, were gentle and respectful to each other and considerate to all the members of their household, including a large number of servants. Instructing his oldest son in how to "acquire pleasant manners," Charles advised him to "try to please everybody you come near, your school-fellows, servants, & everyone."[5] The household staff included a butler and a slew of gardeners along with the men and women who kept the house. Brigades of nannies and nurses ensured that all of their children were carefully attended, even in the years when several of them were very young. The food at Down House was plentiful and good, if a little plain. The Darwins' doctors recommended a bland and simple diet.

Charles was not as physically strong as he had been before his voyage. He worked hard and he loved to walk, but he also spent many hours resting indoors. The house was stocked with books and toys of all sorts, including a wooden slide that fit over the treads of the stairs, turning the staircase into a rainy-day playground. On the quieter side, the children were welcome to read books in Charles's study, tucked under a blanket on the couch while he worked. When he finished his tasks, he would pull out the backgammon game so they could play together.

The biographer Deborah Heiligman points out that Charles's hypothesis that morality was instinctive for humans—that traits such as kindness and honesty were biologically determined—tended, like many of Darwin's scientific theories, to make religion less relevant for him. If the Darwin children were born with good morals, Charles might have wondered, why did they need to learn the moral tenets of a particular religion? Emma saw the question from another angle. For her, the inherent goodness of humans felt like a confirmation of her faith in God, rather than a challenge.

Here, Charles and Emma differed significantly. Emma Wedgwood's Christian faith was a source of great comfort to her. When her dearly loved sister Fanny had died suddenly in 1832, Emma had become convinced that she and Fanny would be reunited in heaven. The two sisters, who had collectively been known as "the Dovelies," had been inseparable until Fanny contracted an infection and died at age 26. Emma would have found the loss unbearable without the concept of the afterlife. Before Fanny died, Emma had been the less religious of the two. After her sister's death, Emma promised herself to be as Christian as her sister had been so she could be sure of a heavenly reunion.[6]

Charles had not been completely sure about Christianity for many years. Before his marriage, his father had urged him to conceal his doubts from Emma, but Charles could not be dishonest. He told Emma everything. He did not describe himself as an atheist—he never saw himself that way. Instead he saw himself as uncertain, doubting. Even Emma's relatively nondogmatic Unitarian strain of Christianity was a little too definite for his taste. How could anyone be certain of an afterlife? And if there were no certain promise of heaven (and no threat of hell), why was God necessary? For Charles, everything pointed toward a vague theism—a universe that might perhaps be suffused with good but did not actually show much evidence of an active, intervening God.

Even before they married, their religious differences had troubled Emma. Charles also seems to have been troubled by his lack of religious conviction. He was never much of a rebel, and he had no desire to offend his Cambridge mentors. He had trained as a priest, after all. Henslow and Sedgwick, the two Cambridge professors with whom he kept close ties, were both quite religious. In the village, Charles supported the parish church and regularly served on charitable committees. He was a close friend of the Reverend John Brodie Innes, who held the appointment of perpetual curate of Downe and lived in the village from 1846 to 1861.[7] The Darwin children were christened in the Church of England and attended Sunday services with Emma. But Charles never went to church. He felt that sitting in the damp cold was bad for his health. Because Emma was a Unitarian, she skipped reciting a few lines from the liturgy that were too Trinitarian for her, quietly closing her mouth and turning slightly to the side. Charles could not mouth any of the service with conviction. Unfailingly honest, he felt it was more ethical to stay away than to participate in a hypocritical way.

There had long been welcoming niches in the Church of England for those who were reluctant to define their beliefs. In the wake of violent disputes between Catholics and Calvinists in the seventeenth century, the Church of England had developed a relatively open attitude to theological differences. By the late eighteenth century, many social circles welcomed nonbelievers. The Lunar Men, for example, were widely respected despite their often heterodox attitudes toward religion. Charles's grandfather Erasmus had been quite open about his atheism, while the other, Josiah Wedgwood, had opted for Unitarianism. In the nineteenth century, Unitarians like Emma were welcome to attend Anglican services and to join their parish churches. The nineteenth-century Church of England made room for a wide range of ideas.

Emma Darwin was just as accustomed to a variety of religious opinions as her husband. She certainly feared that Charles's reluctance to identify as a committed Christian meant he would not be admitted to heaven, but she also admired and valued his sincerity. Before their marriage she had written an ambivalent letter to Charles on the subject. She admitted that it frightened her and made her sad to know that "our opinions on the most important subject should differ widely," though she tried to avoid "melancholy thoughts" about the "painful void" between them. Then she reminded herself that "my reason tells me that honest & conscientious doubts cannot be a sin" and concluded, "I thank you from my heart for your openness with me."[8] One of the things she loved most about Charles was that he was "honest & conscientious."

Although it scared her sometimes, the fact that he was open about his religious uncertainty was admirable and even endearing to her.

Emma Darwin's beliefs would have been just as shocking in mid-nineteenth-century Western Massachusetts as her husband's doubts. Emma Darwin particularly admired Francis Newman, a theologian who argued that almost everything in scripture could be ignored if believers just focused on the character of Jesus. Emma and Charles read and admired Newman's heartfelt spiritual memoir, *Phases of Faith*, which detailed his passage from mainline Anglicanism to evangelical Calvinism toward a nondogmatic theism. Emma liked Newman's emphasis on Jesus, while Charles liked his rejection of dogmatic creed. Both of them were interested in theological ideas. There is a strong contrast between the Darwins and the Dickinsons here. Although the Dickinson Family Library would eventually include a copy of Theodore Parker's *Prayers*, published in 1860, Unitarian theology was frowned on in Amherst. Thanking friends for introducing her to Parker's writing, Dickinson remarked, "I heard that he was 'poison.' Then I like poison very well."[9] It is hard to imagine a member of the Dickinson circle reading a radical theologian like Francis Newman. The Congregationalists of Amherst did not tend to be open-minded about religious matters.

In Darwin's mind, his private theory about the transformation of species was only tangentially related to theology. But long before he published anything about evolution, Darwin also knew that many conservative Christians would find the concept of natural selection antithetical to their religious beliefs because it explained how living things could have developed by chance, without the involvement of a divine creator. Though Darwin hated controversy and hoped to avoid offending people, he was determined to find the truth and make it known. But there was no reason to hurry. Darwin needed to study and think, to try his ideas out on a variety of living creatures, and to share his thoughts with a few colleagues and family members. As it turned out, Darwin would spend two decades deliberating before making his ideas public.

Shortly before his marriage, while he was still living in his depressing rooms on Great Marlborough Street in London, Darwin had started to investigate possible explanations for the transmutation of species. The idea that living things were organized into clearly delineated species was relatively modern—the Swedish botanist Carl Linnaeus's system of biological classification was barely one hundred years old. Yet despite the brief history of the concept, many scientists in the 1830s presumed that biological species were eternal and unchanging. Some thought that they had been created by God within a preordained, hierarchical order. Linnaeus's *Systema Naturae* (and the idea of immutable species) could easily slot into the medieval *Scala Naturae* (the concept of a hierarchical ladder of creation, a Great Chain of Being).

In the mid-nineteenth century, there was not a strong consensus about the transmutation of species. Many professing Christians believed in the concept that would later be known as *evolution*, the idea that species changed over time. By the mid-nineteenth century, most geologists, including religious Christians like Edward Hitchcock, were convinced that the earth was much, much older than biblical accounts indicated and that fossils from previous millennia were often the remains of species that no longer existed. It seemed clear that species went extinct. It also seemed highly unlikely that life on earth was simply a matter of subtraction. Many people, including some committed Christians, found it hard to believe that God had created all the species at once and then arranged matters so that extinct species left fossilized remains while those that survived left no geological trace at all. It was becoming clear that most living creatures belonged to species that had developed more recently than fossils.

Before Charles Darwin, other biologists had proposed that species changed over time, that as some plants and animals went extinct others developed into new life-forms. Though his grandfather Erasmus Darwin was a well-known transmutationist, by the 1840s the concept of evolution—known by then as transformation—was most often identified with Jean-Baptiste Lamarck, the French scientist who argued that plant and animal behavior and experience influenced the traits passed along to offspring. If a certain trait was used frequently, Lamarck thought, the next generation might have a more robust version of that trait. Conversely, if the trait was not used, it was likely to be less robust in subsequent generations. When Darwin was a medical student in Edinburgh in 1826, he and Robert Grant had discussed Lamarckian transformation of species as they walked along the coast, staring out into the North Sea.

Decades later, Darwin was still thinking about the mutability of species. In the spring of 1842, a few months before the Darwins moved to Kent, Charles and Emma and their two children made a long visit to Emma's childhood home, Maer Hall. While he was far away from his *Beagle* specimens and the endless task of sorting and identifying them, with his book on coral reefs complete, and with the second volume about his geological observations on the voyage—the volcano book— yet to start, Charles had the distance he needed to write an overview of his own ideas about how species changed. For the first time, he roughed out a written sketch of his theory of natural selection.

Charles Darwin was certainly not the first person to imagine that species changed and evolved. What was different was that he proposed a new mechanism to explain how species changed. Lamarck had theorized that use or disuse affected the heritability of traits. Darwin argued that some traits conferred advantages in particular environments. He thought that plants or animals that possessed these traits were likely to be healthier and to produce more offspring. Within a few generations, these beneficial traits would come to dominate the population in that place. Darwin's approach was fundamentally materialist, like Erasmus Darwin's and Lamarck's theories of species development: it required no divine intervention or guidance. His version of transmutation was gradualist, like Charles Lyell's version of geology: it did not require any dramatic catastrophes. Finally, his theory focused on population groups, like Malthus and Martineau: it did not center on individual behaviors, experiences, or choices. The thirty-five-page draft of Darwin's theory was vague in places, but it was a good outline. He did not use the word "evolution" to explain his theory. Interestingly, both Lyell and Martineau, his most important London mentors (acknowledged and unacknowledged) used the term in their writings of the period. For both Lyell and Martineau, "evolution" denoted gradual developmental change over generations.[10]

Joseph Dalton Hooker by George Richmond (1855). Wikimedia Commons.

In the beginning of 1844, Darwin explained his theory of natural selection to a new friend, the botanist Joseph Hooker. With Hooker's encouragement he wrote a 231-page manuscript about natural selection. Darwin did not see it as complete, but it was solid enough by the summer of 1844 for him to hire the local schoolmaster to make a clean copy in his elegant handwriting. He kept this draft in his files with a note requesting that Emma make sure it was published if he died before he finished his final draft. The note began, "My dear Emma, I have just finished my sketch of my species theory. If, as I believe that my theory is true & if it be accepted by even one competent judge, it will be a considerable step in science I therefore write this, in case of my sudden death, as my most solemn & last request."[11] He asked her to publish it with the help of a conscientious editor (he hoped Lyell would take it on). Darwin later sent an additional copy to Hooker in London. A few years later, he wrote an outline of his theory for Asa Gray in Massachusetts.

The same year, in October 1844, a book bound in scarlet cloth, *Vestiges of the Natural History of Creation*, was published anonymously. Because Darwin had been circulating his manuscript, some people attributed *Vestiges* to him. It would be decades before the author was identified as Robert Chambers, an Edinburgh publisher and amateur scientist who was a member of the Royal Society of Edinburgh. *Vestiges* sold very well. The little red book offered fascinating summaries of many strands of nineteenth-century evolutionary thinking. However, although the book was sensationally popular, its polemical tone offended some readers.

Religious people were upset by the book's irreverence. God's role in creation was casually dismissed. The anonymous author asked: "How can we suppose that the august Being who brought all these countless worlds into form by the simple establishment of a natural principle flowing from his mind, was to interfere personally and specially on every occasion when a new shell-fish or reptile was to be ushered into existence on *one* of these worlds? Surely this idea is too ridiculous to be for a moment entertained."[12]

Not surprisingly, those who believed that God directly intervened in every event were insulted by the way their ideas were ridiculed. Adam Sedgwick wrote to Charles Lyell in outrage: "If the book be true, the labours of sober induction are in vain; religion is a lie; human law a mass of folly and a base injustice; our labours for the black people of Africa were works of madmen; and men and women are only better beasts!"[13] Sedgwick was deeply offended by the radical, antihierarchical implications of *Vestiges*. If the social world, like the natural world, had been organized by random chance rather than ordained by God, then religion, law, and shared moral codes lost all validity.

More progressive thinkers, like the poet Alfred Tennyson (who had attended Cambridge University at the same time as Charles Darwin), also wondered about the theological and social implications of *Vestiges*. In his popular poem *In Memoriam* Tennyson asked, "Are God and Nature then at strife?"[14] He concluded that although nature was cruel ("red in tooth and claw," as he put it), the natural world was like a staircase that sloped "thro' darkness up to God." Tennyson's poem explained:

> I falter where I firmly trod,
> And falling with my weight of cares
> Upon the great world's altar-stairs
> That slope thro' darkness up to God,
>
> I stretch lame hands of faith, and grope,
> And gather dust and chaff, and call
> To what I feel is Lord of all,
> And faintly trust the larger hope.[15]

Tennyson's "larger hope" was that human beings would gradually progress toward Christian understanding despite the somewhat terrifying power of random chance in biological transformations. Yet few people noticed Tennyson's progressive Christian "altar-stairs" metaphor. Instead readers latched onto the bleak and bloody image of nature—"red in tooth and claw." After *Vestiges of the Natural History of Creation* and Tennyson's *In Memoriam*, evolutionary thought was linked to irreverence, social upheaval, violence, and cruelty.

Darwin was in an impossible position. On one hand, he knew that his theory might be scientifically important, and he thought it was probably true. On the other, he was appalled by the growing association between evolution and values he saw as antisocial. He had confided to Hooker in January 1844 that "I am almost convinced (quite contrary to opinion I started with) that species are not (it is like confessing a murder) immutable."[16] His parenthetical interjection "it is like confessing a murder" was lighthearted, but there was a grain of truth to it. Later that year he approached the subject with more confidence, writing to his Cambridge colleague Reverend Leonard Jenyns, "The general conclusion at which I have slowly been driven from a directly opposite conviction is that species are mutable & that allied species are co-descendants of common stocks. I know how much I open myself, to reproach, for such a conclusion, but I have at least honestly & deliberately come to it."[17] In both of these letters, Darwin referred to evolution's bad reputation but also expressed his growing conviction that species were mutable.

Would Darwin have pressed ahead and published his theory of natural selection in the 1840s if *Vestiges of the Natural History of Creation* had not been published? Perhaps not. He hated conflict, and he knew that his theory would expose him to widespread reproach and upset some of his mentors. John Stevens Henslow and Adam Sedgwick at Cambridge did not like the idea of evolution at all. Charles Lyell was also disposed against it. Though he had entrusted his manuscript to Emma, he knew she also was troubled by his idea that God was not really necessary in the creation of new life-forms.

Just as troubling, when *Vestiges of the Natural History of Creation* was published, theories of evolution lost status in Charles Darwin's network of scientific friends. Scientific readers were offended by what they saw as the book's carelessness about geology and biology. *Vestiges* was as cavalier about natural science as it was about theology and philosophy. At the time, many in the scientific community were trying to develop clear guidelines for a scientific method. They tended to hew to Bacon's idea that scientific logic should be inductive—that general laws should inferred from direct observations. Works based primarily on deductive reasoning (inferred from unexamined common knowledge rather than empirical observation) were often condemned as speculation rather than science. *Vestiges* fell into this category. The author did not base his arguments on his own observations or experiments.

Evolution had long been associated with political radicalism and theological heterodoxy; now it was also associated with the popular press instead of the scientific establishment. *Vestiges* outraged serious scientists because it relied on speculation instead of empirical evidence. Adam Sedgwick savaged the book in the *Edinburgh Review*, commenting, "All in the book is shallow; and all is at second-hand. The surface may be beautiful, but it is the glitter of gold-leaf without the solidity of the precious metal."[18] Darwin wrote that he read Sedgwick's review with "fear and trembling," but he was reassured to find that his own unpublished work on evolution anticipated most of Sedgwick's critiques.[19] Many years later, in the "Historical Sketch" that he added to the third edition of *On the Origin of Species*, Darwin would comment: "The work, from its powerful and brilliant style, though displaying in the earlier editions little accurate knowledge and a great want of scientific caution, immediately had a very wide circulation. In my opinion it has done excellent service in calling in this country attention to the subject, in removing prejudice, and in thus preparing the ground for the reception of analogous views."[20] In 1844, when *Vestiges of the Natural History of Creation* was first published, Darwin had not been so calm. He was not unduly troubled by the fact that *Vestiges* presented "analogous views" to his own, but the fact that the arguments were presented with "little accurate knowledge and a great want of scientific caution" had flustered him.

In 1845 Darwin wrote to his cousin and old college companion, William Darwin Fox: "Have you read that strange unphilosophical, but capitally-written book, the Vestiges, it has made more talk than any work of late, & has been by some attributed to me. – at which I ought to be much flattered & unflattered."[21] He was "unflattered" because the book was unscientific—"unphilosophical," as he phrased it—while on the other hand he was "flattered" because, after all, it was "capitally-written." The great popular interest in the book may have been encouraging to Darwin; it was said that Queen Victoria and her beloved husband read the book aloud to each other.

Darwin knew for certain that he did not want to publish his own version of *Vestiges*. He wanted to base his theory on careful, personally verified observations. He felt he needed to work out every detail to prove—first to himself and then to the world—that there was concrete evidence for his theory.

In 1846, a decade after the *Beagle* voyage ended, Darwin was finally approaching the end of his reports on the expedition. In addition to the popular book about his experiences (*Journal of Researches*, now known as *Voyage of the Beagle*), he had written and published three books about his geological findings on the voyage. He had farmed out his zoological specimens and acted as organizer and editor for a five-volume work on *The Zoology of the Voyage of H.M.S. Beagle*, which included books by Richard Owen on Fossil Mammalia, George R. Waterhouse on

Mammalia, John Gould on birds, Leonard Jenyns on fish, and Thomas Bell on reptiles and amphibians. He had worked hard, and his output was impressive: he had written four books and edited five more in ten years. Darwin had made his name. However, because he had enlisted others to describe his zoological specimens, he had not really established himself as a biologist or a comparative anatomist. Charles Darwin was known as a geologist, not a zoologist, and certainly not a *specific naturalist*—Joseph Hooker's term for a scientist who had fully mapped out a species.[22]

In light of the harsh reception for *Vestiges*, Darwin felt compelled to find more empirical evidence to support his own theory. He needed to establish his credentials—and his expertise—as a biologist. He looked through the few specimens that had not yet been described, searching for a species he could truly master. When he noticed a tiny, mysterious animal that he had collected almost ten years before, on an island off the coast of southern Chile, he got excited.

Darwin had picked up a conch shell perforated with boreholes in 1835 as he strolled along the beach at Lowe's Harbor, in the Chonos Archipelago. It was one of the most beautiful places he had ever been. The view of the Cordillera Mountains in the distance may have made him feel as if he were walking in Humboldt's footsteps. In his diary he wrote, "I cannot imagine a more beautiful scene, than the snowy cones of the Cordilleras seen over an island sea of glass."[23] The shell he carried back aboard the *Beagle* was more peculiar than beautiful. He wondered what sort of animal could have made the holes. When he got it under the microscope, he was amazed. The minute creature—he described it as an "ill-formed monster"[24]—looked exactly like a barnacle and cemented itself into the base of the hole like a barnacle, but it had no shell of its own. At the time, zoologists agreed that all barnacles had shells. If this animal was a barnacle, the entire taxonomy of Cirripedia (the barnacle group) would need to be reconceived. Darwin stored the conch shell and all its tiny residents in a jar of spirits of wine. Along with his ever-expanding collection, it traveled around the world with him, back to England and eventually to his study at Down House. In the autumn of 1846, after he had completed *Geological Observations on South America*, the third and final volume on the geology of the *Beagle* voyage, he pulled the jar from its shelf.

The view from Darwin's windows out across the flat chalky landscape of Kent was very different from the view of the mountains from Lowe's Harbor. The low-angled northern sun illuminated the gold-colored spirits of wine in the large glass jar. There was beauty in its glow, but it was on a very different scale from the bright sun and the glassy sea of the South American beach. In one of her poems about the Cordillera Mountains, Emily Dickinson wrote:

We see – Comparatively –
The Thing so towering high
We could not grasp its segment
Unaided – Yesterday –

This Morning's finer Verdict –
Makes scarcely worth the toil –
A furrow – Our Cordillera –
Our Appenine – a knoll –[25]

For Darwin, the scales were about to shift in the opposite direction, from small to large. The "ill-formed monster" that Darwin had plucked from the beach below the Cordillera Mountains was smaller than the head of a pin. A decade later, it would turn out to have giant implications.

Marine invertebrates had fascinated Darwin since Edinburgh, when he had caught that first, thrilling glimpse of cilia propelling larval *Flustra* across the glass slide of his microscope. The Chilean barnacle—if it was a barnacle—was at least as strange as the Edinburgh *Flustra*. Most barnacles had shells. This miniscule creature, which bored into another animal's shell, was an outlier, on the very edge of the border between barnacles and some other taxonomic group. In order to decide if it was a species of barnacle, Darwin would need to study the entire group of cirripedes. He could use empirical observation of barnacles to explore—and perhaps prove—his ideas about zoology.

At first, Darwin thought the barnacles would be easy. They were small and very portable, so he could gather specimens from contacts far and wide. He assumed that there would not be all that many different species—he thought he could work up the full taxonomy of the group in a relatively short time. But the closer he looked, the more complicated barnacles became. In *Darwin and the Barnacle*, the historian Rebecca Stott describes Cirripedia as "a minute marine monument to mutability."[26] In that respect, barnacles would turn out to be the perfect choice for illustrating the branching process of speciation. Darwin was disciplined and patient. He would eventually manage to construct a comprehensive taxonomy of the group, but it would require many years of work.

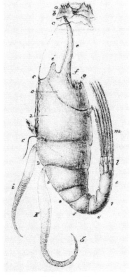

Darwin's barnacle, *Cryptophialus minutus*. From *A Monograph on the Sub-class Cirripedia*. (London: Ray Society, 1854), plate 23.

In the beginning, Darwin loved "doing barnacles." For one thing, he felt like a bit of a barnacle himself. In October 1846 he wrote to his old captain, FitzRoy, "My life goes on like Clockwork, and I am fixed on the spot where I shall end it."[27] In his next letter, he told FitzRoy that he had spent the past "half month daily hard at work in dissecting a little animal about the size of a pin's head from the Chonos Arch. & I could spend another month on it, & daily see some more beautiful structure!"[28] At first, he called the new Chilean barnacle Mr. Arthrobalanus. Later, he would classify it as *Cryptophialus minutus*. He spent his mornings happily peering through his microscope, marveling at the barnacle's anatomical intricacies, cementing himself into the chalky soil of Kent. During those years, barnacles were at the center of Darwin's family life. Visiting a neighbor, one of his young sons was puzzled by the absence of a room filled with the intermingled vapors of spirits and sea creatures. "Where does your father do his barnacles?" he asked.[29]

By 1848 the barnacles had become overwhelming. He wrote to Edward Cresy, complaining, "My work retrogrades, i.e, [*sic*] as I keep on finding out new points, I have to hark back to genera, which I thought

Charles Darwin by R. Burgess (1849). Wellcome Collection.

I had completed. . . . What zeal you have to care about my poor Barnacles, at which most of my friends laugh."[30] By 1850 he had lost sympathy for the "poor Barnacles." He cursed them in a letter to Joseph Hooker: "Confound & exterminate the whole tribe; I can see no end to my work."[31] Two years after that, in October of 1852, he wrote to William Darwin Fox, "I hate a Barnacle as no man ever did before, not even a Sailor in a slow-sailing ship."[32] The barnacles had slowed him down more than he expected. He was painfully aware that his scientific career had become something of a slow-sailing ship.

During the eight years of barnacle study, many changes took place for the Darwin family. Charles's father died in 1848. After this difficult loss,

Charles's health took a turn for the worse. Like Edward Hitchcock, he described his illness as dyspepsia. But in his case, it was more frightening since his mother had died of a stomach ailment. He worried about his own sudden death. For a time the entire family moved to Malvern so that Charles could try Dr. Gully's fashionable water cure. Every day he was rubbed with cold wet cloths, wrapped in cold wet sheets, and doused with cold showers. He gave up tobacco, kept to a plain diet, and walked seven miles a day. It seemed to help a bit; when they returned to Down House, they constructed a cold shower outdoors so that he could keep up the regimen.

It was a great honor when John Herschel invited Darwin to write the essay on "Geology" for his *Manual of Scientific Enquiry*, published in 1849. At the conclusion of his chapter, Darwin offered encouragement to aspiring geologists: "By acquiring the habit of patiently seeking the cause of everything which meets his eye, and by comparing it with all that he has himself seen or read of, he will, even if without any previous knowledge, in a short time infallibly become a good geologist, and as certainly will he enjoy the high satisfaction of contributing to the perfection of the history of this wonderful world."[33]

The mode of scientific inquiry that Darwin advocated was certainly empirical. He urged aspiring geologists to patiently seek "the cause of everything which meets the eye." But he also made room for some speculation, encouraging them to compare each instance with "all that he has himself seen or read." Reading the works of other scientists was an important part of the process. Later, in an encouraging letter to a young naturalist in the distant Malay Archipelago, Alfred Russel Wallace, Darwin would emphatically declare, "I am a firm believer that without speculation there is no good & original observation."[34] For Darwin, speculation and observation worked together.[35]

Perhaps the most striking part of Darwin's essay on "Geology" was its final phrase. Darwin promised that a good geologist could—and certainly would—contribute to the "perfection of the history of this wonderful world." The world described by science was "wonderful," he assured readers, not simply "red in tooth and claw." And the purpose of scientific inquiry was to contribute to the "perfection" of understanding—not to tear things down but to make things more comprehensible.

In the summer of 1849, the Darwin girls fell ill with scarlet fever. Etty and Elizabeth recovered their strength easily, but their oldest daughter, Annie, did not. In January 1850, Charles and Emma's eighth child, Leonard, was born. He was a sturdy child, like most of his siblings. Only Annie seemed fragile and listless. Before the fever, she had loved to pirouette in circles around Charles as he walked his laps on the Sand Walk. She had called it "doing custards."[36] All through the summer of 1850, Annie languished. She stayed tucked up on the sofa most of the time, with little energy for acrobatics. That fall the Darwins tried sea bathing at Ramsgate and then consulted with doctors in London. Nothing seemed to help.

In March 1851 the Darwins decided to take Annie to Malvern for a water cure. Emma was unable to travel because she was pregnant, expecting to give birth in May. Charles made the journey with 10-year old Annie, her little sister Henrietta, their nurse Jessie Brodie, and their governess Catharine Thorley. Once they were established, Charles returned to Down House. But Annie did not get better. Charles was called back in early April. For a few weeks, frantic letters passed between Charles in Malvern and Emma at Down House. Annie passed through a crisis and seemed to rally. But then, on April 23, 1851, she died.

The death of Anne Elizabeth Darwin, Charles and Emma's oldest daughter, was the most "severe grief" they would ever face.[37] A few days after his return from Malvern, Charles wrote a brief description of his "angelic" child that concluded, "We have lost the joy of the household, and the solace of our old age. She must have known how we loved her. Oh, that she could now know how deeply, how tenderly, we do still and shall ever love her dear joyous face! Blessings on her!"[38] Although some scholars see Annie's death as a turning point in Charles Darwin's attitude toward Christianity, there is little evidence for a sudden change in his

Annie Darwin (1849).
Wikimedia Commons.

religious beliefs. He may not have known precisely what he meant by "blessings," but there is no question that he responded to the loss with soft sorrow. He and Emma clung to each other, weeping together.

Many years later, in 1879, in a letter to John Fordyce, Darwin wrote:

> What my own views may be is a question of no consequence to any one except myself. – But as you ask, I may state that my judgment often fluctuates. Moreover whether a man deserves to be called a theist depends on the definition of the term: which is much too large a subject for a note. In my most extreme fluctuations I have never been an atheist in the sense of denying the existence of a God. – I think that generally (& more and more so as I grow older) but not always, that an agnostic would be the most correct description of my state of mind.[39]

In 1851 the word *agnostic* had not yet been coined. Charles Darwin did not have words for his own religious feelings, but he knew that Christianity offered him little solace. He hoped it would be a comfort to

Charles Darwin's study. Wellcome Collection.

Emma. He resumed his routine. Cold showers. Two and a half hours of dissecting tiny shellfish under the microscope. Laps around the garden on the Sand Walk. Afternoons and evenings by the fire playing backgammon or reading the latest novel.

Gradually, life came back to Down House. Horace Darwin was born in May 1851. There were new dogs, new types of pigeons, new plants in the garden. And there were always more barnacles. He may have felt like a "sailor aboard a slow-sailing ship," whose progress was impeded by the confounded creatures, but Darwin doggedly persisted.[40]

CHAPTER 9

Excitement in the Village
Amherst, 1851–1857; Dickinson, Age 20–26

A LITTLE after seven o'clock on the evening of Monday, September 29, 1851, church bells sounded the alarm in Amherst, Massachusetts. The villagers rushed out into the autumn chill, expecting some kind of emergency. Instead of a catastrophe, however, the bells that night announced a "beautiful Phenomenon" (as Emily Dickinson would describe it).[1] The sky was aglow with the northern lights. The red, gold, and pink colors of the aurora were so intense that it took a few minutes for the startled townspeople to realize nothing was on fire.

From 1827 to 1860, there were many chances to see spectacular auroras. The astronomer Denison Olmsted (author of the textbook Dickinson had used at Mount Holyoke) described the northern lights displays during the mid-nineteenth century as "probably among the most remarkable that have ever occurred since the creation of the world."[2] As the scientist Mary Somerville explained, "The disturbing effects of the aurora borealis" were a significant factor in the development of "the theory of electro-magnetism, the most interesting science of modern times."[3] For scientifically inclined observers in Amherst, the aurora was an intriguing electromagnetic mystery, a not-quite-decipherable planetary telegram.

The spectacular aurora of September 29 was visible as far south as Charleston, South Carolina,[4] but Edward Dickinson of Amherst may have been the only person so thrilled by the colors that he rang his town's alarm. To get a sense of what it was like to see one of the great auroras of the mid-nineteenth century, we turn to his daughter's letters. It would be many years before Emily Dickinson wrote a poem about the aurora that she thought worth saving,[5] but the 20-year-old was already an aspiring poet self-consciously exploring the craft of writing. In the autumn of 1851, most of her energy went into her correspondence.

A few nights after the aurora, Emily ate her dinner as quickly as she could, eager to reply to her brother's most recent letter. Longing to connect with Austin, hoping to be interesting and modern, Dickinson wrote energetically. She mocked herself for her obsession with the letters that were flying around her circle of friends, wished that there was some way to send Austin some ripe peaches, hinted at angelic or telepathic modes of communication, and finally told the aurora story:

> There was quite an excitement in the village Monday evening. We were all startled by a violent church bell ringing, and thinking of nothing but fire, rushed out in the street to see. The sky was a beautiful red, bordering on a crimson, and rays of a gold pink color were constantly shooting off from a kind of sun in the centre. People were alarmed at this beautiful Phenomenon, supposing that fires somewhere were *coloring the sky*. The exhibition lasted nearly 15. minutes, and the streets were full of people wondering and admiring. Father happened to see it among the very first and rang the bell *himself* to call attention to it.[6]

The northern lights were particularly thrilling to the Dickinson family because they suggested the possibility of a universal telegraph, with or without wires. Earlier that year, Edward Hitchcock's *Religion of Geology* had described "the Telegraphic System of the Universe,"[7] arguing that many phenomena—including auroras—sent electromagnetic signals across space and time, the way that animals sent tiny galvanic impulses along nerve tissues and telegraphs sent dots and dashes along copper wires. Dickinson's father was eager to bring telegraph lines to Amherst. Emily and Austin were more intrigued by the way the aurora borealis demonstrated the possibility of an alternative, wireless telegraphy. They longed to communicate telepathically.

In 1851, when Dickinson and her group of friends were frenetically writing letters, it had recently been established that neural impulses were electrical and that thoughts and emotions traveled like electrical currents. Because muscular movement was also galvanic, they imagined their handwritten letters to be the result of a mysterious combination of electrical processes. Telegrams were obviously electrical too, while the northern lights were a natural electromagnetic phenomenon that implied the possibility of cutting out the intermediary steps and communicating across great distances—writing without paper, telegraphy without wires.

Leaping from handwritten letters to telegraphy to telepathy (by way of the aurora borealis) did not seem far-fetched at the time. After Byron described an "electric chain" that bound human feelings to natural phenomena, American authors Lydia Maria Child and Nathaniel Hawthorne plotted novels around aurora-like displays that communicated their characters' thoughts and emotions. Such themes were common in the literature of the day.[8] In 1857 Elizabeth Barrett Browning would publish *Aurora Leigh*, a multivolume novel in verse whose visionary heroine was a poet named Aurora.[9]

Dickinson read hungrily during these years. Amherst booksellers offered a surprisingly wide range of materials intended to tempt the students and professors at the college and their intellectual friends and families. She was interested in everything, from popular novels and poetry to essays about science and philosophy. That winter, Dickinson and her friends were all reading Henry Wadsworth Longfellow and Charlotte Brontë. They wove discussions of fiction and poetry into almost every letter, narrating their own lives by alluding to the literature they shared.

One of their favorite books, Charlotte Brontë's *Jane Eyre*, contained a particularly good explanation of how telepathy was imagined in the mid-nineteenth century. Near the end of the novel, the title character receives a telepathic summons from her lost love, Mr. Rochester. Brontë described the sensation as "an inexpressible feeling that thrilled . . . and passed at once to my head and extremities. The feeling was not like an electric shock; but it was quite as sharp, as strange, as startling."[10] Although the notion that distant lovers could call to each other through the ether was, admittedly, right on the edge of witchcraft, Brontë placed it squarely in the realm of science.

The belief that very strong emotion could generate magnetic storms that facilitated telepathic communication accorded with the views of Romantic scientists. Sir John Herschel's *Preliminary Discourse on the Study of Natural Philosophy* explained that electricity was "one of those universal powers which Nature seems to employ in her most important and secret operations." According to Herschel, "This wonderful agent, which we see in intense activity in lightning, and in a feebler and more diffused form traversing the upper regions of the atmosphere in the northern lights, is present, probably in immense abundance, in every form of matter which surrounds us, but becomes sensible only when disturbed by excitements of peculiar kinds."[11] In *Jane Eyre* the peculiar excitement that allowed for telepathic communication was generated by lovers who longed for each other. In the fall of 1851, equally powerful currents of love and longing circulated around Dickinson's inner circle. When the great auroras blazed overhead, Dickinson may have wondered if the heavens were responding to the disturbing and peculiar passions generated by her group of friends.

Frequent letters and occasional telegrams flew between four of them, all in their early twenties: Emily and Austin Dickinson and Martha (Mattie) Gilbert and her sister, Susan (Sue) Gilbert. The Gilbert sisters were known in Amherst as the "twins," though in fact Mattie was Austin's age, a year older than Sue and Emily. Both of the Gilberts were bright and good-looking. Mattie had a sunny disposition, while Sue was often "stormy and unpredictable."[12] In a letter to Sue, Austin commented, "You seem to live in a rather tempestuous latitude, where tis a common thing for a bright day to be overcast with dark clouds—where the conflicts of the various elements are severe."[13] During that year of Romantic introspection and prolific, passionate letters, both Emily and Austin fell for the dark and stormy sister.

In Amherst, Mattie and Emily met almost every day to share the letters each had received from Austin and Sue. When the weather allowed, they read together outdoors, where there was more privacy. Eventually, the group tried an experiment with telepathy. At Thanksgiving, Austin returned to Amherst. Before leaving Boston (with its new telegraph lines), he sent a telegram to Sue in Baltimore to inform her that he and Mattie would expect her virtual presence at a party at precisely 7:00 p.m. on Thanksgiving Day. Afterward, he wrote to Sue, "No one knew that *you* were to be there, but Mat & I." He hoped she had enjoyed the evening even more "than if your material form had been amongst us."[14] The scheme was surprisingly elaborate; in addition to Austin's telegram, Mattie had written a letter of invitation.

Emily's letter to Susan on October 9, 1851, just a few days after her letter to Austin about the aurora, worked out more of the implications of paper letters and telepathic fantasies. It opened with a strangely intimate use of writing paper, announcing "I wept a tear, here, Susie."[15] A faint mark on the manuscript may be an actual tearstain.[16] Declaring "we are the only poets, and everyone else is *prose*," Emily invited Susan to imagine reveling with her in the "native ether" of poetry.[17] Dickinson imagined the "ether" as the highest part of Earth's atmosphere, where the winged creatures of poetry and myth frolicked in the flickering colors of the aurora.

Emily's letters to Susan Gilbert are different from her letters to Abiah Root. The letters to Abiah, which Emily wrote as a teenager, are extraordinary documents, often witty and sometimes passionate, but they are not love letters. The letters to Sue are much more intimate, more impassioned, and more seductive. In *Open Me Carefully*, Ellen Louise Hart and Martha Nell Smith characterize the relationship between Emily and Sue as an "intimacy that cannot be reduced to adolescent flurry."[18]

It is fascinating, if somewhat uncomfortable, to compare Austin's and Emily's love letters to Sue. Emily's are better, of course—more imaginative and more detailed, foreshadowing the great poet she would become—but both lovestruck siblings fantasized about communicating telepathically with her. From Boston, Austin sent telegrams. From Amherst, Dickinson sent her own tears.

Susan Gilbert Dickinson.
bMS Am 1118.99b (29.1),
Houghton Library,
Harvard University.

In December 1851, Emily Dickinson turned 21. The following winter was uneventful. Austin wrote to Sue, "Amherst, the folks all say, has been unusually quiet this winter. – hardly a ride or party or excitement of any sort."[19] One exception to the antisocial winter was the celebration of the start of construction of a railroad on February 6, 1852. Emily's letter to Austin about the railroad echoed her description of the "excitement in the village" on the night of the aurora.[20] This time she wrote, "The grand railroad decision is made, and there is great rejoicing throughout this town and the neighboring.... Every body is wide awake, every thing is stirring, the streets are full of people talking cheeringly, and you really should be here to partake of the jubilee." She concluded enthusiastically, "Nobody *believes* it yet, it seems like a fairy tale, a most *miraculous* event in the lives of us all."[21]

It is hard to tell whether Dickinson's enthusiasm was sincere. She had a sly sense of humor, and she may have been making fun of her father's railroad mania. Around the same time, she was at work on a comic valentine for one of her father's law clerks that poked fun at a wide range of patriarchal figures, from Adam to Ralph Waldo Emerson. A long rhyme that joked about many of the subjects she had studied at Amherst Academy and Mount Holyoke, it opened with Latin: "Sic transit gloria mundi" (Thus the world's glory passes). In seventeen four-line stanzas, Dickinson mocked everything she could think of, including the laws of gravity, the book of Genesis, and the history of the United States. It was a good poem, funny and remarkably energetic. The clerk, William Howland, admired it so much that he sent it to the *Springfield Daily Republican*, a national newspaper with thousands of readers.

On February 20, 1852, Dickinson was surprised to discover her own valentine on page 2 of the paper. The praise from the editors was particularly thrilling. Their note said, "The hand that wrote the following amusing medley to a gentleman friend of ours, as a 'valentine,' is capable of very fine things, and there is certainly no presumption in entertaining a private wish that a correspondence, more direct than this, may be established between it and the *Republican*."[22]

The night before, on February 19, 1852, another great aurora had blazed above Massachusetts. Coincidentally, the page that contained Dickinson's first published poem also included a description of the glorious spectacle: "The exhibition of the Aurora Borealis, last evening, was the most splendid we have seen for many a night. The King of the North shook out shimmering folds of all his banners, red, argentine, and golden."[23]

Reading the *Republican* that afternoon, finding one of her most lighthearted verses printed—and praised—in its pages, Dickinson may have felt haloed by the "red, argentine, and golden" glory of the aurora. Or she may have felt embarrassed, either because she was not ready to publish anything or because "Sic transit gloria mundi" was so slight compared to her other works. She did not mention the poem's appearance to anyone. Publication—without her consent—was upsetting. In the years that followed, Dickinson would remain reluctant to publish.

On the other hand, the editors' invitation to strike up a private correspondence probably thrilled her—she was not at all ambivalent about wanting to connect with people who understood literature.

Despite her shyness she would soon start exchanging letters with Josiah Holland and his wife, Elizabeth, and by the end of the decade she would be in regular correspondence with Samuel Bowles, the editor in chief. Even if Dickinson was troubled by having her relatively nonsensical valentine published, the event heralded her growing sense of possibility.

The entire Dickinson family was riding high. In June 1852 Edward Dickinson attended the Whig Convention in Baltimore; later that year, he was elected to Congress. By June 9, 1853, the railroad to Amherst was complete. Residents of the town of New London packed themselves into the cars and rode to Amherst to celebrate the opening of the line. Emily hid out in the woods and watched the train arrive in town by herself. Her solitude that afternoon was brief. Later that day, Josiah Holland dropped by the Dickinson house and introduced himself in person. He invited Emily and her sister to come to Springfield for a visit. Despite Emily's shyness, she was charmed. A few weeks later Josiah called again, this time bringing his wife, Elizabeth. Their visit was unexpected, but it was a great success. They drank champagne and laughed and talked with giddy pleasure. When the Hollands repeated their invitation, Emily and Lavinia agreed to visit them after commencement. In September 1853, the Dickinson sisters spent night at the Hollands' in Springfield. The easy warmth of the household delighted them both; they would return for a longer visit the following September.

Her friendship with the Hollands was significantly new for Emily. Up to this time, all of her correspondents had been family members or Amherst people. Even at Mount Holyoke, she had shared a room with her cousin. Josiah and Elizabeth Holland were not related to her nor were they from Amherst or connected to the college in any way. They were her first purely literary friends. Josiah was a successful writer. Elizabeth was witty, warm, and bright. Dickinson loved them both.

In addition to Dickinson's new acquaintances, her family's changing circumstances also opened things up. Although her father would only serve in the House of Representatives for a single term, his time as a congressman broadened the family's horizons significantly. They traveled more widely and met a variety of people from outside Amherst. All of them became more independent.

When the rest of her family went to Washington in April 1854, Emily stayed home in the big white clapboard house on Pleasant Street. Susan came to keep her company. John Graves, a Dickinson cousin who was a student at the college, stayed along with them.

There is little information about the weeks that Emily and Susan kept house together in 1854. One letter of Susan's survives. She offhandedly mentions, "I forgot to tell you, I am keeping house with Emily, while the family are in Washington—We frighten each other to death nearly every night—with that exception, we have very independent times."[24] How the women frightened each other is not clear. John Graves later remembered Emily playing "heavenly music" on the piano in the middle of the night.[25] Perhaps Emily's moonlight musical improvisations disturbed Susan's sleep, or perhaps the sexual electricity felt a little scary. One of Dickinson's best-known love poems celebrates "Wild Nights–Wild nights!" The nights that Emily and Susan spent together on Pleasant Street may have offered an opportunity to act on the intense intimacy of their epistolary relationship.

The "very independent times" that Susan described are equally ambiguous. Perhaps the young people enjoyed their solitude—independent from each other and from their families. Susan may have relished a chance to share the freedom from social obligations that Emily took for granted. On a fundamental level, Emily Dickinson had the independence that Sue craved. The Dickinson family was intensely close, and at times it could be suffocating. But neither Emily nor Lavinia had any sense of economic insecurity or marginalization. Marriage would have been likely to diminish the freedom of the Dickinson sisters.

Susan Gilbert was not so lucky. Her parents had died when she was in her early teens; she did not have the option of staying at a parental home. She could teach (and live at school), as she had done in Baltimore; she could make long visits to family members (as she did in Amherst and in upstate New York); or she could marry and set up her own household. Marriage was probably the best of these choices, but she would manage to put it off for two years longer. She reluctantly accepted Austin's proposal while continuing her correspondence with Ned Hitchcock, son of the famed geologist and the other very eligible bachelor in Amherst. Sue wanted independence and she also wanted social success. Marriage was her best chance to secure both.

In April 1854, when they were staying together on Pleasant Street, Emily and Sue may have imagined themselves as characters in *Jane Eyre*. There were many parallels between their experience alone in the house in Amherst and that of the fictional denizens of Thornfield. The nights were full of strange noises and unexplained frights. More significantly, Emily and Sue were caught up in a Gothic love triangle.

The following summer was a jealous season. Although we can only speculate about how the currents of envy circulated, the triangulated friendship seems to have gotten too overheated for comfort. In a letter draft preserved in the archives, Austin wrote angrily to Susan: "As to your deprivation of 'Spiritual converse' with my sister—I Know Nothing." In his hurried draft, Austin asked Susan to "not suspect me of having interfered with your epistolary intercourse with her" and assured Susan that Emily's "choice of friends and correspondents is a matter over which I have never exerted any control." The draft concluded with Austin adopting an arch mock formality: "Knowing therefore that you will not suspect me of having interfered with your epistolary relations & assuring you of my sentiments of regard for yourself—my respect and admiration for the President of the United States and the Gov Gen of Canada—I remain yours truly—Wm A. Dickinson."[26] Susan had accused Austin of interfering with her letters to Emily. He denied it, but whatever the reason, Emily and Susan quarreled. By September, Susan had broken off correspondence with Emily completely. Eventually, they would patch things up, but their letters were never as confiding as they had been before that summer.

The end of 1854 was difficult for the whole Dickinson clan. Edward Dickinson suffered an embarrassing electoral defeat. Susan left town again, this time going all the way to Michigan. Emily became resigned to the end of her friendships with both Susan and John Graves, the cousin who had stayed with them in April. In letters to each she announced that she was putting them in her "box of Phantoms." Her letter

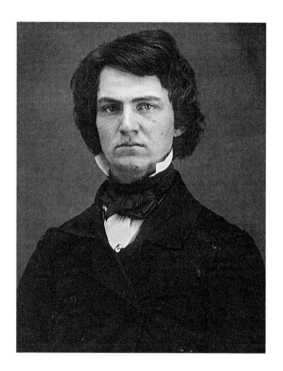

Austin Dickinson (1850). Amherst Archives and Special Collections.

to Susan was particularly sad: "I miss you, mourn for you, and walk the Streets alone – often at night, beside, I fall asleep in tears, for your dear face, yet not one word comes back to me from that silent West. If it is finished, tell me, and I will raise the lid to my box of Phantoms, and lay one more love in; but if it *lives* and *beats* still, still lives and beats for *me*, then say me *so*, and I will strike the strings to one more strain of happiness before I die."[27] As far as we know, Susan did not answer this letter. It may have been hard to match Emily's melodramatic intensity. Their correspondence reached a lull.

Despite her protestations, Emily was also starting to move on. In September 1854, she went on her long visit to the Hollands' in Springfield. In February 1855 (just before Susan and Martha Gilbert returned to Amherst), Emily and Lavinia traveled to Washington, DC, for their father's final (lame duck) session of Congress. They stayed at the Willard Hotel with all the fashionable people. Though Emily commented that her elaborate traveling clothes made her feel like an "embarrassed

Peacock," she enjoyed the trip.[28] The warmer southern weather delighted her. After three weeks in Washington, Lavinia and Emily went to Philadelphia, where they stayed with Eliza Coleman, a cousin whom they had known well in Amherst. During their weeks with the Colemans, Emily somehow managed to secure an introduction to the mysterious Charles Wadsworth, a famously reclusive preacher whose sermons were carried by newspapers across the country.

After 1855 Dickinson grew more determined to connect herself to literary people. She continued to write frequently to the Hollands. She began to write to Charles Wadsworth and eventually to Samuel Bowles and Thomas Wentworth Higginson. She also wrote to her cousins and some childhood friends. Among her old friends, the most notable was Helen Fiske from Amherst Academy, who married twice and gained fame writing under her married name. As Helen Hunt Jackson, Dickinson's childhood friend would become a successful novelist and a prominent advocate for Native American rights. By the late 1850s, Dickinson's circle of correspondents still included some family members, but more and more, she wrote her letters to published writers.

Looking back, Austin recalled that his sister "reached out eagerly, fervently even, toward anybody who kindled the spark."[29] For Dickinson, letters were always an electrifying source of inspiration. She sought out writers with wide-ranging interests—in poetry, to be sure, but also in theology, politics, and science. Dickinson also started to frame her letters differently, pushing the boundaries of conventional writing and including poems. In the years to come, Dickinson's letters would more and more frequently serve as cover letters for her poems.

Late in 1855, the Dickinson family moved from the house on Pleasant Street to the family Homestead in Amherst just around the corner. For Edward, it was a triumphant return to the grand brick house that his father had built on Main Street forty years before. Austin, Emily, and Lavinia had been born at the Homestead, but in 1833, the family had been forced to sell because of financial trouble. After twenty years of effort, their fortunes were restored. As part of the renovation, Edward Dickinson had an elegant, glass-paned conservatory constructed for Emily and Lavinia to use as a greenhouse. For Austin and Susan he went even further, building an entire house—the nineteenth-century Amherst version of a Tuscan-style villa—in the lot beside the Dickinson Homestead. The Dickinson compound included the two houses (the Homestead and the Evergreens), a full barn, extensive gardens, and an apple orchard. Across the street were the Dickinson meadows.

Emily Dickinson rambled the fields with her dog Carlo, cultivated plants in the gardens and greenhouse, watched birds and butterflies, and tended the animals in the garden and the barn. Indoors was a (sometimes overwhelming) load of housework. When she could get out of the kitchen and away from her sewing, Emily practiced piano and read everything she could get her hands on—the latest books, the daily papers, and the monthly journals that arrived in a constant stream. For the first time in her life, she had a room to herself. Late at night, early in the morning, between chores in the kitchen, while she was working outside—whenever she could—she wrote poems. In the privacy of her room, she revised them, copied them, and made them tighter and brighter.

In 1855 Hitchcock opened the Appleton Cabinet, a new fossil museum larger than the Octagon, just a few blocks away from the Homestead. There Emily could contemplate the world's largest collection of dinosaur tracks or page through the ingenious "stone books" that Hitchcock had fashioned by prying apart layers of sedimentary rock so that students could examine changing fossil impressions across the centuries. In June 1856, a ten-ton boulder scored with glacial striations was excavated from in front of the Dickinson's house.[30] A crew of enthusiastic geology students rolled the giant stone up the hill and brought it to rest in front of the old Octagon.

In Edward Hitchcock's Amherst, a twenty-thousand-pound boulder was not just a rock—it was a telegram from the deep past of the earth. Judging from the composition of the stone, Hitchcock determined that it had once formed part of the ledge of a nearby mountain, Mettawampe (Mount Toby). Glaciers had carried the boulder ten miles south and deposited it in the rich soil where humans would eventually come to live. The marks on the surface of the stone told a story of great planetary forces working over thousands of years. Hitchcock's community shared his vivid sense of the tiny, infinitesimal smallness of their own quotidian

Amherst lives, measured on the scales of astronomical space and geological time. Glacial erratics and fossil footprints—and the mineral-rich soil of Amherst—were geological messages from the deep reaches of time. Changing angles of sunlight, the unchanging stars, and—occasionally—the flaring colors of the great auroras illuminated the vastness of planetary space.

For Hitchcock the implications of nineteenth-century science were primarily theological. His 1851 book *The Religion of Geology and Its Related Sciences* had summed up what was at stake for him. For Dickinson the stakes were different. The intersection of science and religion interested her, but the intersection of science and emotion interested her more. When her thoughts turned to geology, she speculated about the feelings of stones ("How happy is the little Stone") and volcanoes ("Volcanoes be in Sicily"). When she thought about astronomy, she wondered about the ether. If consciousness—thought and emotion—was somehow electromagnetic, could it travel across space? When she considered infinity, her thoughts leapt beyond the God of Christian orthodoxy. Her questions—poetic and scientific—were about consciousness and emotion: Could she find a way to express how infinity felt?

In a letter to Elizabeth Holland, Dickinson compressed her questions about the infinite and the infinitesimal into a brief phrase, calling them "problems of the dust." Dickinson's mother had fallen ill when they moved back to the Homestead. On the one hand, Dickinson feared her mother might die—death was the large (infinitely large) problem of the dust. On the other, she was frustrated that housework and care work took so much time away from her writing and thinking—the literal (if infinitesimal) dust drove her to distraction. Finally, there was the geological/theological question that troubled Edward Hitchcock: "It is said that the body laid in the grave is ere long decomposed into its elements, which are scattered over the face of the earth, and enter into new combinations, even forming a part of other human bodies."[31]

In her January 1856 letter, Dickinson wrote, "I often wish I was a grass, or a toddling daisy, whom all of these problems of the dust might not terrify."[32] Contemplating the greensward of the Dickinson meadow, its tall grass splashed with daisies, offered Dickinson some solace, but it also raised more questions. What would it feel like to be a blade of grass? Walt Whitman had offered one approach to the question in *Leaves of Grass*, published in 1855, but Dickinson was not interested in Whitman.[33] She preferred to imagine being grass, being a daisy, being dust, in her own way.

A few years after her letter to Holland, in 1862, Dickinson would copy a poem about that wish into a carefully stitched booklet. "The Grass so little has to do" begins:

The Grass so little has to do,
A sphere of simple Green –
With only Butterflies, to brood,
And Bees, to entertain –

Yet as it turned out, Dickinson's concept of a "sphere of simple Green" was not so simple. Subsequent verses outline the life cycle of a grass, through day and night, and eventual death, finally concluding:

The Grass so little has to do,
I wish I were a Hay –[34]

In the context of Dickinson's letter to Holland, this poem may be suggesting that Dickinson would rather be dead than do housework. There is something comical about wanting to be "a Hay." On the other hand, there was also an edge of complicated grief in Dickinson's meditation on the hayfield. Not even grass or a daisy could completely evade the "problems of the dust." Years later, in 1871, Dickinson would comment to Holland that "Darwin does not tell us" why death, "the Thief ingredient accompanies all Sweetness."[35] Long before Darwin published *On the Origin of Species*, Dickinson wondered how death shaped the natural world.

Though Whitman failed to inspire her, Dickinson would soon find a role model. In 1857 Elizabeth Barrett Browning published *Aurora Leigh*, an eleven-thousand-line narrative poem as long as a novel.[36] Dickinson was enthralled by the way Barrett Browning managed to bring modern nineteenth-century approaches to science and philosophy into her work without abandoning her idealism or her emotional warmth. The combination of science and emotion was pure magic.

Elizabeth Barrett Browning was a poet, a political radical, a widely respected intellectual, and the author of a best-selling novel in verse. She was also famous for escaping her repressive father and eloping with Robert Browning, another widely respected poet. Dickinson was particularly drawn to Barrett Browning's generous skepticism when it came to scientific and religious orthodoxies. In a letter about her rigid and difficult father, Elizabeth Barrett Browning insisted that "the evil is in the system."[37] Her father's "system" was patriarchal and white supremacist (proslavery) as well as highly controlling. In contrast, Elizabeth Barrett Browning turned away from "system" in general. As she put it, she cared "very little for most dogmas and doxies."[38] Barrett Browning was politically engaged, and she often spoke out against slavery, colonialism, and sexism, but she did not fall strictly into line with any organized ideology. As Kate Field would write in the *Atlantic Monthly* in 1861, Barrett Browning "scorned to take an insular view of any political question."[39]

When it came to nature and natural philosophy, the young Elizabeth Barrett Browning had hesitated about joining the antimaterialist faction that Samuel Taylor Coleridge and the Romantic poets had represented in an earlier part of the century. An early poem gleefully pushed back against the Romantic poets' version of mystical paganism, declaring, "Pan is dead!"[40] However, Barrett Browning also refused to accept materialist models that drained the natural world of spiritual significance.

The heroine of *Aurora Leigh* was a brilliant and successful woman poet. Throughout the book, Aurora discussed art, politics, and science with confident pleasure. The poet-heroine described herself as

> . . . creeping in and out
> Among the giant fossils of my past,
> Like some small nimble mouse between the ribs
> Of a mastodon.[41]

This paleontological metaphor is cheerful and a little humble (she likens herself to a "small nimble mouse"). In this conception, the past may be dead, but the poet is very much alive, at play within its fossilized structures. She is nothing like a paleontologist, though it is clear that she understands nineteenth-century science very well, Aurora does not see sciences such as paleontology or geology as depressing. To the contrary, she imagines her vocation as an epic poet in the warmest of geological terms. Her goal is to express "the full-veined, heaving, double breasted Age"—to capture nineteenth-century thought in "the burning lava of a song."[42]

Aurora's suitor Romney has a very different response to modern science. "My soul is grey," he tells her.

> ... Observe, – it had not much
> Consoled the race of mastodons to know,
> Before they went to fossil, that anon
> Their place would quicken with the elephant;
> They were not elephants but mastodons;[43]

Romney, an uneasy aristocrat who longs for social justice, presumes that narratives of evolution and extinction forecast the eventual extinction of humans. Aurora admires Romney's egalitarian and humanitarian impulses, but she is repelled by his ideologically systematic approach because it strikes her as not only hopeless but also loveless. Wondering if Romney is capable of love, she muses:

> A pagan, kissing for a step of Pan
> The wild-goat's hoof-print on the loamy down,
> Exceeds our modern thinker who turns back
> The strata ... granite, limestone, coal, and clay,
> Concluding coldly with, "Here's law! where's God?"[44]

When she compares Romney's modern, scientific way of understanding the world to a goat-worshipping pagan's, she concludes that paganism is preferable.

It would be a mistake to read the character of Aurora simply as a mouthpiece for Barrett Browning. The novel-length poem makes room for a wide variety of voices. Barrett Browning represents Romney as a "modern thinker," whose views are shaped by the social theories of

Malthus and Fourier. At first Aurora rejects Romney and his cold "dogmas and doxies," but in the course of the story, his soul becomes less gray. He warms up and becomes less rigid in his thinking. Ultimately, she embraces him. The poem concludes with Aurora and Romney's marriage, a glorious union between poetry and science.

Dickinson loved Barrett Browning's intellectual ambition and her remarkable range, which stretched from small, perfectly crafted sonnets to her sprawling eleven-thousand-line novel in verse. The critic Páraic Finnerty explains that reading Barrett Browning changed Dickinson, in part because it "altered the way she heard and saw the natural world."[45] Picking up on the "burning lava" in *Aurora Leigh*, Dickinson used volcanic metaphors to describe her poetic vocation, most famously in "Volcanoes be in Sicily," in which the speaker of the poem is a "Vesuvius at Home."[46]

In *Aurora Leigh*, Barrett Browning declared that "Poets should / Exert a double vision." Dickinson responded with her own poems about seeing from multiple perspectives. In "The Admirations and Contempts—of Time," Dickinson expanded on Barrett Browning's idea of "double vision" by offering her own concept of "compound vision," which was exponentially more complex than Barrett Browning's.[47] Where *Aurora Leigh* argued that poets should combine close and distant perspectives, "The Admirations and Contempts—of Time" asked readers to look through an "Open Tomb" and imagine how death could expand perspective.

In her best-known poem about Elizabeth Barrett Browning, "I think I was enchanted," Dickinson described reading the poet's work as magical. Describing the effect, she wrote:

> And whether it was noon at night –
> Or only Heaven – at noon –
> For very Lunacy of Light
> I had not power to tell –[48]

Her description of reading Barrett Browning likened the experience to an aurora's "noon at night." The aurora reference was no accident. Barrett Browning's roman à clef featured a great woman poet named

Elizabeth Barrett Browning by Elliot & Fry after Macaire (1858). National Portrait Gallery, London.

Aurora whose life story echoed *Jane Eyre*. Dickinson had never quite been able to imagine herself as a novelist, much less a governess. But Elizabeth Barrett Browning convinced her that she might imagine being a great poet. Reading *Aurora Leigh* helped her understand her own vocation.

As Dickinson considered the significance of the aurora for Barrett Browning, her thoughts may have turned to her memory of seeing the northern lights in Amherst or to the fact that her first published poem had been heralded by an aurora. By 1857 Emily Dickinson knew that she wanted to be the next Aurora Leigh. She longed to press nature and God together, to express the "Lunacy of Light," to find the magical words that could join the material world to the infinite world. As the 1850s drew to a close, she spent her time writing more and more, sharing her poems with her correspondents, and carefully revising.

In England, Charles Darwin was also writing. His daily routine was similar to Dickinson's, though he spent a bit more time in the garden and a bit less time dusting. They were both passionate observers of the natural world that surrounded their homes. They both worked ceaselessly. Both spent their most productive years in seclusion with their families, but neither Darwin nor Dickinson lived a wholly solitary life. Their large, close-knit family groups provided lots of emotional interaction. They also corresponded with wide (and sometimes overlapping) circles of friends and professional acquaintances.

By the 1850s Charles Darwin's voyaging years were far behind him. He rarely ventured from his home outside London. Yet although he was increasingly reclusive, Darwin corresponded regularly with scientific friends and acquaintances, including Charles Lyell, Joseph Hooker, Thomas Huxley, John Stevens Henslow, Asa Gray, and William Whewell. His decades of research had provided him with data that upheld his ideas about the transmutation of species. Like Dickinson, he was hesitant about publishing. He preferred to circulate manuscript drafts among his network of friends.

CHAPTER 10

On the Origin of Species
Downe, 1858–1860; Darwin, Age 49–51

CHARLES DARWIN was at least as enthusiastic about his mail as Emily Dickinson. He angled a mirror outside his study at Down House so that he could watch for the post without going to the window. When it arrived he would often drop what he was doing and hurry to find out what the mailbag held for him. There was usually quite a bit. In addition to his regular correspondence, Darwin was in the habit of writing to scientists all over the world to ask for information and to build his own network. Measured by the letters that have been collected and published, Darwin wrote ten times as much mail as Dickinson. Though his prose was more conventional, in many respects Darwin's letters resembled Dickinson's. None of his surviving letters are as frankly erotic as Dickinson's could be, but they are often quite expressive in other ways. He was a warm and witty correspondent who kept up a voluminous correspondence with family members. His letters to scientists, like hers to writers, contained a mix of friendship and professional effort lightened by humor.

Darwin finished the barnacle books in 1854. As it turned out, he published more than twelve hundred pages of barnacle taxonomy: two large illustrated volumes on nineteenth-century barnacles (*Living Cirripedia*), both cone shaped (*sessile*) and stalked (*pedunculated*), and two smaller accompanying volumes on fossil barnacles, both sessile and pedunculated. After the final barnacle book was published, he described his plan for the next winter's work to Joseph Hooker, "I have been frittering away my time for the last several weeks in a wearisome manner, partly idleness, & odds & ends, & sending ten-thousand Barnacles out of the house all over the world. – But I shall now in a day or two begin to look over my old notes on species."[1] With the ten thousand barnacles cleared from his house, Darwin was eager to turn toward the more speculative theoretical work he had always loved. His "old notes on species" had not faded during his barnacle years. They retained their power.

Darwin had been awarded a Royal Medal for Natural Science in honor of the barnacle books in 1853. He hoped that he had earned credibility as a "specific naturalist," a person who had systematized the taxonomy of a particular species. But his overwhelming feeling was relief to be finished with Cirripedia. In March 1855, Darwin wrote gleefully to William Darwin Fox, "Far the greatest fact, about myself is that I have at last quite done with the everlasting Barnacles."[2] Yet although Darwin's study at Down House was no longer encrusted with barnacles, he was still sailing slow.

As part of his species research, he decided to try breeding pigeons. He joined two pigeon fancier's societies in London and began gathering as many varieties of pigeons as he could. He hoped to be able to offer an account of variation under domestication that would be as exhaustive as his barnacle work had been. Perhaps he would publish twelve hundred pages on pigeons. From 1854 to 1858, for nearly four years, Darwin worked very slowly. It seemed likely that the next phase of his species project would take even longer than the barnacle books.

In the spring of 1858, Darwin was working quietly at home, as he had been for decades. His health was stable, though his stomach troubles—and his regimen of cold showers—continued. He and Emma had been married for nineteen years. Their oldest child, William, was almost 18 years old. Their youngest—Charles Waring Darwin, born December 6, 1856—was approaching 18 months. Both Charles and Emma were about to turn 50, and they did not expect to have any more children. Little Charles was the baby of the family. They usually referred to him as *Baby*. He was a remarkably good-tempered child, but he had not started to walk or talk. He seemed to be developing more slowly than their other children. They thought he might be more physically vulnerable than his older siblings, though he was also extraordinarily sweet. The Baby spent hours snuggled in his parents' laps. Charles and Emma treasured Baby's company, but they were anxious about his future.

The Darwins always worried about illness. In the preantibiotic era, even the slightest infection could turn serious. Before their marriage, each of them had experienced sudden death in their families. Charles's mother and Emma's sister had both died within a few days of contracting seemingly minor illnesses. Later, as a married couple, they had shared the desolation of their daughter Annie's death in 1851. Now, in the spring of 1858, disease threatened the family again. Willie's school had an outbreak of measles. Etty fell ill and was diagnosed with diphtheria, which was very dangerous. Some neighboring children in the village were rumored to have scarlet fever. Charles and Emma were beside themselves with worry. In June, just as Etty started to recover, baby Charles got sick. Soon, Charles and Emma's youngest son was very ill.

In the midst of all the illness and anxiety, on June 18, Darwin received a devastating letter. It came from Alfred Russel Wallace, a self-taught biologist who supported himself by collecting specimens for more established scientists. Darwin had exchanged a few letters with Wallace years before, asking him to forward specimens from the Malay Archipelago and encouraging him in his work. He had even exhorted Wallace to speculate freely, arguing, "Without speculation there is no good & original observation."[3] Now, shockingly, this marginal figure had written a paper outlining a theory of evolution very similar to his own. Darwin had been working painstakingly on gathering evidence for natural selection for more than twenty years. He was in danger of being scooped.

Wallace's parcel arrived at a particularly bad time. Darwin forwarded the manuscript to Charles Lyell immediately, with a note: "Your words have come true with a vengeance that I shd. be forestalled. You said this when I explained to you here very briefly my views of 'Natural Selection' depending on the Struggle for existence. – I never saw a more striking coincidence. If Wallace had my M.S. sketch written out in 1842 he could not have made a better short abstract! Even his terms now stand as Heads of my Chapters."⁴

Darwin was upset. In the same note he commented, "So all my originality, whatever it may amount to, will be smashed." But he consoled himself by thinking of his barnacles and pigeons and all the examples he had gathered. He hoped "my Book, if it will ever have any value, will not be deteriorated; as all the labour consists in the application of the theory."⁵ Wallace's brief sketch was not fully worked out. It was only a few pages long, and it was primarily speculative with very few concrete examples. The fact that his logic was substantially the same as Darwin's was both troubling and oddly reassuring. Darwin felt the loss of precedence more keenly than he had expected. At the same time, however, he could see that Wallace's independent formulation helped to validate the ideas. Like Darwin, Wallace took the concept of gradual, incremental change from Lyell and the concept of focusing on population groups instead of individuals from Malthus. Because Wallace started from the same theoretical underpinnings, much of his terminology was strikingly similar. Darwin's analysis of the situation was right. His originality was "smashed," but his theory was independently validated. The value of his concrete application of the theory—the pigeons and barnacles, dogs and finches—remained.

Darwin had started his transmutation notebooks nearly twenty years before. He had shared his species manuscript with Joseph Hooker more than ten years before, and he had explained or alluded to his thinking in many letters and conversations over these decades. Even so, it took Wallace's brief outline to jolt him toward publishing. When he realized that Wallace's somewhat hasty sketch was publishable and that it would be ungenerous of him not to help Wallace publish it, it also became glaringly obvious that his own sketches from previous years were equally publishable.

Darwin's fear of being a bad sport—taking advantage of a weaker, less well-established rival—warred with his fear of being preempted yet again. He worried about his own reputation. At the same time, however, he also felt protective of the theory. In 1844 the anonymous author of *Vestiges of the Natural History of Creation* had stolen Darwin's thunder, but the greater damage had been that *Vestiges* made evolutionary thought scientifically disreputable. Alfred Russel Wallace had some standing as a naturalist, but he was not very well positioned to defend evolution. If the concept was attributed solely to Wallace, would it sink down into Wallace's obscurity?

On June 25, Darwin wrote to Lyell. He was considering publishing his own brief sketch—perhaps a copy of the letter outlining his theory that he had sent to Asa Gray in 1857, which was roughly equivalent to Wallace's sketch in length and detail. But he was worried that doing so would make him look mean-spirited. He commented, "I would far rather burn my whole book than that he or any man shd. think that I had behaved in a paltry spirit." He did not know what to do. "I am worn out with musing," his letter continued. "I fear we have [a] case of scarlet-fever in House with Baby. – Etty is weak but is recovering." The letter concluded, "My good dear friend forgive me. – This is a trumpery letter influenced by trumpery feelings." His final postscript ended with a hasty promise: "I will never trouble you or Hooker on this subject again."[6]

The following day, however, Darwin wrote once more, expressing his great confusion. He asked Lyell to consult with Hooker and decide on his behalf. "I have always thought you would have made a first-rate Lord Chancellor; & I now appeal to you as a Lord Chancellor." As for Darwin, he had more important things to worry about at home. The children were ill. He told Lyell, "Etty is very weak but progressing well. The Baby has much fever but we hope not S. Fever. – What has frightened us so much is, that 3 children have died in village from Scarlet Fever, & others have been at death's door, with terrible suffering."[7] The Darwins, Charles and Emma, were much more concerned about the Baby's fever than they were about Charles's book projects. Unfortunately, their fears were justified. On June 28, 1858, baby Charles Waring Darwin died.

The next day, on June 29, 1858, Charles Darwin wrote to Joseph Hooker. His letter is worth reading in entirety, in part because it expresses his sorrow in such moving terms and in part because it shows Darwin's level of confusion. In the letter he tells Hooker that he "cannot think" and promises to write again, "as soon as I can think."

My dearest Hooker

You will, & so will Mrs Hooker, be most sorry for us when you hear that poor Baby died yesterday evening. I hope to God he did not suffer so much as he appeared. He became quite suddenly worse. It was Scarlet-Fever. It was the most blessed relief to see his poor little innocent face resume its sweet expression in the sleep of death. – Thank God he will never suffer more in this world.

I have received your letters. I cannot think now on subject, but soon will. But I can see that you have acted with more kindness & so has Lyell even than I could have expected from you both most kind as you are.

I can easily get my letter to Asa Gray copied, but it is too short. –

Poor Emma behaved nobly & how she stood it all I cannot conceive. It was wonderful relief, when she could let her feelings break forth –

God Bless you. – You shall hear soon as soon as I can think.

> Yours affectionately |
> C. Darwin[8]

One striking aspect of Darwin's letter to Hooker on June 29 is the moment when Darwin instinctively turned toward overt religious language. Unable to think, he invoked God three times ("I hope to God," "Thank God," "God Bless you"). He also described his emotions using religious language ("It was the most blessed relief to see his poor little innocent face resume its sweet expression").

It is not surprising that Darwin reached for this kind of language at the moment of his son's death. Many people turn to religious metaphors in extreme circumstances without becoming devoutly religious. It would be a mistake to read Darwin's letter as evidence of some sort of momentary bedside conversion to Christianity. Invoking God when his son died may have been little more than a gesture toward a familiar cultural script. On the other hand, it is fair to note that this is not a letter from a hardened antitheist.

There is no question that Darwin was distraught when he wrote to Joseph Hooker. Nonetheless, it is quite striking that his plan for publishing his work on evolution was effortlessly (perhaps inextricably) intertwined with expressions of love, loss, and friendship framed in terms of God's blessings. If we believe that Darwin's later description of his religious judgment was accurate, that indeed his attitude fluctuated, then the moment when he decided to publish his species theory—or to allow his dear friends to publish it on his behalf—seems to have been a moment when Darwin felt less distant from God. The letter to Hooker weaves between grief for the Baby and plans for the publication, thanking Hooker and Lyell for their advice and advocacy and promising to get his letter to Asa Gray copied. For a moment, his authorial anxieties predominate, as he remarks, "It is too short." But then his letter turns back to Emma and to their shared sorrow. He closes with "God Bless you. – You shall hear as soon as I can think."

In the following weeks, Charles Lyell and Joseph Hooker took over and managed the scientific publication questions while Charles and Emma focused on their family. The Darwins went to the Isle of Wight to grieve. Lyell and Hooker decided to add Wallace's paper, alongside one of Darwin's, to the program of a July meeting of the Linnean Society in London. Neither Wallace nor Darwin would be present. Both papers would be read aloud at the same session so that both authors' papers would have the same publication date.

After the papers were jointly presented and published in the proceedings, Lyell and Hooker urged Darwin to put aside the extremely long, encyclopedic work that he had been laboring over for decades and publish a quick, book-length "abstract" of the theory. He turned it around in a year. In November 1859, Darwin published *On the Origin of Species*.

The first edition sold out the day it was published. It was a modest volume, bound in green cloth, unlike the bright red *Vestiges*. The color was very fitting. Where *Vestiges of the Natural History of Creation* was confrontational, *On the Origin of Species* was conciliatory. *Vestiges* implied that nature was red in tooth and claw. In contrast, *Origin* likened the natural world to a "great tree" covered in "green and budding twigs."[9] According to Darwin, the "great Tree of Life" continually blossomed with "ever branching and beautiful ramifications."[10]

It was an elegant, accessible work presenting solid evidence with clear logic. Darwin himself was widely respected—his wealth gave him independence and social authority while his long record of steady research and his network of elite scientist friends gave him scientific credentials. Evolutionary ideas had been circulating for many years, but Darwin's theory of natural selection was the clearest, most compelling explanation yet.

The page facing the title page of *On the Origin of Species* had two epigraphs, one from William Whewell's Bridgewater Treatise, the other from Francis Bacon's *Advancement of Learning*. Bacon is often credited with laying the foundations for empirical science in the seventeenth century. In the early nineteenth century, Whewell played a similar role. The Bridgewater Treatises were a series of books commissioned by the Earl of Bridgewater in which prominent scientists argued that science accorded with Christianity. Whewell's volume, *Astronomy and General Physics: Considered with Reference to Natural Theology*, was an empirically based argument for Christian natural theology.

Both of Darwin's epigraphs declared that empirical observation of the material world was compatible with religious observation. The quotation from Bacon concluded by asserting that no one "can search too far or be too well studied in the book of God's word, or in the book of God's works; divinity or philosophy; but rather let men endeavour an endless progress or proficience in both."[11] Thus, even before Darwin introduced his own argument, he asked his readers to try to move away from thinking of empirical science and religion as opposed to each other. He claimed to aim for Bacon's ideal: "proficience in both."

The "Introduction" began modestly: "WHEN on board H.M.S. 'Beagle,' as naturalist, I was much struck with certain facts in the distribution of the inhabitants of South America, and in the geological relations of the present to the past inhabitants of that continent. These facts seemed to me to throw some light on the origin of species—that mystery of mysteries, as it has been called by one of our greatest philosophers."[12] The philosopher whom he quotes in his first sentences is John Herschel, whose *Preliminary Discourses* on the scientific method had inspired Darwin as an undergraduate. In an 1836 letter to Charles Lyell, Herschel had praised Lyell for daring to consider "that mystery of mysteries the replacement of extinct species by others."[13] Although there is an almost biblical resonance to Herschel's phrase, it is actually very different from the phrase it echoes: "Holy of holies." No particular set of beliefs is required to describe something as a mystery.

By starting with the "mystery of mysteries," Darwin continued to appeal to both religion and science. As he put it, "I am well aware that scarcely a single point is discussed in this volume on which facts cannot be adduced, often apparently leading to conclusions directly opposite to those at which I have arrived."[14] This had long been his experience. Half a lifetime before, he and Emma had drawn opposite theological inferences from their children's innate kindness. Either it meant that God was beneficent or that God was irrelevant. The core arguments of *On the Origin of Species* were similarly double-edged.

Readers had questions. What did evolution mean? In the first edition Darwin wrote, "There is grandeur in this view of life, with its several powers, having been originally breathed into a few forms or into one; and that, whilst this planet has gone cycling on according to the fixed law of gravity, from so simple a beginning endless forms most beautiful and most wonderful have been, and are being, evolved."[15] But who was breathing life into the planet? In the 1860 edition, he amended his sentence, changing "breathed" to "breathed by the Creator."[16] His revision begged the question: Did he believe in a divine Creator, or was the Darwinian "Creator" an impersonal force along the lines of gravity? Was it possible to believe in God and in natural selection? Or had Darwin replaced God with chance?

Darwin resolutely refused to declare for either side. He enlisted many orthodox Christians to support his concept of natural selection while he also welcomed the support of the less orthodox. According to Adam Gopnik, Darwin "sensed that his account would end any intellectually credible idea of divine creation, and he wanted to break belief without harming the believer, particularly his wife Emma, whom he loved devotedly and with whom he had shared, before he sat down to write, a private tragedy that seemed tolerable to her only through faith."[17] Gopnik concludes that *On the Origin of Species* was a "triumph of style" because it gently made space for nonbelief without ridiculing—or even opposing—religious belief.[18] Gopnik is certainly right to describe Darwin's book as a "triumph of style." *On the Origin of Species* is clear, persuasive, and engaging. It is studded with vivid examples that make it remarkably fun to read. However, in my view Darwin's invocation of "mystery"—his claim that the origin of species is the "mystery of mysteries"—is not simply a subtle strategy for deflecting attention from Darwin's nonreligious attitude toward the natural world. *On the Origin of Species* is framed in terms of wonder and mystery because Darwin saw the earth as enchanted and enchanting.

This is not to say that Charles Darwin shared his wife Emma's belief in God. Most of the time, he was not particularly religious. He objected to religious certainty, and he was offended by the idea that one particular sect might know the only correct version of God. Darwin was equally skeptical of the concept of *revelation*, the belief that God could (and did) directly reveal the truth to his favorites. The strict division between being pro-Christian and anti-Christian seemed as untenable to him as the division between variations of life-forms. Darwin was always interested in the intermediate cases—the *Flustra*, marine invertebrates right on the edge of kelp; the barnacles, crustaceans very close to mollusks; and eventually, the *radicles*, emerging root tips of plants capable of something like thought, "receiving impressions from the sense organs and directing the several movements" of the growing roots.[19] Darwin believed in change and variation much more than he believed in an unchanging, unchangeable divine order, but he never accepted the premise that refusing to adhere to a single religious orthodoxy made him an atheist, much less a nihilist.

Darwin's friend Thomas Huxley, the biologist and fierce polemicist, would soon coin the term "agnostic" to describe the refusal to adhere to a single definition of the divine. As Huxley explained it:

> When I reached intellectual maturity and began to ask myself whether I was an atheist, a theist, or a pantheist; a materialist or an idealist; a Christian or a freethinker; I found that the more I learned

and reflected, the less ready was the answer; until at last, I came to the conclusion that I had neither art nor part with any of these denominations, except the last. The one thing in which most of these good people were agreed was the one thing in which I differed from them. They were quite sure they had attained a certain "gnosis"—had, more or less successfully, solved the problem of existence; while I was quite sure I had not, and had a pretty strong conviction that the problem was insoluble.[20]

Huxley probably saw Darwin as an agnostic. In his review of *On the Origin of Species* in the London paper the *Times*, he praised Darwin for "doubt which so loves truth that it neither dares rest in doubting, nor extinguish itself by unjustified belief."[21] But Darwin hesitated about Huxley's terms too. He would have preferred not to define his religious beliefs at all.

In subsequent years Darwinism has often been defined as anti-Christian. Sometimes Darwinian thought is identified as agnostic, but it is more frequently associated with atheism or even with nihilism. The struggle for existence that Darwin describes in *On the Origin of Species* is often considered cruel and meaningless, and many people think of Darwin's world as completely amoral, with no place for any type of ethics. These understandings of the ramifications of natural selection are central to the disenchantment narrative and to the great disciplinary divorce, the sudden, stark separation of science from philosophy and the arts. After Darwinism took hold, many people began to think of science as anathema to the humanities. Some even began to think of science as inhumane, perhaps antihuman.

The responses to *On the Origin of Species* in 1860 were just as varied as later reactions. Some critics disliked the evolutionary premise so much that they attacked the book without reading it carefully. Some of Darwin's most fervent supporters were just as quick to praise his work—and equally likely to misinterpret it. These readers were not entirely to blame. The book hedged on many of its central questions. It avoided discussion of human beings as cautiously as it avoided mentioning God. Few readers were able to share Darwin's restraint. The implications of natural selection were horrifying to some readers and thrilling to others, but few could resist discussing them. Whether they hated Darwin's book or loved it, everyone was interested.

Like so many of his intellectual and emotional responses, Darwin's reaction to the reception of *On the Origin of Species* was ambivalent. On one hand, he was delighted that the book had such a large audience. Emma wrote to William, "It is a wonderful thing the whole edition selling off at once.... Your father says he shall never think small beer of himself again & that candidly he does think it very well written."[22] Shortly after, Darwin wrote to Charles Lyell with profuse thanks: "What a grand, immense benefit you conferred on me by getting Murray to publish my Book." He was amazed by the popular interest, reporting that an acquaintance "heard a man enquiring for it at *Railway Station*!!! at Waterloo Bridge.... The Bookseller said he had not read it but had heard it was a very remarkable book!!!"[23]

Darwin was proud of his accomplishment and pleased by the initial reception of the book. He wanted his theory to make an impact. But he had always been shy and afraid of controversy. Now, at 50 years old, he was thrown into the public eye. He confessed to Lyell that it was hard to process the "plethora of Reviews."[24] Though he was happy about his book's wide popularity, he was hurt by the critiques. A few of his old friends turned against him. He thought the review by Adam Sedgwick, who had mentored him in geology, misrepresented his ideas in "abusive terms."[25] Richard Owen, the paleontologist who had collaborated with him on the *Beagle* fossils, wrote an even harsher review. Darwin thought it "extremely malignant, clever & ... very damaging." Darwin felt Owen's review was "atrociously severe," "very bitter," and sneering, full of misquotations and "scandalous" misrepresentations. He commented, "It is painful to be hated in the intense degree with which Owen hates me."[26]

Darwin had feared these reactions. He had seen the way Coleridge and the Romantic poets responded to his grandfather's Darwinism, and he had watched in distress as the Oxbridge establishment anathematized the evolutionary ideas advanced in *Vestiges of the Natural History of Creation*. He had been reluctant to publish because he did not want to be drawn into a controversy between science and religion. In part,

Darwin's refusal to enter the debate about whether evolution was anti-Christian was a simple reflection of his character and his circumstances. He hated intellectual disputes, and his health was fragile. Fighting upset his stomach. More important, though, was Darwin's profound discomfort with the false binary.

For Darwin, *atheist* was an insult. Although he was certain that plant and animal species developed over millions of years by means of natural selection and he was well aware that the evolutionary account did not square with Milton's *Paradise Lost* (much less with Genesis), Darwin was ambivalent. He may have balked at the single-minded certitude of Christian orthodoxy, but he still wanted the solace of religion—if not for himself, then at least for Emma. In 1879 he would write, "It seems to me absurd to doubt that a man may be an ardent theist and an evolutionist."[27]

For himself, Darwin believed in multiplicity, in variation, in "endless forms, most beautiful and most wonderful."[28] His was a positive stance. Huxley's emphasis on doubt and unknowing struck Darwin as slightly wrongheaded. It was not that natural selection prompted Darwin to doubt the Christian God but rather that the wonder, beauty, and mystery of the natural world enchanted him so deeply that he had little interest in theological debate.

In England there was great controversy over whether Darwin's book had atheist implications. For months, the papers were filled with articles, reviews, and sermons. The new University Museum at Oxford had not even installed its collections when the British Association for the Advancement of Science held a special meeting in the empty space, inviting Samuel Wilberforce, bishop of Oxford, to debate Thomas Huxley about Darwin's theory. Richard Owen, Darwin's friend turned nemesis, helped the bishop prepare. Darwin himself coached Huxley.

Darwin did not go to Oxford. Though he often avoided conferences, his absence was noticeable since *On the Origin of Species* was the central topic of the entire conference. Perhaps Darwin's dread of controversy made him ill. His stomach troubles were always exacerbated by anxiety. At any rate, Darwin's ill health prevented his attendance. He wrote to Hooker, "My stomach has utterly failed; & I cannot think of Oxford; on Thursday I go for week of water-cure."[29] Then he skipped the meeting.

The 1860 debate was one of the most dramatic events in the history of the BAAS. Bishop Wilberforce ridiculed Darwin and Huxley and defended biblical creation. Huxley, who would later be known as *Darwin's Bulldog*, staunchly defended natural selection. When Wilberforce mocked him, asking if his grandfather or his grandmother was an ape, Huxley whispered, "The Lord hath delivered him into mine hands!" He concluded his reply to Wilberforce by declaring that he would rather have an ancestor who was an ape than an ancestor who resorted to ad hominem mockery as Wilberforce had done.[30] Huxley was a better debater, and most people agreed that he won the contest. But Huxley's oratory did not change many minds. Since *Vestiges*, the conviction that evolution was not only un-Christian but anti-Christian had hardened. If it was difficult for Christian believers to give serious consideration to Huxley's arguments, it was even harder for them to think about Darwin's.

Wilberforce's invocation of human ancestry was a clear illustration that he was not really discussing Darwin's book. Darwin did not write about human beings at all in *On the Origin of Species*. When he finally did explain his thoughts about the evolution of humans, Darwin made it very clear that he did not think humans were descended from apes. Instead, he argued that humans and apes shared a common ancestor. The frequent caricatures of Darwin as a monkey or an ape irritated him, not because he disliked these primates but because it was such a misreading of his theory.[31]

Attacks from old friends like Adam Sedgwick and Richard Owen hurt, but support from figures like Herbert Spencer was even more upsetting. Those who adopted his ideas sometimes used them in startling ways to justify political and economic ideas from capitalist exploitation to outright anarchy. At times he was as distressed by his advocates as by his adversaries. Spencer, who coined the phrase *survival of the fittest*, broadened evolution to include social change as well as biological change. Janet Browne comments, "Much of what was ultimately attributed to Darwin was the result of philosophical shifts expressed in one form or another by Spencer."[32] Darwin kept his distance from Spencer's developing ideas, which came to be known as *social Darwinism*.

Caricature of Darwin as an ape (1874). From *The London Sketch Book* 1, no. 4, Wellcome Collection.

For Darwin himself, "survival of the fittest" was only part of the story. It was true that *On the Origin of Species* had emphasized violent struggle: the book's subtitle ended with "The Struggle for Life."[33] His next theoretical books, *The Descent of Man* and *The Expression of the Emotions*, would focus equally on the other side of natural selection: sexual attraction and emotion. In *The Descent of Man*, Darwin would explain the importance of beauty, arguing that animals whose traits were aesthetically pleasing to potential partners were more successful in reproduction. In his book on emotions, he would explore a wide range of psychological states—not only anger, fear, and hatred but also affection, joy, and even love.

With the benefit of hindsight, twenty-first-century readers can see that Darwin's theories were capacious. For Darwin, life was shaped by many forces that often seemed to contradict each other. Living things evolved in response to both love and hatred, both altruism and greed, both friendship and violence, both beauty and truth.

In 1860, however, public perception of Darwin was one-sided. Most opponents and supporters of evolution thought that Darwin's theory implied that nature was fundamentally brutal. Whether they liked his ideas or not, many readers tended to fixate on the violence of natural selection. Few paid attention to the invocations of mystery and wonder in *On the Origin of Species*. None knew of Darwin's interest in "Joy, High Spirits, Love, Tender Feelings, Devotion." He would not explain his ideas about these emotions until 1872, in chapter 8 of *The Expression of Emotions*.

As Darwin reeled from the controversy over *Origin*, he thought about how to explain his theory more fully. He had started sketching out his ideas about sexual attraction and the expression of emotions around the same time he sketched out natural selection in his private notebooks. Alfred Russel Wallace's letter had pushed him to publish his account of natural selection. Though he had not really wanted to make his ideas public, and he certainly hated the attacks, there were unexpected pleasures in fame. He enjoyed it more than he expected. He would always be a shy person—in fact, in *The Expression of Emotions* he would do his best to analyze that "odd state of mind."[34] Nonetheless, in future he would be quicker to overcome his own shyness. *On the Origin of Species* was a best seller, but it did not quite succeed in convincing readers of "the beauty and infinite complexity of the coadaptations between all organic beings."[35] Darwin would need to try again.

CHAPTER 11

If You Saw a Bullet

Amherst, 1857–1861; Dickinson, Age 26–31

EMILY DICKINSON entered her creative zenith about the same time as Darwin. Through her early twenties, she had put at least as much energy into her social life as into her writing. She had made visits to Springfield, Philadelphia, and Washington, DC, as well as Monson and Boston. She was a brilliant writer, but the literary quality of her letters of the early 1850s was somewhat incidental. She had written them to strengthen social and emotional bonds, which were never easy for her. Dickinson's literary talent did not always help her to communicate. Her unmodulated verbal play—and her wild ideas—sometimes scared people off.

After Susan Gilbert married Emily's brother in 1856 and all of them moved into the family compound in Amherst, there were fewer letters to write. There was little reason for Emily to leave the family grounds. As she approached her thirties, she became more reclusive. More and more, she turned her attention to writing.

Because Emily and Lavinia lived at the Homestead with their parents while Sue and Austin were on the far side of the garden in a separate house, the poet's days were quieter than days in the Darwin household. But Charles Darwin's routines were strikingly similar to Emily Dickinson's. He walked on the Sand Walk while she walked in the Dickinson meadows. They both had Newfoundland dogs, whose companionship they treasured. Both of them worked in their gardens with keen botanical curiosity and focus. Both wore loose, casual clothes most of the time. Darwin defaulted to dark colors, while Dickinson often wore a white wrapper—an informal dress that did not require corsetry, unlike the day dresses most women wore at the time. Darwin and Dickinson both liked fashionable paisley shawls in rich, bright colors. Both of them wrote prolifically during the late 1850s and early 1860s, and for both the philosophical and psychological implications of natural science remained deeply fascinating.

From 1858 onward, Dickinson wrote furiously, sometimes composing two or three poems a day. She was no longer obsessed with valentines. Relatively few of her poems were about romantic love. More often, she wrote about the natural world, perception, and poetry itself. Immensely talented, supported by her family, unfettered by marriage, she probably would have written poetry no matter what. As it turned out, however, Dickinson was living through the intellectual earthquake of Darwinism and the political earthquake of the Civil War.

Dickinson's period of sustained poetic effort started around 1858, just as Darwin started drafting *On the Origin of Species*. Before that, from 1856 to 1858, there is a gap in the Dickinson archives.

For Dickinson, 1857 should have been a good year. Her family was prosperous. Her brother and her best friend had married each other and moved next door. From the outside, Austin and Sue's new house was quite showy. The interior of their new house—the Evergreens—was equally luxurious and modern. Across the yard the renovation of the Dickinson Homestead also signaled the family's wealth. Edward had not only managed to buy back the house his father had lost but also built a new kitchen, a few extra rooms for servants, and the glassed-in conservatory for Emily and Lavinia. However, despite her ostensible good fortune, by 1857 the poet had entered a "time of troubles."[1]

It is not clear what the trouble was, precisely. One probable issue was her mother's illness. Emily Norcross Dickinson, the poet's mother, became an invalid around the time their family moved back to the Homestead. She was incapacitated for years. Nothing seemed to help, not even the water cure in Northampton. Another factor was the change in the

Emily Dickinson's writing table. Photograph by Jerome Liebling, Getty Images.

poet's relationships with Sue and Austin. When two of the people she loved most in the world married each other, Dickinson might have felt she had lost them both. Her beloved Susan Huntington Gilbert was gone forever, transformed into a wife—into Mrs. Austin Dickinson. Some scholars speculate that Dickinson might have had a secret love affair around this time, but there is little evidence to support the conjecture. It does seem clear that she had no desire for the self-erasure she associated with wifehood. Whatever the case, Dickinson did not leave a record of what was happening in her personal life. There are no surviving letters from 1857.[2]

The relationship between Dickinson and her sister-in-law and sometime love, Susan, was complicated. Emily and Susan never fully recovered the intimacy of the early 1850s, but they were not estranged. A constant flow of handwritten notes and shared books, flowers, and other small gifts circulated between the two houses. Susan read many of Dickinson's poems and sometimes offered editorial suggestions. Dickinson valued Sue's opinions, though she tended to trust her own judgment more than any editor.

Susan's marriage distanced her from Emily. Perhaps jealousy played a role. Emily might have had trouble accepting that Susan had chosen marriage to Austin over romantic friendship with her. The bigger issue, however, was that Susan's social role—and her interests—changed when she became a married woman. She started to care about housekeeping, which Emily abhorred. In time Susan would become something of a society hostess. Her house was elaborately decorated with richly upholstered furniture and fashionable wallpaper, and the walls were hung with valuable paintings. Even the gardens were fancy: Austin and Susan consulted the landscape architect Frederick Law Olmsted about what to plant.

Another great difference was their attitude toward groups of people. Susan entertained on a lavish scale. Over the years many cultural luminaries, including Ralph Waldo Emerson, Frances Hodgson Burnett, and Harriet Beecher Stowe, were guests at the Evergreens. Emily rarely attended Susan's parties, though she would sometimes venture over to dinner when old family friends such as Sam Bowles or Otis Phillips Lord were the only guests.[3] Bowles, the editor of the *Springfield Republican*, was exhilarating company who could always be relied on for literary gossip and sparkling good humor. In comparison, Judge Lord was a little awkward. Closer in age to Emily's father, Edward, than to Emily, Lord graduated from Amherst College in 1832. Like Edward Dickinson, he was active in Whig politics. He held a number of different offices in the Massachusetts state government, eventually becoming a judge on the Massachusetts Supreme Judicial Court. He had started out as a friend of Edward's, but in later years Lord was Susan and Austin's frequent guest. In her *Annals of the Evergreens*, Susan recounted his ability to recite poetry from memory. Though Emily was reluctant to attend parties with strangers, she liked to spend time with friends who shared her interest in literature.

Beyond the Dickinson family, the late 1850s were an anxious time across the United States. It was becoming clear that the compromises between slaveholding states and free-soil states would be difficult to sustain. In March 1857 the Supreme Court ruled against Dred Scott, an enslaved African American man who argued that he and his family (wife Harriet and daughters Eliza and Lizzie) should be granted freedom from enslavement because they had lived in Illinois and Wisconsin, where slavery was illegal. In a sweeping decision, Chief Justice Roger B. Taney declared that people descended from Africans were ineligible for United States citizenship and that the Missouri Compromise, which had balanced free states and slave states since 1820, was invalid. Many people doubted that the United States could stay united in the aftermath. The *Dred Scott v. Sandford* decision destabilized westward migration since it was no longer certain that any western territories would become free-soil states. This uncertainty undermined the railroad companies. Soon, crashing railroad stocks led to widespread financial losses. The Panic of 1857 started a global economic depression that lasted for three years.

In Amherst, Edward Dickinson's railroad company fell victim to the panic. After a hasty reorganization, the Amherst and Belchertown line survived, but many of its stockholders lost a substantial portion of their investments.[4] Luckily, Edward Dickinson was a bondholder rather than a stockholder. Though he did not lose his own money, it was a close call. The Dickinsons were aware that their wealth and social position were somewhat precarious. The family had lost their home once. They could lose it again.

The mood of precarity was widespread. Around the world, many people lost money and property. Feelings of panic went beyond finance. Dred Scott's loss—of freedom, citizenship, even legal personhood— was more terrifying than any monetary loss. In 1858, in Edwardsville,

Illinois, a candidate for the U.S. Senate, Abraham Lincoln, explored the implications. He warned, "The logical consequences of the *Dred Scott* decision" would not stop with the "dehumanization" of Black people that "made it forever impossible for [them] to be but as the beasts of the field." Once Dred Scott had been stripped of his humanity, Lincoln told his predominantly white listeners, "the demon you have roused will . . . turn and rend you." He urged those who acquiesced to slavery to "familiarize yourselves with the chains of bondage, . . . you are preparing your own limbs to wear them."[5] The critic Priscilla Wald has explained that Lincoln used "uncanny black subjects—enslaved persons and free descendants of Africans—to emblematize the fate of any subject without a Union that guaranteed liberty for all its subjects."[6] After *Dred Scott v. Sandford*, personhood had become precarious for everyone.

The Panic of 1857 spread at lightning speed, thanks to the telegraph wires that linked small towns like Amherst with Boston, New York, and Washington, DC. Most people believed that the United States would either dissolve or enter into war. Edward Dickinson's congressional term had concluded in 1855, but he and his family were still very politically engaged. Like many others, they took politics personally. In the following years, Emily Dickinson's writings would repeatedly collapse the personal and the political. She would return again and again to the questions of selfhood and dehumanization that the *Dred Scott v. Sandford* decision brought to the national stage. For her, the family dramas—her father's financial uncertainty, her mother's fragile health, Sue's transformation to a society wife—seem to have echoed the national dialogue. Eventually, in letters and poems, she would bring the panic home.

There are straightforward explanations for the skimpy archive of Dickinson papers from 1857. With the entire family (including Susan) living in the same compound, there were fewer occasions for long letters. Dickinson's mother's illness—and the elaborate routines of cold showers and wet linens required by the water cure—probably took much of Dickinson's time. The poet and her sister were overwhelmed by their new housekeeping duties. The most mundane (and most likely) reason that so few papers survive is that Dickinson cleaned her desk and organized her papers in 1858. At that time she made careful copies of the documents she wanted to preserve. No one knows what she discarded.

In the absence of writings from 1857, Dickinson's reading looms large. For the first part of the year, nothing changed. She continued to read the Amherst papers and the *Springfield Republican*. She had access to bookshops in Boston as well as Amherst, and she continued to exchange books with friends and family members. Then, in November 1857, a new publication was launched in Boston, and the Dickinson family subscribed immediately. From that time forward, the *Atlantic Monthly* would be Emily Dickinson's mainstay.

The original publisher of the *Atlantic Monthly*, Moses Dresser Phillips, was the principal owner of Phillips, Sampson, and Company, a leading Boston publishing house. A few years before he had rejected Harriet Beecher Stowe's *Uncle Tom's Cabin* because he wanted to avoid the political controversy over slavery. This was a terrible business decision. Phillips and his partners missed out on the extraordinary profits generated when the book sold more than a million copies in the first year of release.

By 1857 Phillips had decided that the days of nonpartisanship were over. The *Atlantic Monthly* would be an ambitious journal with a decidedly antislavery stance. As he planned the magazine, Phillips gathered a group he described as a "brilliant constellation of philosophical, poetical, and historical talent,"[7] including Ralph Waldo Emerson, Henry Wadsworth Longfellow, Oliver Wendell Holmes, and James Russell Lowell. Soon, Lowell would be appointed editor. Although it was a Massachusetts publication, they named it the *Atlantic Monthly* because they hoped it would draw national or even transatlantic readership. Phillips boasted that his early contributors were "known alike on both sides of the Atlantic, and . . . read beyond the limits of the English language."[8]

The *Atlantic Monthly* offered Emily Dickinson a connection to the sort of intellectual community she had longed for. She pored over each issue, reading hungrily. She was passionately interested in all the topics the magazine covered—"Literature, Science, Art, and Politics."[9] The quality and scope were inspiring. Reading the *Atlantic Monthly* enlarged the literary horizon for Dickinson. It made her confident that Massachusetts writers could hold their own in any company. The next year, in 1858, Dickinson would begin her most ambitious literary projects, the "Master" documents and the fascicles. Both were experimental and both were private.

The "Master" documents include three prose pieces addressed to a figure whom Dickinson calls "Master." In *Writing in Time: Emily Dickinson's Master Hours*, the Dickinson scholar Marta Werner suggests that a few brief pieces that use the word are related to the three longer pieces. The "Master" documents are not conventional poems or letters. The longer pieces are written in highly artistic prose that frequently crosses into poetry. These documents were not made public until 1955. When they were first published, they were read as letters and known as the three "Master letters." Most mid-twentieth-century critics thought of them as artifacts of a relationship with a man who was probably married. Samuel Bowles and Charles Wadsworth were mentioned most frequently as possible lovers, though there was little agreement about the identity of Dickinson's supposed "Master." More recently, critics have questioned whether the "Master" documents are personal letters or some other kind of writing. Most prominently, Werner argues that it is much more accurate to describe them as literary experiments than as private correspondence. At any rate, the "Master" documents were unconventional writings that addressed—and often questioned—the concept of mastery.

By the summer of 1858, Dickinson had also started work on the booklets that scholars call *fascicles*. At first these booklets contained neat final copies of poems. Later the texts were looser and included many alternate words and variant lines. To make a fascicle, she copied poems onto a few sheets of loose paper. She carefully folded the sheets, then stacked the folded sheets, poked holes near the fold, and sewed them together. The grouping of poems onto sheets of paper within the fascicles created fascinating intertwining dialogues. Once her poems were copied this way, she usually destroyed earlier drafts. The literary scholar Alexandra Socarides has contrasted Dickinson's laboriously constructed fascicles with the more common nineteenth-century collections of poetry copied into prebound blank books.[10] Dickinson's technique for making booklets was rarely used for portfolios of poetry or for diaries, though the method was a common way of preserving sermons at the time. Dickinson's booklets differed from popular commonplace books, autograph albums, and scrapbooks. For one thing, she did not include works by anyone else. For another, she did not pass them around to her friends. As far as we know, she did not circulate the fascicles at all.

In the preface of this book, I have discussed the first leaf of the first fascicle. Although the fascicle project would go on to encompass all the topics covered in the *Atlantic Monthly* (literature, science, art, and politics), it is notable that it began with flowers. Her fascicle project was related to the herbarium-making of her school days. At the beginning, the fascicle's central questions seemed to be about literature and science, particularly poetry and botany. If the "Master" documents addressed human mastery from below—from the perspective of someone with a master—the fascicle projects took a slightly different angle, eschewing mastery altogether and writing from the perspective of someone on the same level as the rest of the natural world. In both projects Dickinson identified with natural phenomena, particularly birds, insects, and flowers.

Eventually, there would be three passionate pieces of prose addressed to a mysterious entity whom the author called "Master." These texts veered from specificity to vagueness in a way that made them seem more like fantasies than first drafts of actual letters. The editor Marta Werner calls them Dickinson's "Master experiments."[11] Werner explains, "This experiment is not a language experiment only, but one, laid bare by the presence, either explicit or implicit, of an address, in writing to another: not to a particular other—though it may originate with the image of someone actual in the world . . . but to another describable only in writing and perhaps only fully real in the time of composing."[12] Dickinson was experimenting with audience as well as with language.

Dickinson's family did not participate in her private experiments. Sometimes she shared copies of her poems with Sue or with other correspondents, but none of them realized the extent to which her days were filled with linguistic experimentation. She sent frequent letters to people she knew (and occasional ones to strangers). She often included poems in her correspondence. But she kept many of her writings private. Hundreds of poems were uncirculated. Many were addressed to readers whom she secretly imagined. By their allusions we can identify some of her works as conversations with figures such as the scientists Isaac Newton, John Herschel, and Urbain Le Verrier, and the writers John Keats, Ralph Waldo Emerson, Charlotte Brontë, and Elizabeth Barrett Browning. A few seem aimed straight at her neighbor, the geologist Edward Hitchcock, who had been president of Amherst during her student years. If the "Master" documents had started as a conversation with a particular person, they changed when Dickinson made private copies for herself. The copy of the "Dear Master" letter that Dickinson made in early 1858 was, like the fascicles, uncirculated. The later documents seem wilder, even less likely to have been addressed to a person Dickinson knew.

According to Marta Werner's time line, Dickinson wrote most of the "Master" documents after July 1860, when *On the Origin of Species* was reviewed in the *Atlantic Monthly*. By then the conflict over slavery had come to a head in the United States. These seemingly separate events were surprisingly related to each other—Darwin's theory was immediately taken up by abolitionists as scientific evidence that neither Dred Scott nor any other Black people could be denied humanity. Both the start of the Civil War and the publication of Darwin's theory would have profound implications for Dickinson's thinking about mastery.

The reception of *Origin* in Britain and across Europe was often at odds with Darwin's genteel progressive politics. Unlike most of his British and European colleagues, Darwin did not think "survival of the fittest" was the only implication of his theory. In contrast, the response to his theory in the United States aligned perfectly with his political values. Although *Origin* had not directly discussed human beings, Darwin's theory of natural selection offered a scientific rationale for the abolition of slavery since it explained how variations could occur within a single species. Darwin had always been a passionate abolitionist. He hated slavery. When New England abolitionists began to use his book to argue for abolishing slavery, Darwin was thrilled.

In part, Darwin's opposition to slavery came from his family history. Most of the Lunar Men had been abolitionists. In 1787 his grandfather Josiah Wedgwood's pottery had created an antislavery medallion depicting a kneeling, chained man sculpted in relief on a background emblazoned, "AM I NOT A MAN AND A BROTHER?" Two generations afterward, *On the Origin of Species* offered a resounding affirmation of the shared humanity of different races. If Darwin's theory of natural selection fostered uncertainty about God, when it came to race, it offered moral clarity.

A cartoon in the May 18, 1861, edition of *Punch* pointed out the connection between Darwin's theory and his grandparents' abolitionism. Though England had abolished slavery before Darwin was born, many British colonies and former colonies continued to organize their economies around enslavement, and Britain continued to reap the profits. Darwin did not personally witness the slave system or meet any enslaved people until he traveled to South America aboard the *Beagle*. Before then, as a medical student in Edinburgh, Darwin had studied taxidermy with John Edmonstone, who had been born into enslavement in Guyana. He respected Edmonstone and made good use of what Edmonstone taught him. When he got to South America and learned that the captain of the *Beagle*, Robert FitzRoy, supported slavery, he was outraged. In 1832, after a quarrel about slavery, Darwin left the ship, determined to take passage home. Fitzroy apologized to Darwin, and the two men patched things up well enough for Darwin to continue the voyage, but their relationship was deeply damaged by their differing views on the slavery question. From then on Darwin was personally committed to the abolition of slavery. In the early 1860s, Darwin wrote to the American botanist Asa Gray, describing slavery as a "hopeless curse."[13]

In New England, abolitionist-leaning intellectuals were desperate to get their hands on Darwin's book. Ralph Waldo Emerson wrote home

Wedgwood antislavery medallion (1787). British Museum.

from a lecture tour through Indiana, complaining about "prairie mud" and adding, "I have not yet been able to obtain Darwin's book which I had depended on for a road book.... It has not arrived in these dark lands."[14] It did arrive in Massachusetts in December 1859, very soon after it was published. Asa Gray, to whom Charles Darwin had sent an outline of his theory in 1857, got one of the very first copies. In Concord, which congratulated itself on being enlightened, Emerson's friends gathered to discuss the two most exciting developments of the winter: the publication of *The Origin of Species* and John Brown's raid on Harper's Ferry.

In *The Book that Changed America*, the literary historian Randall Fuller has described the first Massachusetts discussions of Darwin's

"Monkeyana" cartoon, *Punch*, May 18, 1861. Wellcome Collection.

book. On January 1, 1860, Asa Gray's cousin, Charles Loring Brace, went to a small dinner party in Concord. Though Emerson was out of town, Bronson Alcott and Henry David Thoreau came. They gathered at the home of Franklin Benjamin Sanborn. After dinner the four men pored over the copy of *On the Origin of Species* that Brace had borrowed from Asa Gray. Darwin's book was a welcome distraction from the cloud of anxiety after Harper's Ferry. John Brown had been convicted of treason and executed in Virginia. The same fate impended over the six men, popularly known as the "Secret Six," who had raised the money for his raid. Sanborn, the host of the dinner, was one of the Secret Six. Fearing arrest, Sanborn would flee to Canada twice that winter.

Sanborn's coconspirators were a mix of industrialists and intellectuals. Another member of the Secret Six was the dashing Thomas Wentworth Higginson, a writer whose articles were often featured in the *Atlantic Monthly*. A man of great courage (with a saber scar on his chin from his 1854 fight to prevent Anthony Burns from being returned to enslavement under the Fugitive Slave Law), Higginson was the only one of the six conspirators who refused to flee. Later that winter he would give a public lecture in Concord that used examples drawn from Darwin to bolster the fight against slavery. In 1860 Higginson was known primarily as a radical Unitarian who wrote essays on women's equality, abolition, and the outdoors. Later he would serve in the Union Army, an officer of the 1st South Carolina Volunteers, the first Black regiment in the U.S. armed forces. He would correspond with—and eventually visit—both Emily Dickinson in Amherst and Charles Darwin in Kent. In the winter of 1860, however, all that was in the future. Thomas Wentworth Higginson and Henry David Thoreau gave their attention to *On the Origin of Species*. The book seemed to hold out the promise of a new philosophical approach to the natural world.

Gray was the first person to review *On the Origin of Species* in the United States. In the July 1860 issue of the *Atlantic Monthly*, Gray explained that Darwin showed that "the Negro and the Hottentot" were "blood-relations" of the white readers of the *Atlantic Monthly* (a view that Gray thought was supported by both "reason" and "Scripture").[15] Gray argued that reason and scripture also supported the theory of natural selection. In Gray's view, species change was divinely guided, and Christianity was perfectly compatible with evolutionary thought. He summed up the ramifications by quoting Shakespeare: Darwin's arguments implied that nature had made "the whole world kin."[16] This essay was probably the first place Emily Dickinson learned of Darwin's ideas about evolution.

In some ways Darwin's thinking would have sounded familiar to Dickinson. As Gray put it, "Investigations about the succession of species in time, and their actual geographical distribution over the earth's surface, were leading up from all sides and in various ways to the question of their origin."[17] As a resident of Amherst and an alumna of Mount Holyoke and Amherst Academy, Dickinson was well versed in such subjects. Gray's description would have made perfect sense to her, particularly when he explained: "The theory hitches on wonderfully well to Lyell's uniformitarian theory in geology, – that the thing that has been is the thing that is and shall be, – that the natural operations now going on will account for all geological changes in a quiet and easy way, only give the time enough, so connecting the present and the proximate with the farthest past by almost imperceptible gradations, – a view which finds large and increasing, if not general, acceptance in physical geology, and of which Darwin's theory is the natural complement."[18]

Like many of her contemporaries, Dickinson was fascinated by the philosophical implications of this new combination of geology and biology. What did it mean to say "that the thing that has been is the thing that is and shall be?" Had Darwin discovered *how* "nature makes the whole world kin?" And if everyone was family, what did that mean for mastery?

Even before Dickinson learned about Darwin, she had been writing about the natural world in ways that challenged hierarchies. One thrilling aspect of Darwin's theories was that they helped Dickinson clarify her own thinking. Another was that her imagined connection with the writers featured in the *Atlantic Monthly* began to deepen. Dickinson had long been able to imagine herself in conversation with Emerson, Longfellow, and Lowell. Now she began to imagine herself in conversation with Asa Gray and his friends: around Boston, his fellow contributors to the *Atlantic Monthly*; in Britain, Charles Darwin.

Dickinson had always liked uncertainty in natural science, just as she liked uncertainty in religion. Now, Darwin's ideas about death and creation combined with her own sense of loss and her powerful poetic ambition. Her experimental writing emphasized the magical kinship that Darwin's writing implied. Chris Gosden reminds us that "magic sees spirits in the land, considers how people and animals are related, and tries to understand transformations around birth and death."[19] In the early 1860s, Dickinson was experimenting with magic in all her writing. She enjoyed the "mysterious incalculable forces" that Max Weber thought nineteenth-century disenchanters had banished. In the texts we know now as the "Master" documents, she addressed the concept of mastery directly, and then she gleefully dismantled the theory that "all things ... can be mastered" (as Weber phrased it).[20]

In late summer or early autumn of 1860, she penciled a brief note, "Mute – thy coronation –," that Werner considers a "Master" document.[21] Dickinson did not copy this piece into a fascicle. "Mute – thy coronation –" describes the moment when the "Master" quietly becomes a king. It is written from the perspective of a small being—perhaps a bee, perhaps a piece of paper—folded into the beloved master's robes. Other than this short piece (about thirty-five words long), she would not return to the "Master" documents until 1861.

Abraham Lincoln was elected president of the United States in November 1860. By the time he was inaugurated in March 1861, seven slave states had seceded to form the Confederacy. Dickinson wrote the second long "Master" document, "Oh' did I offend it," in the first months of 1861. This prose composition, written from the point of view of a daisy or a person named Daisy, expressed pain at being distanced from the "Master." It filled four sides of a sheet of stationery folded like a book.[22] Later that spring she composed or copied "A wife – at Daybreak – I shall be –," a poem that Werner includes among the "Master" documents. A few years later, Dickinson would copy a "A wife – at Daybreak" into one of the fascicles. The longer prose piece, "Oh' did I offend it," is about the pain of separation, while "A wife – at Daybreak" is about the pain of union—the loss of identity that comes with marriage. Both "Master" documents from early 1861 are ambivalent meditations on love, death, and domination, shot through with longing. They are intimate documents, saturated with extraordinary emotion. At the same time, their overarching themes—the relationship between a "Master" and an enslaved person, the perils of union and separation—are cued to the political circumstances of the Civil War.

On April 12, 1861, Confederate troops fired on Fort Sumter, marking the formal beginning of the Civil War. In Amherst the Dickinson family cautiously supported the war effort. Sue was pregnant. Edward and Austin were busy trying to encourage their neighbors to volunteer as Union soldiers. Emily was anxious and lonely, worried about the Amherst College men who were eager to join the army, worried about the ordeal of childbirth facing Sue, fearful that her intense need for love would drive everyone away from her, as it seemed to have done with Sue and perhaps with another person she imaginatively addressed as "Master."

The summer of 1861 was eventful. On June 19, 1861, Sue gave birth to her first child. She and Austin named him Edward, after his grandfather. He would be called Ned. Ten days later, on June 29, 1861, in Florence, Italy, Elizabeth Barrett Browning would breathe her last. Within a few weeks, on July 21, 1861, the Union Army would suffer its first major defeat at the Battle of Bull Run. After Bull Run, many young men from Amherst enlisted in the 21st Massachusetts Infantry Regiment. As they waited to go south, the college students and other local men drilled on the Amherst Green, learning to operate rifles and bayonets. There was no escape from the noise: drums, marching, target practice.

Sometime that summer, Dickinson wrote the last of the "Master" documents. It begins with a strange conceit. A dead bird speaks, only to lie about his pain:

Master.
   If you saw a bullet
hit a bird – and he told you
he was'nt shot – you might weep
at his courtesy, but you would
certainly doubt his word –
One drop more from the gash
that stains your Daisy's
bosom – then would you believe?[23]

Breathless, unpunctuated, ecstatic, the document moves from a bird pierced by a bullet to a bleeding daisy, and then on to a silent dialogue between volcanoes: "Vesuvius dont talk – Etna – dont – They one of them – said a syllable – a thousand years ago, and Pompeii heard it, and hid forever – She could'nt look the world in the face, afterward – I suppose – Bashfull Pompeii! 'Tell you of the want'—you know what a leech is, dont you – and remember that Daisy's arm is small – and you have felt the Horizon – hav'nt you – and did the sea – never come so close as to make you dance?"[24]

The speaker describes Italian volcanoes as conscious beings, capable of thought, emotion, and speech. She imagines Pompeii as a shamed female, hiding forever because "she couldn't look the world in the face, afterward." Although she invokes sexual shame, the purpose of this passage may be to move beyond embarrassment. Directly after she describes Pompeii as bashful, she answers an interlocutor who has asked her to explain her desires by comparing the question to a leech that could potentially unleash a flow of blood or sap. It is not clear that this "Daisy" is a human being. She might be a flower. Shortly after describing her arm as small, she speaks of feeling the horizon and dancing with the sea. This literary experiment puts readers in the realm of magic geology, where geological features are as laden with emotion (and consciousness) as biological creatures, plant and animal. It gives voice to a dead bird and expresses a heady mix of pain and desire.

When Dickinson wrote "Master. / If you saw a bullet," Darwin was still hesitant to make explicit his view that human beings were not at the apex of a hierarchy but merely branches on the great green tree of life, kin to other living things. At the time Dickinson's experimental work would have shocked him. In the early 1860s, her poems and letters expressed dizzying sexual longing and powerful, polymorphous love for a surprising variety of people, plants, animals, and things. She concluded "Master. / If you saw a bullet" with a couplet: "No Rose, yet felt myself a'bloom / No Bird – yet rode in Ether." The natural world intoxicated her.

"Master. / If you saw a bullet" is beautiful and moving, but it is also puzzling. It is hard to follow. Many readers are disturbed by the strange energy of the piece and overwhelmed by questions about context. It is almost impossible to read the piece without making conjectures about the intended recipient. Indeed, Dickinson might have imagined addressing an editor (like Sam Bowles) or a spiritual leader (like Charles

Wadsworth), as twentieth-century scholars have hypothesized. She might have been thinking of a philosopher or scientist, perhaps a figure like Charles Darwin, whose work *On the Origin of Species* reframed death as a creative force and encouraged readers to look closely at small variations and to question the separation between human and nonhuman beings.

These possibilities do not preclude each other. "Master. / If you saw a bullet" is extraordinarily open-ended. Werner's suggestion that the "Master" documents are experiments of address is helpful here. Around the time Dickinson wrote "Master. / If you saw a bullet" in summer 1861, she had probably learned of Elizabeth Barrett Browning's death. Dickinson identified with Elizabeth Barrett Browning. Alongside the possibility that the later "Master" documents address a Darwinian scientist, we must also consider the possibility that they address a poet. Dickinson might have imagined Elizabeth Barrett Browning as her "Master." She might also have identified so closely with Barrett Browning that she imagined Robert Browning as her own "Master."

For Dickinson, Barrett Browning's death was a profoundly moving event. Before then she had admired the poet, and even aspired to emulate her. After Barrett Browning died, Dickinson's conception of her own poetic vocation changed in a Darwinian way. Barrett Browning modeled clear-eyed, multiperspectival empirical observation and a willingness to tackle modern scientific thought about the natural world, but she was certainly not a Darwinist. Because Barrett Browning died a year or so after the publication of *On the Origin of Species*, none of her poetry directly discussed Darwin. She did not share Darwin's view that death was a powerful creative force that shaped evolution. In this respect, Dickinson's interest in the meaning of death aligns more closely to Darwin than to Barrett Browning.

In the months after Barrett Browning died, Dickinson gathered as much information about her as she could. She collected three photographs of the poet and eventually secured a photograph of her tomb in Florence. In an 1862 letter, Dickinson would connect her poetic vocation to Elizabeth Barrett Browning, remarking, "I sing, as the Boy does by the Burying Ground – because I am afraid."[25]

The literary critic James Guthrie has argued that Dickinson approached the literary marketplace as Darwinian competition.[26] From this perspective the evolutionary logic of the survival of the fittest may help explain why Barrett Browning's death was so inspiring to Dickinson. Dickinson hoped to take Barrett Browning's place, just as the elephants had replaced the mastodons in *Aurora Leigh*. Intensely ambitious, Dickinson thought it might be possible to devise a new species of poet.

Dickinson wrote at least three poems about Barrett Browning. The summer after Barrett Browning died, Dickinson wrote a letter to Samuel Bowles, who was traveling to Italy. "Dear friend –" she said, "Should anybody where you go, talk of Mrs. Browning, you must hear for us – and if you touch her Grave, put one hand on the Head, for me – her unmentioned Mourner."[27] With this she offered a complex identification of herself as Barrett Browning's possible successor and "unmentioned Mourner." A year later she sent her sister-in-law, Susan, a copy of another poem about Barrett Browning's death. "Her – last Poems –" concluded:

> Nought – that We – No Poet's Kinsman –
> Suffocate – with easy Wo –
> What – and if Ourself a Bridegroom –
> Put Her down – in Italy?[28]

Here, the narrator identifies with Robert Browning. She admits that she is "No Poet's Kinsman" but imagines herself as Barrett Browning's "Bridegroom," burying his dead wife in Italy.

Another poem on Barrett Browning's death, "I think I was enchanted," described Barrett Browning's poems as "Tomes of Solid Witchcraft" that transformed the natural world into a "Titanic Opera" and filled the speaker with "Divine Insanity." The poem concluded by declaring that even though Barrett Browning had died, her magic would last forever:

> Magicians be asleep –
> But Magic – hath an element
> Like Deity – to keep –[29]

This declaration of faith in the abiding magic of poetry was an important turning point in Dickinson's career. She would always be skeptical of absolutist claims about God, but Barrett Browning's works filled her with a sense of the possibilities of poetry. They showed her that it was possible to express something real and true about the natural world in poetry as well as in science. And, just as importantly, Barrett Browning offered a vision of nature that was empirically accurate without being dull or mechanical. *Aurora Leigh* showed that a nimble poet could play among the fossils without becoming a cold, gray soul.

Even so, the leap from *Aurora Leigh* to enchantment is surprising. There are no sorceresses mumbling magic spells in *Aurora Leigh*. The novel in verse is realistic rather than supernatural. Why did Dickinson describe this highly intellectual realist as a "Magician"? It might be helpful to recall archaeologist Chris Gosden's observation: "The forces defined by science find echoes in magic's insistence that spirits animate the world."[30] In Barrett Browning, Dickinson found a poet who embraced modern scientific thought and confidently insisted that the natural world was animated with meaning. She had been inspired—even enchanted—by Barrett Browning. In 1861 she was powerfully affected by grief at the death of the poet she had never met.

All this came together in September 1861 in the pages of the *Atlantic Monthly*. Browsing through this issue of the magazine, Dickinson would have found all her preoccupations gathered into one volume. Many of the articles looked back at the grim news of the summer. Charles Eliot Norton's article on "The Advantages of Defeat" argued that "Bull Run was in no true sense a disaster; that we not only deserved it, but needed it; that its ultimate consequences are better than those of a victory would have been. Far from being disheartened by it, it should give us new confidence in our cause, in our strength, in our final success."[31] Despite his fervor, it was hard to contemplate the carnage without anxiety. In "Bread and the Newspaper," Oliver Wendell Holmes discussed the emotional costs of the ceaseless barrage of war news, which he described as "disturbances of the nervous system as a consequence of the war excitement in non-combatants."[32] He wrote: "When any startling piece of war-news comes, it keeps repeating itself in our minds in spite of all we can do. The same trains of thought go tramping round in circle through the brain, like the supernumeraries that make up the grand army of a stage-show. Now, if a thought goes round through the brain a thousand times in a day, it will have worn as deep a track as one which has passed through it once a week for twenty years. This accounts for the ages we seem to have lived since the twelfth of April last."[33] The image resonated with Dickinson. In her poem "I felt a Funeral, in my Brain," the poet would describe how "Mourners treading to and fro / kept treading, treading" through the brain.[34]

Although Dickinson had known about all the deaths at Bull Run and about the death of Elizabeth Barrett Browning since July, the topics were still painful. The same issue of the *Atlantic Monthly* contained Kate Field's obituary, "Elizabeth Barrett Browning," which was hard to read and equally hard to stop thinking about. The pages are scissored out of Dickinson's copy of the magazine.[35]

The remaining pages of the September 1861 issue of the *Atlantic Monthly* offered some relief. A brief review of George B. Prescott's *History, Theory, and Practice of the Electric Telegraph* might have taken Dickinson back to her enthusiasm for the aurora of 1851, ten years before. There was another installment of Harriet Beecher Stowe's *Agnes of Sorrento*, a historical novel set in Renaissance Italy. And then there was an unsigned article on "My Out-Door Study." It is not certain that Dickinson would have known that the essay was by Thomas Wentworth Higginson, the dashing abolitionist and Darwinian literary man. But in the face of defeat and death, she could not avoid recognizing a kindred spirit.

"My Out-Door Study" describes searching for a "realm of enchantment beyond all the sordidness and sorrow of earth" and finding inspiration in "a bird, a bee, or a blossom, beside [the] homestead door."[36] Living in her own homestead, deeply fascinated by flowers, insects, and birds, Dickinson must have thrilled with recognition. Here was someone who thought the way she did about nature and about literature.

Around this time, Dickinson seems to have stopped writing letters addressed to a "Master." She was done with that experiment. As 1861 came to a close, she would start to imagine a different kind of correspondence. She wanted kinship, not mastery. She needed a friend.

Dickinson had started her first fascicle with an invocation:

In the name of the Bee –
And of the Butterfly –
And of the Breeze –[37]

Now she had found a writer who felt the same deep, somewhat mystical connection between poetry and science that she did. The author of "My Out-door Study" asked, "How many living authors have ever attained to writing a single page which could be for one moment compared, for the simplicity and grace of its structure, with this green spray of wild woodbine or yonder white wreath of blossoming clematis?" He declared, "A finely organized sentence should throb and palpitate like the most delicate vibrations of the summer air."[38] He celebrated how everything was connected: "The same mathematical law winds the leaves around the stem and the planets round the sun. The same law of crystallization rules the slight-knit snow-flake and the hard foundations of the earth. The thistle-down floats secure upon the same summer zephyrs that are woven into the tornado. The dew-drop holds within its transparent cell the same electric fire which charges the thunder-cloud."[39]

In the face of defeat, death, and loneliness, Dickinson longed for connection. She knew that she could find solace in her sense of kinship with the nonhuman world. Now she began to hope that she could find other people who shared her beliefs. Maybe she was not alone. In the pages of the *Atlantic Monthly*, she had found a stranger who could see that the world was green with magic.

CHAPTER 12

Wild Experiment

Downe and Amherst, 1860–1862

DARWIN THOUGHT of *On the Origin of Species* as a brief abstract of the great theory he had been working on for decades. After he published it, he planned to get to work on the larger book right away. But the response to *Origin* was overwhelming. Over the next few months, he spent countless hours consulting his circle of scientific allies about how to defend his theory—and himself—from the widespread attacks. His stomach troubles flared up, perhaps because of anxiety. The summer after he published *Origin*, his daughter Etty fell ill with scarlet fever. Wandering on the heath in Sussex during her illness, Darwin noticed some strange-looking plants, which he identified as *Drosera rotundifolia*. He was fascinated and "gathered about a dozen plants."[1]

While Emma tended to their sick child, Darwin spent his days writing letters and experimenting with his *Drosera* plants. When Etty began to recover, he wrote to Charles Lyell, "All this dreadful illness for last six months (& that wicked dear little *Drosera*) has made any progress in my larger Book almost nothing."[2] He was not complaining. Now that his daughter was on the mend, it was a relief to take a break from the public debates over evolution. He had developed a strong emotional attachment to "that wicked dear little *Drosera*." In the same letter, he confessed, "at this present moment I care more about *Drosera* than the origin of all the species in the world."[3]

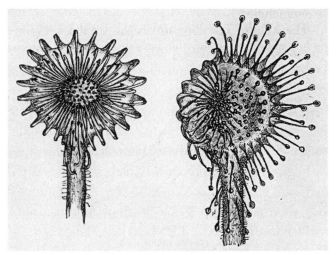

Drosera rotundifolia by George Darwin. Charles Darwin's caption for the leaf on the right reads "Leaf (enlarged) with the tentacles on one side inflected over a bit of meat placed on the disc." From Charles Darwin, *Insectivorous Plants* (London: John Murray, 1875).

Drosera rotundifolia, known as the common sundew or the round-leaved sundew, grows in swamps and boggy places. The leaves of the sundew are edged with delicate reddish stalks that glisten with clear, sticky droplets so they look like they are drenched in morning dew long after sunlight has dispersed all traces of moisture from the surrounding plants. When an insect lands on a leaf, it is trapped by the flypaper effect of the sticky secretion. Then the red tentacles fold around it, capturing it completely. Once caught, the insect is digested, its nutrients absorbed directly into the plant.

Before Darwin, botanists had observed that sundews produced a liquid that acted as an insect adhesive. When Darwin's grandfather Erasmus described *Drosera* in *The Botanic Garden* in 1789, he compared the clear droplets to diamonds:

> A zone of diamonds trembles round her brows;
> Bright shines the silver halo as she turns;
> And as she steps, the living lustre burns.[4]

In a note, Erasmus had likened the "viscous material" of the *Drosera* to earwax, explaining it as a defense mechanism that prevented "small insects from infesting the leaves. As the ear-wax in animals seems to be in part designed to prevent fleas and other insects from getting into their ears."[5] Many years later, his grandson Charles had begun to suspect that the *Drosera* was not merely defending itself. He thought the plants might be carnivorous—that they might capture insects in order to supply themselves with elements that were unavailable in the nutrient-poor soil of the bogs. In a letter to botanist Daniel Oliver, he relayed an "extraordinary fact. – I have placed over & over again minute atoms of paper, stick, cinders meat flies &c on glands of *single* hairs of *Drosera* & they always became inflected."[6] The plants never let go of the insects or the other tiny bits of meat, but when he tried feeding his *Drosera* bits of his own hair and toenails, they first clasped, then rejected them (once they determined that the offerings were indigestible).[7]

Emma Darwin commented, "At present he is treating *Drosera* just like a living creature, and I suppose he hopes to end in proving it to be an animal."[8] She was right. He wrote to Hooker, "The leaves are first rate chemists & can distinguish even an incredibly small quantity of any nitrogenised substance from non-nitrogenised substances."[9] He told Lyell he was perfectly sure that *Drosera* was "far more sensitive to a touch than any nerve in the human body."[10] These observations astonished him, but they also scared him. In three separate November 1860 letters to Charles Lyell, John Lubbock, and Asa Gray, he repeated similar sentiments: "I am frightened and astounded"; "I have got actually frightened"; "I am frightened at my results & must retest them."[11]

Why was Darwin frightened? First, the concept of carnivorous plants is just plain scary. Before Darwin, no one had suspected it was possible for plants to eat animals. In later years, Arthur Conan Doyle, H. G. Wells, and countless screenwriters would write stories about malevolent plants devouring innocent animals.[12] For Darwin, however, there was a second, even more frightening aspect to his research. He found it extremely "painful to be hated in the intense degree" with which his fiercest critics hated him.[13] Men like Adam Sedgwick, Richard Owen, Louis Agassiz, and Samuel Wilberforce attacked him in the press with disturbing frequency.

Now he had stumbled on evidence that plants might possess consciousness. He was certain that if he published anything that blurred the dividing line between plants and sentient animals, his critics would pounce.

And yet, despite his fears, Darwin was enthralled. He wrote to Daniel Oliver at Kew, begging for Venus flytraps (*Dionæa muscipula*): "If I could get one or two plants of *Dionæa* I would experimentise on them; ... It is so important to me ... that I should extremely much wish to ... purchase (at almost any price) one or two or three of the young plants."[14] By the end of 1860, he was determined to continue experimenting with sundews and flytraps. He loved to "experimentise." But he had also decided to keep his work private for the time being. He would not publish his book on *Insectivorous Plants* for fifteen years.

Darwin's constant *experimentising*—the term he used for fiddling around with experiments—involved his entire household. According to the biologist James T. Costa, at Down House, "sheets of damp paper stuccoed with frog eggs lined the hallway ... pigeons cooed boisterously in a dovecote in the yard, row upon row of glass jars with saltwater and floating seeds filled the cellar."[15] Outdoors, "malodorous pigeon skeleton preparations permeated the air.... There was a terrarium of snails with suspended duck feet, heaps of dissected flowers, and the fenced-off plots in the lawn where the grass was carefully scraped away to study struggling seedlings."[16] In later years, when he became interested in earthworms, he constructed a device he called a worm stone that used a millstone to measure how the action of worms changed the level of soil in his garden. He built elaborate paper mazes for worms and subjected them to puffs of smoke and other stimuli. He even tried to determine their reactions to piano melodies. Darwin's experiments were a constant source of amusement for his busy family. Everyone participated, including the servants and the houseguests. But these experiments were not always public. Darwin was reluctant to expose his ideas to critics.

Nonetheless, he kept experimenting. In a springtime letter to his friend Joseph Hooker in 1863, Darwin begged him to send orchid pods, lichens, and mosses. He wrote, "*For love of Heaven favour my madness ... I am like a gambler, & love a wild experiment.*"[17] In a striking similarity, Dickinson's letters a few years later, in April 1875, to her sister-in-law,

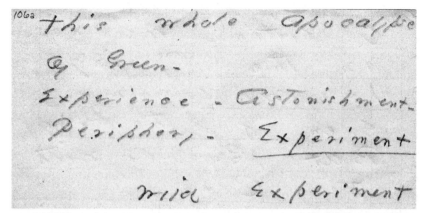

Emily Dickinson's "wild experiment." "A little madness in the Spring" manuscript (verso). Amherst College Archives and Special Collections.

Susan, and to Elizabeth Holland, enclosed copies of a poem that parallels Darwin's language:

> A little madness in the Spring
> Is wholesome even for the King
> But God be with the Clown
> Who ponders this Tremendous scene
> This whole Experiment of Green
> As if it were his own –[18]

According to the editor Ralph W. Franklin, a surviving "worksheet reveals a determined attempt to find a reading for two words in line 5. . . . On she went, testing nouns until 'Experiment' emerged and she tried it with a new adjective, 'wild.'"[19]

Cristanne Miller lists the alternatives for line 5:

> [This] gay • bright • *quick* • whole • swift – • fleet • whole [legacy of Green]
>   [This] fair *Apocalypse* of Green – • This whole Apocalypse of Green – • [This whole] Experience – • Astonishment • Periphery – • *Experiment* •
>   wild Experiment [of Green].

The similarities between Darwin's and Dickinson's messages are notable. They used the same words—madness and wild experiment—to characterize their passion for plants in springtime. Of course, there are also some significant differences. Darwin's hasty postscript is about experimentation, but it is not—in itself—experimental writing. Dickinson's worksheets and multiple alternatives and variants show that she was much more focused on word choices than Darwin. In her case, the wild experiment was primarily literary.

Perhaps the most significant connection between Darwin's and Dickinson's versions of experimentation is that they approached the natural world with a startling combination of ecstasy and humility. The growth of plants in springtime was thrilling to both. They were possessed by the urge to experiment. Darwin did not propose or assume that his experiments with orchids and insectivorous plants would result in sober rational mastery. He linked the experiments themselves to wildness and madness. Dickinson was even more explicit about avoiding mastery. "God be with the Clown –," she wrote,

> Who ponders this Tremendous scene
> This wild Experiment of Green
> As if it were his own –[20]

For Dickinson, as for Darwin, the experimental approach to nature and the natural world was as distant from the rational mastery that Max Weber would ascribe to disenchantment as it was from the divinely ordained "dominion over the fish of the sea, and over the fowl of the air, and over the cattle, and over all the earth, and over every creeping thing that creepeth upon the earth" in Genesis.[21] As an alternative to both the disenchanted and the religious approach to the natural world, Darwin explained his own attitude: "The most humble organism is something much higher than the inorganic dust under our feet; and no one with an unbiassed mind can study any living creature, however humble, without being struck with enthusiasm at its marvellous structure and properties."[22] Like Dickinson, Darwin approached nature with enthusiasm and a sense of wonder, without a sense of superiority. Instead of helping them to master the green world, Darwin and Dickinson's experiments tended to equate mastery with madness.

Like Darwin, Dickinson was intrigued by questions about plant consciousness. Many of her poems and other writings become easier to understand if we read them as part of a conversation with Darwin and his interlocutors. Dickinson's works are experimental, scientifically precise, and intently focused on the natural world. They invoke wonder and insist on ecological humility. They tend to attribute sentience to nonhuman beings.

Darwin would publish five monographs that patiently documented plant movement and consciousness. He built slowly toward the idea that his "wicked dear little *Drosera*" was "an animal."[23] It took twenty years. Darwin published *The Fertilisation of Orchids* (1862), *On the Movements and Habits of Climbing Plants* (1865), *Insectivorous Plants* (1875), *The Different Forms of Flowers on Plants of the Same Species* (1877), and *The Power of Movement in Plants* (1880). In the last of these books, Darwin would conclude that plants were capable of thought: "It is hardly an exaggeration to say that" the tip of a plant's roots "acts like the brain of one of the lower animals; the brain being seated within the anterior end of the body, receiving impressions from the sense-organs, and directing the several movements."[24] Darwin's decades of carefully documented experiments convinced him that plants and animals were kin to each other and that consciousness had evolved from the very roots of the tree of life.

Dickinson's experimentation was different. In some respects, describing Dickinson's writing projects as *experiments* harks back to the early Romantic poets. In the 1802 preface to *Lyrical Ballads* (the collection of poems by Wordsworth and Coleridge that had been supported by the Wedgwood's patronage), William Wordsworth had described the book as an "experiment" and predicted that future poets would help to transfigure "what is now called science," into "a dear and genuine inmate of the household of man."[25] Wordsworth's prediction did not come true.

By the 1860s, when Dickinson and Darwin were occupied with their experiments, science was shifting away from private households and into professional laboratories. Darwin's and Dickinson's private at-home experimentation was much more unusual in the 1860s than it would have been half a century earlier. Further, although Dickinson's poetic experimentation was shaped by Darwin, she had no thoughts of making Darwinian science more accessible to the "household of man." Like Darwin, Dickinson kept her ideas private, sharing them only with a small circle of friends and allies. Her poetry did not propose to endear Darwin's ideas to the public.

Another distinction between Dickinson's experiments and those of the Romantics is their differing attitudes toward the relationship between human and nonhuman. Like Darwin, Dickinson often anthropomorphized, attributing personhood to nonhuman beings. The critics George Levine and Gillian Beer contrast Darwin's tendency to anthropomorphize with the tendency of thinkers like William Wordsworth and Ralph Waldo Emerson to approach the natural world anthropocentrically, as if nature were a reflection or projection of the human spirit. As Levine puts it, Darwin "recognized not spiritual but literal kinship" between humans and nonhumans.[26] Beer explains that although Wordsworth was "an imaginative inspiration to Darwin," Wordsworth's anthropocentrism made "problems for Darwin."[27] Unlike Wordsworth and Emerson, Darwin did not believe that human thoughts and emotions were central to the natural world. Although he decentered the human, he did not dismiss human feelings or ideas. It was just that he tended to approach nonhumans with humility. He avoided imposing human values or desires on nature. Instead, Darwin used his imagination and his emotions—his values and desires—to find out as much as possible "about how we are all connected."[28]

Dickinson had the same impulse. She often personified plants, sometimes identifying with them (when she imagined being a grass plant or a daisy), at other times attributing different qualities to them. In her botanical riddles, common plants like the mayflower (*Epigaea repens*, trailing arbutus) or the purple clover (*Trifolium pratense*) expressed their personalities.[29] Twentieth-century critics sometimes dismissed

Dickinson's flower poems as anthropomorphizing and sentimental. More recently, the ecocritic Christine Gerhardt has argued that Dickinson's "emotionally charged attention" to particular plants is laudable because it allows her to frame "human-nature relationships" with a nonanthropocentric ethical stance.[30] For Dickinson, as for Darwin, anthropomorphism opened up for humble affection between interconnected species.

Dickinson's portrait of the purple clover, "There is a flower that Bees prefer –," describes the plant as a happy warrior, bravely blossoming in a Darwinian world:

> She doth not wait for June –
> Before the World be Green –
> Her sturdy little Countenance
> Against the Wind – be seen –
>
> Contending with the Grass –
> Near Kinsman to Herself –
> For privilege of Sod and Sun –
> Sweet Litigants for Life –[31]

Dickinson often identified herself with the grass. When she depicted clover and grass as "Near Kinsman" contending for survival, the broader implications were clear. Dickinson believed that plants were related to each other, that humans and plants were related, and that many species struggled to survive. Personifying the brave clover was a form of botanical or philosophical thought experiment. If the clover was a form of person, perhaps a human might take courage from understanding that human beings were a form of clover. For Dickinson, the leap from plants to animals was irresistible. She identified with both.

Dickinson felt strong kinship with the hills, trees, and animals of Amherst, and her poems were full of plants and animals. In "Of Bronze – and Blaze –," a poem about the aurora borealis, Dickinson declared, "My Splendors, are Menagerie –."[32] In the next line, she went on to express humility, admitting that her poetry could not quite compete with the aurora. The wonders of the natural world "entertain the Centuries" long after the poet's body had returned to grass.

> My Splendors, are Menagerie –
> But their Competeless Show
> Will entertain the Centuries
> When I, am long ago,
> An Island in dishonored Grass –   some – [Island]
> Whom none but Daisies, know –   [but] Beetles –[33]

In the last line of this poem, Dickinson provided an alternative word, substituting "Beetles" – for "Daisies." Linking flowers to beetles points toward natural history in general and gestures in the specific direction of Darwinian science. In childhood, Charles Darwin's boyhood passion for beetles had marked him as a naturalist. His beetlemania persisted throughout his life. In 1859 the *Entomologist's Intelligencer* featured a brief entry about rare beetles attributed to "Darwin, Darwin, and Darwin"— Franky (10), Lenny (8), and Horace (7). Their father surely had a hand in his children's entomological exploits.[34]

Alternating between beetles and butterflies, flowers and grasses, Dickinson held on to a greener, more Darwin-like vision of naturalism, resisting the bleak assumptions that were increasingly seen as scientific. Most of her peers were adopting Tennyson's "red in tooth and claw" understanding of nature as bloody, violent, and probably meaningless. In contrast to such conflictual understandings of the natural world, Dickinson's choice to end "Of Bronze – and Blaze –" with all-knowing grass, daisies, and beetles was profoundly affiliative. Dickinson was developing a scientifically informed vision of nature that was not "red in tooth and claw." Her world, like Darwin's, blazed with green. Her poetic experiments were glorious—often intoxicating—because they contained a diverse and vibrant "Menagerie."

"Of Bronze – and Blaze –" is just one example of hundreds of Dickinson poems that express kinship with nonhuman beings. Dickinson frequently mentioned beetles in poems and letters. She used "Daisy" even more often. The poet identified with daisies not only because she loved flowers and botany but also because "Daisy" was a common nickname for someone small and innocent, just as *Carlo* was a common moniker for a big dog. (In Dickinson's reading, "Daisy" was used in both

David Copperfield and *Little Women* as a nickname for young, innocent children; *Reveries of a Bachelor* and *Jane Eyre* both featured dogs named "Carlo.") Dickinson often called herself Daisy. Her beloved Newfoundland dog was named Carlo.

Dickinson's world was shaped by her reading—about science, art, and politics, as well as literature. She understood Darwin's importance, but she would never call herself an axiomatic "Darwinist." Instead, she imagined herself as an experimenter. She wrote:

> Experiment escorts us last –
> His pungent company
> Will not allow an Axiom
> An Opportunity –[35]

Dickinson avoided axioms as assiduously as she avoided dogmatisms. She was not a joiner. Instead she was an original thinker, a prolific and profound creator whose work encompassed Darwin's views of the natural world. Darwin's thought bolstered her sense of kinship with nonhuman beings. He helped her to envision death as a transformative and ultimately creative force. He helped her to understand the deep interconnections between living and nonliving beings. Her sense of animated materialism was a view of life that had evolved within a Darwinian world.[36] Without Darwin, Dickinson would probably have been limited to the natural theology espoused by her teachers, who believed that the natural world revealed supernatural truth. But for Dickinson, Darwin reversed the terms. Magic and wonder emerged from the material world and revealed truths about nature.

Throughout their lives, Dickinson and Darwin were fascinated by botany, and they approached their gardens and greenhouses with passionate—and often inventive—scientific curiosity. Beyond their shared loved of botanical experiment, their projects sometimes diverged, though the underlying questions remained very similar. Darwin did all kinds of natural history experiments. He built elaborate devices to determine the aesthetic preferences of earthworms and the reproductive capacities of frogs. He carefully tested the relationships between pollinating insects and pollen-rich plants. At times he focused his experimental attention on human emotions.

Outside the garden and greenhouse, Dickinson's experiments were primarily literary. She often stretched the conventions of prose and poetry to make room for her questions, observations, and hypotheses about the same wild tangles of animals, plants, and emotions that fascinated Darwin.

One of the most confusing aspects of the mid-nineteenth century debates over evolution is that Darwinian thought was often described as materialist. In some ways the label made sense. One of the most decisive moments in the history of modern science was Coleridge's objection to materialist natural philosophy at the BAAS meeting in 1833 and Whewell's responding invention of the term "scientist." After that, scientists and artists were imagined in opposition to each other. In the years that followed, the arts were more and more identified with idealism, while the sciences were frequently framed as materialist. Although the dichotomy is oversimplified, it is not completely wrong. Compared to Coleridge, Darwin was certainly a materialist. However, his version of materialism was surprisingly focused on nonmaterial aspects of life—the interactions and relationships that animated the biological world. Although he is primarily remembered as a biologist, Darwin's work also helped to launch the fields of ecology and psychology.

In the mid-nineteenth century, there was much debate over the question of whether personality—soul—was rooted in material physiology or whether it was wholly nonmaterial. Some scientists condemned materialism altogether. For example, in *The Religion of Geology* (1851), Edward Hitchcock rebuked materialists, whom he defined as those who "deny the existence of the soul, and regard it as a function of the brain."[37] Others tried to find material evidence for their conception of the soul. Darwin's critic Richard Owen sought anatomical proof that Darwin was wrong to think that humans had evolved from other animals. He claimed that the human brain was the only brain that possessed a hippocampus and that the existence of this exclusively human organ proved that humans alone were endowed with souls and separate from animals.

Once again the debate reached a crisis at a meeting of the BAAS. In a dramatic confrontation at the 1862 meeting, Thomas Huxley dissected an ape's brain in front of a spellbound audience. When Huxley brandished the ape's hippocampus in triumph, a young poet in the audience, George Meredith, concluded that "all thinking men" had to be Darwinists.[38] If ape brains were no different from human brains, there was no material basis for thinking that human beings were intrinsically separate from other animals.

It should come as no surprise that George Meredith witnessed Huxley's dramatic dissection at the BAAS. Though poetry and science were increasingly imagined as opposing endeavors, many mid-nineteenth-century poets grappled with evolutionary thought in their works. As discussed in chapter 8, Tennyson's interest in the topic was evident long before *On the Origin of Species* was published. Indeed, Tennyson's grasp of scientific thought was widely celebrated. In 1865 he was made a fellow of the Royal Society—a rare scientific honor for a poet.[39] But Tennyson, like the other members of his generation (including Elizabeth Barrett Browning and Robert Browning) was too far along in his career to structure his poetry around Darwin's ideas.

In *Darwin's Bards*, John Holmes explores poets' responses to Darwin. He makes a distinction between poets who were born at about the same time as Darwin, who first absorbed evolutionary theory from Robert Chambers's *Vestiges of the Natural History of Creation*, and younger poets, whose understanding of evolution was actually shaped by Darwin's theory. According to Holmes, members of Tennyson's generation shared a pre-Darwinian conception of evolution. Even when they referred directly to Darwin, they rarely seem to have been particularly interested in natural selection. For them, the name "Darwin" was just another term for evolution, which they understood in a Larmarckian way that was closer to the views of Erasmus Darwin than those of his grandson Charles. It would fall to the next generation—the poets born from the late 1820s to the 1840s—to try to imagine Darwinist poetry. As Holmes explains it, "Of the poets of this generation, George Meredith and Thomas Hardy engaged most profoundly and persistently with Darwinism."[40]

In 1862, the same year that he watched Huxley dissect the ape's brain, Meredith published a volume of poems that included a series of sonnets on *Modern Love*. In "Modern Love XXX," Meredith danced between materialism and idealism, describing human beings as "First, animals; and next / Intelligences at a leap." Meredith's sonnet begins:

> What are we first? First, animals; and next
> Intelligences at a leap; on whom
> Pale lies the distant shadow of the tomb,
> And all that draweth on the tomb for text.
> Into which state comes Love, the crowning sun:
> Beneath whose light the shadow loses form.
> We are the lords of life, and life is warm.
> Intelligence and instinct now are one.[41]

Meredith followed his observation that "life is warm" with the joyous assertion that "Intelligence and instinct now are one." This bright version of the Darwinist view of human beings was surprisingly rare among Victorian poets. Meredith, who was an optimist by temperament, hoped that modern Darwinist poetry would give voice to what he described as the new "disenchanted harmony."[42]

In contrast, Thomas Hardy was drawn to the bleak aspects of Darwinism. He described himself in his autobiography as "among the earliest acclaimers of *The Origin of Species*."[43] From the start Hardy reveled in the gloomy implications of Darwinism. In 1866 he wrote "Hap," a sonnet about his loss of faith in God and his profound conviction that life was governed solely by chance:

> If but some vengeful god would call to me
> From up the sky, and laugh: "Thou suffering thing,
> Know that thy sorrow is my ecstasy,
> That thy love's loss is my hate's profiting!"

Then would I bear it, clench myself, and die,
Steeled by the sense of ire unmerited;
Half-eased in that a Powerfuller than I
Had willed and meted me the tears I shed.

But not so. How arrives it joy lies slain,
And why unblooms the best hope ever sown?
– Crass Casualty obstructs the sun and rain,
And dicing Time for gladness casts a moan. . . .
These purblind Doomsters had as readily strown
Blisses about my pilgrimage as pain.[44]

In this poem Hardy wallowed in Darwinist despair to the point of making it perversely pleasurable. He declared with emphatic clarity that "joy lies slain." He replaced the vengeful god of Moses with "Crass Casualty" and "dicing Time," freakish figures of chance whom he described as "purblind Doomsters." Hardy's extravagant, almost gleeful language was very different from that of his friend and fellow Darwinist George Meredith.

Always a pessimist, Hardy embraced a dark version of Darwinism based on the mid-nineteenth-century assumption that Darwinism signified a great spiritual and emotional loss. However, according to George Levine, there is an astonishing "love of life manifest in [Hardy's] meticulous descriptions of the nonhuman world." Levine argues that Hardy "was more in love with life, and in Darwinian ways, than most typical Darwinian readings suggest." For Levine, Hardy's view of the world is suffused with Darwinian enchantment.[45]

Although Meredith's and Hardy's responses to Darwin offer useful contrasts to Dickinson, it is not likely that Dickinson read their poetry of the 1860s. *Modern Love* was published in 1862, but it is hard to imagine a member of the Dickinson household purchasing the volume. Reviewers had labeled his book as immoral because it described the dissolution of a marriage. Later the Dickinson library would include at least one of Meredith's novels. There is no evidence that Dickinson read Hardy's novels or his poetry. As for Hardy's "Hap," it was not published until 1898, so she never had the chance to read it.

Dickinson had other opportunities to read about the hippocampus debates and the larger question of the relationships between the human brain and the soul. The most popular account of the debate was published in *The Water-Babies*, a children's book by an Anglican clergyman. The author, Charles Kingsley, was an old-school parson-naturalist who supported Darwin and became a personal friend of Darwin's. His books were favorites in the Dickinson household too. The Dickinson family copy of *The Water-Babies* is preserved in the Emily Dickinson collection at Harvard University.

Pointing out the nonsensical elements of the anatomical debate, Kingsley altered the term from *hippocampus* to *hippopotamus*. He wrote: "Nothing is to be depended on but the great hippopotamus test. If you have a hippopotamus major in your brain, you are no ape, though you had four hands, no feet, and were more apish than the apes of all aperies. But if a hippopotamus major is ever discovered in one single ape's brain, nothing will save your great-great-great-great-great-great-great-great-great-great-great-greater-greatest-grandmother from having been an ape too."[46] Kingsley's comically absurd "great hippopotamus test" poked fun at debates about Darwinian science that had grown bitterly polarized.

A few months after the 1862 BAAS meeting, Dickinson wrote "The Brain – is wider than the Sky –." Her twelve-line poem addressed the debate about consciousness and materialism that had preoccupied the scientific community for decades.

> The Brain – is wider than the Sky –
> For – put them side by side –
> The one the other will contain
> With ease – and You – beside –
>
> The Brain is deeper than the sea –
> For – hold them – Blue to Blue –
> The one the other will absorb –
> As Sponges – Buckets – do –
>
> The Brain is just the weight of God –
> For – Heft them – Pound for Pound –

And they will differ – if they do –
As Syllable from Sound –[47]

"The Brain – is wider than the Sky –" was a clever response to the debates in Cambridge. Though it celebrated the brain, it did not exactly deny the existence of the human soul or of God. Instead it turned toward the stuff of poetry, concluding that the brain and God "differ – if they do / As Syllable from Sound –." It was no accident that the poem ended with syllables and sounds. For Dickinson, poetic language offered a way to give weight to the biological world without foreclosing the world of ideas and ideals.

Like Thomas Hardy's "Hap," "The Brain – is wider than the Sky –" would remain unpublished until the 1890s. This was bad luck for Charles Darwin. Dickinson and Hardy were among the poets who best understood his work, but he would never have the chance to read their poems. Before he became a publishing scientist, he had loved poetry. Perhaps if he had been able to read "The Brain – is wider than the Sky –" it would have sparked his interest in poetry again. In the 1860s, however, both Dickinson and Hardy were keeping their poetry private. Darwin was doing the same thing, confiding the thoughts that "frightened and astounded" to a small circle of friends. He would not release his next "large book" for more than ten years. He would wait even longer—until after *The Descent of Man* (1871) and *The Expression of the Emotions* (1872)—to release *Insectivorous Plants* (1875).

CHAPTER 13

Melody or Witchcraft?
Amherst, 1862–1866

ON THE last night of 1861, Emily Dickinson wrote a melancholy letter to her cousin Louisa Norcross. It was a lonely New Year's Eve. Dickinson had expected her cousin to visit, but Louisa did not come. The weather was cold and sleety. There was nothing to do but write letters, sew, and worry about the war. Another young Amherst soldier had died. The widowed Mrs. Adams had gotten a telegram about the death of her son, Sylvester, just a few months after her other son died. "Dead! Both her boys!" Dickinson wrote. "Poor little widow's boy, riding to-night in the mad wind, back to the village burying-ground where he never dreamed of sleeping! Ah! the dreamless sleep!" "'Happy new year' step softly over such doors as these!"[1]

The telegram announcing Sylvester Adams's death came from Frazar Stearns, a junior officer in the Amherst infantry regiment, the Massachusetts Twenty-First. Frazar was the 21-year-old son of William Stearns (who had succeeded Edward Hitchcock to become the fourth president of Amherst College). As a student, Frazar had loved the physical sciences, particularly chemistry. He had hoped to complete his undergraduate degree in 1863, but the war interrupted. When his favorite chemistry professor, William S. Clark, enlisted and became the commanding officer of the Amherst regiment,[2] Stearns followed, leaving Amherst to serve as Clark's adjutant.

Dickinson worried about Frazar Stearns more than she worried about other Amherst volunteers. She knew him personally. In the letter to Loo, she wrote, "Frazar Stearns is just leaving Annapolis. His father has gone to see him to-day. I hope that ruddy face won't be brought home frozen."[3] Unfortunately, her grim vision would become reality. Ten weeks later, on March 14, 1862, Frazar Stearns died on the battlefield.

Afterward, in late March, Dickinson wrote to Samuel Bowles: "Austin is chilled—by Frazer's murder—He says—his Brain keeps saying over 'Frazer is killed'—'Frazer is killed,' just as Father told it—to Him. Two or three words of lead—that dropped so deep, they keep weighing."[4] In this letter Dickinson chose to use "killed" rather than the more usual "dead." On top of accentuating the violence, the choice avoided rhyming "dead" with "lead" and emphasized the link between Austin ("chilled") and Frazar ("killed"). Notably, Dickinson located the repeating "words of lead" in her brother's "Brain," tying language (and thought) to material physiology in a particularly gruesome way. Austin did not have a physical bullet in his brain, but Dickinson described him as being repeatedly pierced by "words of lead." Though the poet and her brother were far from the battlefield, Frazar Stearns's death hurt them physically.

Dickinson wrote to her Norcross cousins to tell them about the funeral. On New Year's Eve, she had imagined Sylvester Adams riding the wind back to their village. Now she pictured Frazar Stearns making the same ghostly ride. "Just as he fell, in his soldier's cap, with his sword at his side, Frazer rode through Amherst. Classmates to the right of him, and classmates to the left of him."[5] She was alluding to Tennyson's poem "The Charge of the Light Brigade," which describes soldiers riding to their slaughter with "Cannon to the right of them, / Cannon to the left of them."[6] Since Frazar was a much-loved Amherst student, she pictured him surrounded by classmates instead of cannons. After describing his crowded funeral, she concluded, "Austin is stunned completely. Let us love better, child, it's most that's left to do."[7]

The previous summer, Dickinson had drafted the last of the long "Master" documents that she preserved. In "Master. / If you saw a bullet," violence is eroticized, while mastery is a source of fascination. Now, less than a year later, she was thinking differently about bullets. In the spring of 1862 Dickinson (like her brother) was chilled and stunned. Resolved to "love better," the poet turned her attention to caring for the people, animals, and plants closest to her. She also turned to poetry. She worked intently, writing new poems, copying older poems, revising. She was done with the "Master" experiments, but she would keep working on the fascicles for years. Around this time she began to wonder if she should publish. Was it possible that her poetry could bring solace to strangers?

In the April 1862 issue of the *Atlantic Monthly*, Thomas Wentworth Higginson published "Letter to a Young Contributor," an essay that encouraged young women to become professional writers.[8] Higginson was best known for being an abolitionist activist. He was also an advocate for women. In February 1859 he had published a robust defense of women's education in the *Atlantic Monthly*, ironically titled "Ought Women to Learn the Alphabet?"[9] In addition, Higginson was an early champion of Darwin whom he mentioned in many of his *Atlantic Monthly* essays about the outdoors.

Many of Higginson's essays struck a chord with Dickinson. Some of those that cite Darwin, like "The Life of a Bird," go unmentioned in her correspondence. Others, like "The Procession of the Flowers," are explicitly discussed in her letters. Of all Higginson's writings, it was "Letter to a Young Contributor" that changed Dickinson's life most significantly. Dickinson seems to have felt that Higginson was writing directly to her. She had never met him, but nonetheless she felt compelled to respond.

On April 15, 1862, Emily Dickinson wrote a short, brilliant, and peculiar letter to Thomas Wentworth Higginson. Nearly three decades later, Higginson would describe his correspondence with the poet for the *Atlantic Monthly*. He recalled, "On April 16, 1862, I took from the post office in Worcester, Mass., where I was then living, the following letter: . . . The letter was postmarked 'Amherst,' and it was in a handwriting so peculiar that it seemed as if the writer might have taken her first lessons by studying the famous fossil bird-tracks in the museum of that college town."[10] Something about that envelope reminded Higginson of geology. Before he knew her, before he had read any of her poems or become her friend and literary mentor, her handwriting reminded him of "fossil bird-tracks."

> Mr Higginson,
>
> Are you too deeply occupied to say if my Verse is alive?
> The Mind is so near itself – it cannot see, distinctly – and I have none to ask –
> Should you think it breathed – and had you the leisure to tell me, I should feel quick gratitude –
> If I make the mistake – that you dared to tell me – would give me sincerer honor – toward you –
> I enclose my name – asking you, if you please – Sir – to tell me what is true?
> That you will not betray me – it is needless to ask – since Honor is it's own pawn –[11]

Folded inside the brief letter was another smaller envelope containing a card upon which Dickinson had written her name in small, careful script. She also included four poems: "Safe in their Alabaster Chambers," "The nearest Dream recedes unrealized," "We play at Paste," and "I'll tell you how the Sun rose."[12] It was obvious (both from the letter asking if her verse was "alive" and from the poems themselves) that poetry was a matter of life and death to Dickinson. Although the strange poems she sent discussed death, they were among the most vivid works Higginson had ever encountered. He responded encouragingly, and the two struck up a correspondence.

Dickinson's letters to Higginson were peculiar, but they were much more restrained than the "Master" texts had been. When he responded positively to her first letter and asked her about herself, she wrote a second letter that translated some of the wild imagery of "Master. / If you saw a bullet" into more socially acceptable language. She told Higginson that she sang "as the Boy does by the Burying Ground – because I am afraid –" then explained, "For Poets – I have Keats – and Mr and Mrs Browning."[13] Keats was important to her, but the Brownings meant more. Their lives and their work helped her to make sense of herself as a poet.

In the same letter, Dickinson declared that she was not a churchgoer. She explained that she felt a profound kinship with the nonhuman natural world: "You ask of my Companions Hills – Sir – and the Sundown – and a Dog – large as myself, that my father bought me – They are better than Beings –."[14] Dickinson described these nonhuman "Companions" (hills, sundown, a dog) as "better than Beings." She commented that frogs and insects were better musicians than humans ("the noise in the Pool, at Noon – excels my Piano"). She had read Higginson's essays in the *Atlantic Monthly* carefully. She suspected that he would sympathize.

Perhaps most significantly, she asked Higginson if he could teach her to be a poet. She wrote, "I would like to learn – Could you tell me how to grow – or is it unconveyed – like Melody – or Witchcraft?"[15] This strange, musical, magical letter professed humility toward Higginson and toward the nonhuman world, but it also expressed startling self-assurance.

Higginson wrote back. He kept answering. He paid attention. Dickinson had found a friend who would connect her to Darwin and to the literary world. Her experiments with "Witchcraft" would succeed.

When Dickinson chose Thomas Wentworth Higginson to be her literary mentor, she chose a celebrated writer and editor who often wrote about the natural world. Like Ralph Waldo Emerson, he had attended Harvard Divinity School, become a Unitarian, and eventually left the priesthood in order to devote himself to writing. But his writing was different from Emerson's. At times he was more explicit about his radical politics. What made him even more different was that his writing about the natural world was profoundly empiricist, far more scientific than Emerson's would ever be. In the 1830s Emerson had longed to be a naturalist. In the 1860s Higginson actually managed it. He wrote about birds and flowers with scientific precision. In 1872 he would visit England and befriend Darwin. In the early 1860s, when Dickinson reached out to him, he was one of the only people in the United States who was publishing work that combined literary ambition with Darwinian natural science.

Thomas Wentworth Higginson (1862).

If anyone could understand her poetry, Dickinson hoped it might be him. In some ways Higginson lived up to her expectations. He praised her writing. From the very beginning, he was able to read it in the right context: as emerging from the science-saturated environs of Amherst, Massachusetts, where a woman's letters might recall fossil bird tracks. But at first Higginson did not think that Dickinson's poetry was publishable. He thought her work was too unconventional for the *Atlantic Monthly*. He did not appreciate its philosophical and scientific acuity.

According to Higginson, the second group of poems she sent included one of her most celebrated poems, "A Bird, came down the Walk –."[16] Higginson admired the poem, but he never commented on how it might be related to his own essay on "The Life of a Bird" or how it echoed Darwin's descriptions of bird life.

"A Bird, came down the Walk –" is profoundly Darwinian. In the beginning of *On the Origin of Species*, Darwin had written, "We behold the face of nature bright with gladness, we often see superabundance of food; we do not see, or we forget, that the birds which are idly singing round us mostly live on insects or seeds, and are thus constantly destroying life."[17]

The lyrical, ecological, and endlessly generative language of Darwin's book set the stage for Dickinson's work. Indeed, it is not hard to imagine that in the moment before Dickinson's "A Bird, came down the Walk –," it perched in an "entangled bank" very much like the one Darwin described in the last paragraph of *On the Origin of Species*: "It is interesting to contemplate an entangled bank, clothed with many plants of many kinds, with birds singing on the bushes, with various insects flitting about, and with worms crawling through the damp earth, and to reflect that these elaborately constructed forms, so different from each other, and dependent on each other in so complex a manner, have all been produced by laws acting around us."[18] Darwin was probably thinking about a hillside near his house in Kent that is now known as the Orchis Bank. That small slope is still rich with a diverse collection of plants and animals, but it is not unique. There are places like it all over England. Indeed, similar tangles of wildly diverse yet deeply connected plants and animals can be found all over the world.

What was new was the way that Darwin described his "entangled bank." He was not the first to write about nature with specificity; to the contrary, many writers used more specific botanical or zoological details in their descriptions of nature. Shakespeare's description of the "bank where the wild thyme blows, / Where oxlips and the nodding violet grows, / Quite over-canopied with luscious woodbine, / With sweet musk-roses and with eglantine," for example, is far more detailed about the plant life on an imaginary bank than Darwin's description.[19]

Darwin's description was compelling and unique because of the way he imagined the relationships between different forms of life. Everything was interconnected. The birds, bushes, insects, and worms—even the damp earth—were "different from each other, and dependent on each other." In *Darwin's Backyard*, James T. Costa describes the way Darwin imagined the "entangled bank" as a "lucid ecological vision" that "helped give rise to the field of ecology."[20]

Importantly, the relationships that Darwin described were not always peaceful. To the contrary, many of the "laws" that Darwin described were laws of competition, struggle, even violence. "Thus, from the war of nature," Darwin wrote, "from famine and death, the most exalted object which we are capable of conceiving, namely, the production of the higher animals, directly follows."[21] Without exalting the violence, Darwin showed that struggle directly produced complex and variegated forms of life that were entangled with one another.

Darwin's observation—that "we do not see, or we forget" that birds are "constantly destroying life" is the central conceit of Dickinson's poem.

> A Bird, came down the Walk –
> He did not know I saw –
> He bit an Angle Worm in halves
> And ate the fellow, raw,
>
> And then, he drank a Dew
> From a convenient Grass –
> And then hopped sidewise to the Wall
> To let a Beetle pass –
>
> He glanced with rapid eyes,
> That hurried all abroad –
> They looked like frightened Beads, I thought,
> He stirred his Velvet Head.
>
> Like one in danger, Cautious,
> I offered him a Crumb,
> And he unrolled his feathers,
> And rowed him softer Home
>
> Than Oars divide the Ocean,
> Too silver for a seam,
> Or Butterflies, off Banks of Noon,
> Leap, plashless as they swim.[22]

Dickinson's description of the bird echoes Darwin. The critic Kaitlin Mondello explains that "Both Darwin and Dickinson attempt to shift the romantic notion of the songbird away from the 'Sweetness' of nature to acknowledge instead the brutal side of existence."[23] Dickinson's poem starts with the raw struggle for existence—the moment when the bird eats a worm. She sets the bright-eyed bird in a complex natural world, coadapted with dew and grass and beetles. She offers a concrete description of the bird's bright eyes and his awareness. And then, in the last six lines, the poem takes off. The bird becomes a metaphorical boat, rowing home through the sky. Its flight is soft—its wings in the air are softer than oars in the ocean. The bird's flight is like the flight of "Butterflies, off Banks of Noon."

Both birds and butterflies were central to Dickinson's poetic world, just as they were central to Darwin's scientific world. In *Voyage of the Beagle*, Charles Darwin had repeatedly mentioned "large and brilliant butterflies, which lazily fluttered about."[24] He told a story about the ship being overwhelmed by flocks of butterflies: "One evening, when we were about ten miles from the Bay of San Blas, vast numbers of butterflies, in bands or flocks of countless myriads, extended as far as the eye could range. Even by the aid of a glass it was not possible to see a space free from butterflies. The seamen cried out 'it was snowing butterflies,' and such in fact was the appearance."[25] When Dickinson referred to "butterflies, off Banks of Noon," her language evoked the large flocks of tropical butterflies that she had never seen for herself. Travelers' accounts like *The Voyage of the Beagle* allowed her to imagine them.

Perhaps Dickinson hoped that Thomas Wentworth Higginson would notice the elusive references to Darwin in her letters and poems. When Higginson asked her for a photograph in July 1862, Dickinson replied with a description of herself: "I am small, like the wren; and my hair is bold, like the chestnut bur." These similarities were not mere literary devices. Dickinson believed that her likeness to the wren and the chestnut was meaningful. Around that time, observing an American chestnut, Dickinson wrote, "There are two Ripenings –."[26] First, there is the visible ripening: the growth of round, fuzzy green burs on the tree. But then there is "a homelier maturing," a secret "process in the Bur." When she likened her auburn hair to a chestnut bur, she invoked the glossy red-brown of the hidden kernel. By the mid-1860s,

Dickinson felt that she was ripening into a mature poet. Though she probably intended to keep the secret to herself, she could not resist giving Higginson a few hints.

At times Dickinson's communications were cryptic, full of riddles and codes. However, it is not likely that Dickinson thought of her interest in natural science—or in Darwin's ideas—as a dark secret. To the contrary, she compressed her language and layered her allusions to give her words more force. She certainly hoped that the words she chose would resonate for him. She had known that he admired Darwin in April, when she sent "A Bird, came down the Walk –." In July she sent Darwinian images of a wren and a chestnut in lieu of a photograph. In August, knowing that Higginson admired Elizabeth Barrett Browning, she concluded her letter by offering him an image of the woman poet they both liked. "Have you the portrait of Mrs. Browning? Persons sent me three – if you had none, will you have mine?"[27] Subtly, she invited Higginson to visualize her as a Darwinist successor to Barrett Browning.

Dickinson hoped that Higginson would recognize her immense talent. She might also have wanted him to help her publish her poems, but she was reluctant about exposing herself to the public. Since the Higginson correspondence started shortly after Dickinson had completed her "Master" documents, some of the early letters bear traces of Dickinson's longing for a masterful figure to instruct her. But within a few months, the relationship seems to have changed. As they exchanged more letters, the correspondence became more like a conversation between equals.

In the summer of 1862, Higginson enlisted. He joined the Massachusetts Fifty-First as captain. Through the fall he spent long weeks at Camp John E. Wool in Worcester, learning military skills. Later he would recall his early days in the army as blissful. It was as if "one had learned to swim in air, and were striking out for some new planet."[28] Finally, in November 1862, Higginson was invited to lead the first regiment of Black soldiers in the U.S. Army, the First South Carolina Volunteers.[29] Dickinson learned that he had gone to war shortly after she read his essay about natural selection in Massachusetts, "The Procession of the Flowers." She wrote to him again, addressing her letter "Dear Friend."[30]

Dickinson was very worried about Higginson's exposure to danger. At the same time, she did not want to burden him with her terror. She tried to keep her tone light. "I should have liked to see you before you became improbable," she wrote. She feared that his survival was improbable—even if the letter reached his regiment, it was very possible that Higginson would be dead by then. She desperately hoped he would survive. Her thoughts tangled together. At times her worry felt like prayer: "I trust you may pass the limit of war; and though not reared to prayer—when service is had in Church for Our Arms, I include yourself – I, too have an Island, whose 'Rose and Magnolia' are in the Egg, and it's 'Black Berry' but a spicy prospective, yet as you say, 'fascination' is the absolute of clime."

In this sentence, Dickinson leapt from church to an enchanted imaginary island where rose, magnolia, and blackberries flourished. The literary historian John Shoptaw has suggested that Dickinson might have been quoting phrases from a lost Higginson letter describing his experience in South Carolina. We cannot know exactly what such a letter from Higginson might have said, though Shoptaw points out that Higginson's memoir, *Army Life in a Black Regiment,* published in 1869, described his Sea Island camp with similar botanical references: "This picket station was regarded as a sort of military picnic by the regiments stationed at Beaufort, South Carolina; it meant blackberries and oysters, wild roses and magnolias."[31] Later in the memoir, Higginson observes that "to those doing outpost duty on an island, however large, the mainland has all the fascination of forbidden fruit, and on a scale bounded only by the horizon."[32] The "Island" in Dickinson's letter is similar.

It is also possible that Dickinson was responding to his recent *Atlantic Monthly* essay, "The Procession of the Flowers," which she mentioned in the letter. The essay described the sequence of blossoming throughout the calendar year in Massachusetts. It opened by remarking "To a watcher from the sky, the march of the flowers of any zone across the year" is a beautiful spectacle.[33] As the focus narrowed in on Massachusetts, Higginson asked, "Is it by some Darwinian law of selection that the white Hepatica has utterly overpowered the blue?" Discussing environmental changes around Boston, the essay took an elegiac tone, mourning the flowers that had disappeared from Eastern Massachusetts. Higginson remarked that "nothing in Nature has for me a more fascinating interest than these secret movements of vegetation, – the sweet blind instinct with which flowers cling to old domains until absolutely compelled to forsake them."[34] He referred to geological discussions of "successive epochs of heat which led the wandering flowers along the Arctic lands" and remarked, "These humble movements of our local plants may be laying up results as important, and may hereafter supply evidence of earth's changes upon some smaller scale."[35] Throughout, Higginson framed his specific local botanical observations geologically, in terms of epochal climate change.

Dickinson's puzzling remark "'fascination' is the absolute of Clime" was at least partly a response to Higginson's discussion of species loss and climate change. Higginson had claimed that there was nothing more fascinating in nature than the way that flowering plants were "absolutely compelled" to forsake their habitat. Dickinson turned his statement back on him, her improbable correspondent who had forsaken New England for somewhere in the South, within or beyond "the limit of War." In a strained attempt at consolation, she implied that his disappearance from Massachusetts—perhaps his possible extinction—was fascinating because he existed for her in the clime of the absolute: he was one of the greats.

Throughout, the letter changed frames of reference with startling agility, moving from Dickinson's Newfoundland dog Carlo (whom she called her "Shaggy Ally"), to death, friendship, flowers, eggs, brambles, biology, geology, extinction, absolutism, and climate. It continued,

> I was thinking to-day – as I noticed, that the "Supernatural"
> was only the Natural disclosed –
>   Not "Revelation" – 'tis – that waits,
>   But our unfurnished eyes –[36]

Where Christian scripture offers Revelation as a vision of apocalypse, an end of the world, Dickinson's letter offered "Not 'Revelation.'" For Dickinson, "Revelation" held a double meaning: a moment of insight or divine disclosure (when God reveals the truth) and apocalypse (the end of the world as depicted in the Book of Revelation). When Dickinson prefaced revelation with an emphatic "Not," she implied that she and Higginson shared a world without a comprehensible, divinely explained purpose that was also, somehow, a world without end—without a final, demarcated death knell. With these lines, Dickinson offered a small but possibly profound comfort to her Darwinian mentor.

A short time after Higginson received her letter in South Carolina, James Fields published a collection of Higginson's *Atlantic Monthly* essays on natural history. Higginson was thrilled when a copy of his first book, *Out-door Papers*, arrived in camp. Dickinson was almost as excited as the author. Years later, she recalled, "It is still as distinct as Paradise – the opening your first Book – It was Mansions – Nations – Kinsmen – too – to me. . . . I had long heard of an Orchis before I found one, when a child, but the first clutch of the stem is as vivid now, as the Bog that bore it."[37] In Higginson's pages, she found a kinsman whose descriptions of the natural world were shaped by his reading of Darwin. Higginson loved wild orchids and boggy wetlands. He knew that this passion connected him to Darwin. Before she wrote to him, he may not have been aware that it also connected him to Dickinson. In *White Heat*, biographer Brenda Wineapple notes that Dickinson, "like Higginson, discovered self-renewal in natural recurrence, which was not revelation, resurrection, or religious cant."[38] Even so, it was "distinct as Paradise."

Higginson's essays in *Out-door Papers* quoted Darwin enough to make it clear that Higginson considered himself a committed Darwinist. However, despite her interest, Dickinson was not committed in the same way. She approached natural science in general (and Darwinism in particular) with engaged appreciation rather than ideological inflexibility. The debate between the intellectual factions of the mid-nineteenth century—the Darwinists (who did not believe that the natural world was a revelation of the divine) and the anti-Darwinists (who insisted that nature *was* a revelation of the divine)—interested her. However, Dickinson was reluctant to join either team. At times her poetry echoed Darwin's language and thought. Darwin gave Dickinson a new perspective, which she appreciated as "Another way – to see –."[39] Yet as much as she valued his ideas, she did not subscribe to Darwinism any more than any other creed. In this respect Dickinson was more Darwin-like than the Darwinists. Like Darwin's, her judgments fluctuated. She refused to pin down her beliefs. As she had told Abiah, she loved "to buffet the sea."[40]

In many cases, Emily Dickinson's insistent open-mindedness can seem heroic. For example, her attitude toward Christianity was remarkably brave. She refused to bend to the pressures of her community and join the church, but she also continued to take their ideas seriously and remained profoundly interested in religious questions. Dickinson's commitment to exploring scientific questions from all sides was equally admirable. Her reluctance to choose—between different words or between different ideological stances—is a central feature of her poetry.

In other cases, Dickinson's refusal to decide can be frustrating. Her attitude toward the Civil War—and particularly toward the abolition of slavery—is hard to pin down. In a letter to Higginson, she wrote, "War feels to me an oblique place."[41] This is the clearest statement of her stance on the war. Her attitudes toward race and racism were equally oblique. Vivian Pollak's "Dickinson and the Poetics of Whiteness" argues that although Dickinson's work is "not wholly successful in freeing itself of racism," her poems push against racial hierarchy insofar as they render race "semantically unstable."[42] Dickinson's somewhat equivocal attitudes toward racial justice, the abolition of slavery, and the Civil War are particularly unsatisfying when considered in contrast to her friend Higginson, who was unequivocally committed to fighting racism and slavery.

Because she was a woman, Emily Dickinson did not have to decide whether to fight in the Civil War. She did not have the opportunity—or the responsibility—of voting for Lincoln or any members of Lincoln's Republican Party. There is no way to know how she would have voted or whether she would have served as a soldier. When her father was invited to run for lieutenant governor as a Republican in 1861, he refused.[43] When her brother was drafted into the army, he hired a substitute to go in his place.[44] Their reluctance to participate in the war is a matter of public record. Dickinson's feelings about their behavior are not.

On the other hand, Dickinson supported Higginson unequivocally. When she learned that he had been wounded in battle, she wrote immediately, saying that she cared more about his health than her own. Higginson was first injured in the summer of 1863. He recovered enough from his injuries to continue for almost a year, but he was disillusioned with how the army treated Black soldiers. He thought the decision to send Shaw and the Black soldiers of the Massachusetts Fifty-Fourth to their deaths was not only senseless but also very likely rooted in racism. Shaw died on the battlefield. Shaw's adjutant, Wilkie James, the 17-year-old younger brother of William and Henry James, was severely injured. James was transported back to Massachusetts on a stretcher, delirious with pain and fever. His recovery would take a year and a half. In the meantime, descriptions of Wilkie James's suffering circulated around Massachusetts.

According to Wineapple, by the spring of 1864 Higginson's "faith in the military as a great equalizer" had waned.[45] He was as idealistic as he had ever been, but he was more and more ambivalent about the U.S. Army. He resigned his commission and left the Union Army on May 14, 1864. Higginson's decision to resign sheds light on Austin Dickinson's decision to hire a military substitute. Around the same time, on May 23, Austin signed a contract with a person named Frank Paine, paying him $500 to take his place in the army. In *Patriotism by Proxy*, literary historian Colleen Glenney Boggs explains that Paine "was most likely a formerly enslaved black man."[46] This was a terribly complicated transaction. Boggs describes draft substitutes as "the literal and metaphorical battleground on which collective belief structures and belief structures of the collective play themselves out against wartime formulations of nationalism and patriotism."[47] On one hand, antiracist white people in Massachusetts and across the Union believed that paying Black men—particularly formerly enslaved Black men—to serve as battlefield soldiers was an important step toward equality. On the other side, when wealthy white men like Austin Dickinson paid substitutes to become cannon fodder in their stead, the equation's racist implications were unavoidable. Although the transaction was legal and relatively frequent, Austin had made an unholy bargain.

By 1864, the third year of the war, it was hard to hold fast to any clear-cut political principles. Like the English Civil War before it—indeed like every conflict described with that oxymoronic construction—the American Civil War was a morass of contradictions. As the war dragged on, Dickinson wrote to her Norcross cousins, "Sorrow seems more general than it did, and not the estate of a few persons, since the war began; and if the anguish of others helped one with one's own, now would be many medicines."[48] Everyone was grieving because of the war, but the general anguish was small comfort.

Dickinson may have been one of the only people in 1864 whose thoughts turned from the shared grief of the American Civil War to the singular grief of Elizabeth Barrett Browning's death three years before. Perhaps she was reminded of Barrett Browning when, in 1864, Robert Browning published *Dramatis Personae*. At any rate, Dickinson's letter on grief leapt directly from the war to the Brownings. The next sentence of her letter to her cousins remarked, "I noticed that Robert Browning had made another poem, and was astonished – till I remembered that I, myself, in my smaller way, sang off charnel steps."[49]

A charnel house is a place to store exhumed bones. Dickinson's declaration that she "sang off charnel steps" is usually understood either as a general reference to the fact that death inspired her or as a reference to the fact that her most prolific years of writing were the years of the Civil War. Few scholars have commented on the strange juxtaposition of Dickinson's "charnel steps" with Robert Browning's latest publication.

The first few years after Elizabeth Barrett Browning died were Dickinson's most prolific. If she felt surprised or even vaguely ashamed by her longing to take Barrett Browning's place in the literary pantheon, the fact that Barrett Browning's widowed spouse was also writing prolifically would have been a comfort. *Dramatis Personae* was formally innovative and intellectually ambitious. It engaged more directly with Darwin than any of Elizabeth Barrett Browning's works. It included Robert Browning's great anti-evolutionary monologue, "Caliban upon

Setebos," in which the character of Caliban gives full-throated expression to post-Darwinian despair.

Dickinson's response to the "charnel steps" was similar to Robert Browning's in that proximity to death inspired them to write. But Dickinson seems to have felt more excitement than depression. She did not share Robert Browning's hopelessness. To the contrary, she remarked, "Every day life feels mightier, and what we have the power to be, more stupendous."[50]

With that mighty power coursing through her, Dickinson put aside her fears about publishing. According to the literary scholar Karen Dandurand, 1864 was "exceptional in the history of Dickinson's publications. It was the only year in which more than one poem appeared."[51] Dickinson published five poems in the spring of 1864. (Because two of them were republished in numerous newspapers, the five poems account for ten separate instances of publication.) Scholars estimate that at this point in her life, Dickinson had written about five hundred poems. The five she decided to release were representative of her interests: observing the sky ("Blazing in Gold – and"), botany ("Flowers – Well – if Anybody"), bird-watching ("These are the days when the Birds come back –"), natural theology ("Some – keep the Sabbath – going to Church –"), and death ("Success is counted sweetest").

Three of these poems were published anonymously in the Brooklyn *Drum Beat*. The archival scholar Mike Kelly explains, "*The Drum Beat* was edited by R. S. Storrs, Jr. and published just thirteen issues during the duration of the Long Island Fair for the Benefit of the U.S. Sanitary Commission, 22 February to 5 March, 1864."[52] Though it was short-lived, the *Drum Beat* recruited an impressive roster of writers, including Oliver Wendell Holmes, William Cullen Bryant, and Louisa May Alcott. Dandurand argues that Dickinson's *Drum Beat* poems "must be seen as her contribution to the Union cause."[53] It seems significant, however, that the purpose of the fair was not to raise funds for the army but instead to raise money for the Sanitary Commission—to provide medical supplies for wounded soldiers. If Dickinson could not decide how she felt about sending soldiers to war, she knew for certain that she supported caring for the wounded in the aftermath.

At the start of this book, I discussed the opening lines of "Flowers – Well – if anybody." The poem started by explaining the paradoxes and contradictory implications that made botany so fascinating to nineteenth-century thinkers.

> Flowers – Well – if anybody
> Can the extasy define –
> Half a transport – half a trouble –
> With which flowers humble men:
> Anybody find the fountain
> From which floods so contra flow –
> I will give him all the Daisies
> Which upon the hillside blow.⁵⁴

After describing how the "extasy" of botany could "humble men," Dickinson compared her own appreciation of flowers to the way that butterflies know them. The poem concluded:

> Butterflies from St Domingo
> Cruising round the purple line –
> Have a system of aesthetics –
> Far superior to mine.⁵⁵

Like the "Butterflies off Banks of Noon" in "A Bird Came down the Walk," these "Butterflies from St. Domingo" recall the flocks of tropical butterflies that Darwin had described in *The Voyage of the Beagle*. In these final lines, she suggested that the way insects understand blossoms is superior to familiar human aesthetic understanding. Dickinson's biological humility was strikingly similar to Darwin's. She rejected the idea of a hierarchy in which a human poet is superior to a butterfly. To the contrary, she presented the complex beauty of a hillside blooming with daisies or a cloud of butterflies as superior to human systems of thought. Darwin's naturalist observations led him to a similar conclusion.

Perhaps Dickinson chose the poem for *Drum Beat* because of its antislavery implications. Santo Domingo was the site of the Haitian Revolution against enslavement and colonization. By locating itself there, the poem implied that there was a political, abolitionist dimension to studying butterflies in the tropics. Darwin had traveled to the equator as a naturalist, and his travels had convinced him that slavery was deeply wrong.

The logic of "Flowers – Well – if anybody" becomes clear when we read its discussion of botany and entomology, abolitionism, and aesthetics in conversation with Darwin. The same can be said for Dickinson's other 1864 publications. As Dickinson and her community struggled to make sense of the horrific violence that had engulfed them, Dickinson's poems offered readers a vision of an interconnected natural world where death was as closely related to new life as beauty to truth. As she explained to her cousins, she sang "off charnel steps."

After that exciting spring when Dickinson published her poems, Higginson returned alive from the war, and Austin made definitive arrangements to avoid the battlefield, Dickinson's health took a turn for the worse. She developed an eye problem that required her to spend months in a darkened room at her cousins' home in Boston. Her doctor told her she was not allowed to read. She was permitted to write a little bit, but only with a pencil. She hated not being able to go for long walks outside. She was also upset that her dog Carlo stayed in Amherst. She missed him terribly during her medical treatment in Boston.

Of all her companions, human and nonhuman, Dickinson's Newfoundland dog, Carlo, was her closest companion during these years. In an 1863 letter to Thomas Wentworth Higginson, she had described him as her "Shaggy Ally."[56] Later she described him as her "mute confederate."[57] When she was in Amherst, Carlo accompanied her on long walks through the meadows and woods around her house. He was better company than any of her family.

Now, in Boston in the autumn of 1864, Dickinson was more alone than she had ever been.

Charles Darwin loved dogs, and he was as interested in them as Dickinson. In his book *The Expression of the Emotions*, his detailed discussion of the ways dogs expressed themselves preceded his discussions of other animals (including humans). The number of illustrations of dog emotions was second only to illustrations of human emotions. He was so taken with the contrast between the expressive posture of a "hostile" dog and "the same dog in a humble and affectionate frame of mind" that

Illustrations of "A dog approaching another dog with hostile intentions" (*top*) and "The same dog in a humble and affectionate frame of mind" (*bottom*). From Charles Darwin, *The Expression of Emotions* (London: John Murray, 1872).

he added duplicate illustrations of two separate dogs in these contrasting postures.[58]

One of Darwin's letters said, "I most heartily subscribe to what you say about the qualities of Dogs: I have one whom I love with all my heart."[59] He wrote hundreds of letters about dogs. (There are 235 listed on the Darwin Correspondence Project website.) The Darwins had quite a few dogs over the years, including a Newfoundland dog named Bobby (Charles Darwin's own childhood name). *The Expression of the Emotions* included a description of the way that Bobby the dog showed his feelings of "piteous, hopeless dejection" whenever Darwin turned toward his hothouse to check on experiments instead of striking out for a longer walk.[60] In 1870 Emma described Bobby: "He looked so human, lying under a coat with his head on a pillow, and one just perceived the coat move a little bit over his tail if you spoke to him."[61] Bobby's posture—and his demeanor—echoed Darwin's. Like Bobby the dog, Charles often huddled under a coat or shawl in the parlor. Both man and dog were notable for their "humble and affectionate frame of mind."[62] Of course, since Dickinson did not know Darwin personally, she had no idea that he loved Bobby as much as she loved Carlo.

In 1864 and 1865, Dickinson spent many months in Boston for medical treatment. She was miserable—she described her boarding house room as "Prison" and characterized her treatment as great "woe – the only one that made me tremble."[63] Because she was not allowed to read, it was hard for her to keep up with news of the war. She did not learn that Higginson had been injured until months afterward. Both recuperated slowly through 1865.

The Civil War ended even more gradually. On April 9, 1865, Confederate general Robert E. Lee surrendered at Appomattox. Six days later, on April 15, 1865, President Lincoln was killed. Throughout the following summer, Confederate troops surrendered across the South. The official proclamation declaring the end of the Civil War would not come until August 1866. It had been a devastating conflict; Dickinson's and Higginson's complicated grief at this time was widely shared.

And then there was the simple grief. In January 1866, Carlo died of natural causes. Dickinson wrote to Higginson, sending a terse, two-word letter: "Carlo died." In a slightly longer postscript, she asked plaintively, "Would you instruct me now?"[64] In the moment of loss, she might have genuinely wanted advice from Higginson, but the question also implied that she suspected Higginson was no longer able to guide her. Dickinson enclosed a poem about cricket songs that described their music as "an unobtrusive Mass" or a "spectral canticle." In her memorial to Carlo, Dickinson turned again to insects, birds, and the grass:

Further in Summer than the Birds
Pathetic from the Grass
A minor Nation celebrates
Its unobtrusive Mass

No Ordinance be seen
So gradual the Grace
A pensive Custom it becomes
Enlarging Loneliness.

Antiquest felt at Noon
When August burning low
Arise this spectral Canticle
Repose to typify.

Remit as yet no Grace
No Furrow on the Glow
Yet a Druidic – Difference
Enhances Nature now[65]

Though she was full of sorrow, the last lines of "Further in Summer than the Birds" expressed a comforting thought. Carlo's death might somehow connect her to the distant world of beings that were no longer alive. There was some solace in the idea that

>... a Druidic – Difference
>Enhances Nature now[66]

Later that year, when summer had finally come to Amherst, the "Druidic – Difference" she had anticipated did not quite materialize. She wrote to Higginson about her grief for Carlo once more: "Nature, seems it to myself, plays without a friend."[67] Dickinson's walks had grown shorter after Carlo's death. She was staying closer to home. As she explained, "I explore but little since my mute Confederate, yet the 'infinite Beauty' – of which you speak comes too near to seek. To escape enchantment, one must always flee."[68] Without her canine companion, Dickinson was alone with the natural world. Enchantment was inescapable.

CHAPTER 14

Mutual Friends

Downe and Amherst, 1866–1882

BY THE LATE 1860S, Charles Darwin and Emily Dickinson patterned their days in similar ways. For the most part, their daily habits were typical for wealthy men and women of the era. Most of their friends and relations read the latest novels and journals, spent hours on their correspondence, enjoyed their gardens, and went for long walks or rides with animal companions. Darwin and Dickinson, however, also avoided many social interactions, and their immediate circles noticed the extraordinary amount of time they spent writing. Both were prolific and surprisingly ambitious.

Although the pace of life slowed somewhat for both after 1865, neither Darwin nor Dickinson stopped working. From the mid-1860s onward, Darwin continued his great, interconnected projects. Eventually, he published ten more books that built on the foundation of *On the Origin of Species*. With similar dedication and focus, Dickinson kept writing poetry. After 1864 she wrote roughly one thousand new poems. Their most astonishing years lay behind them, but by any other standards, Darwin's and Dickinson's later decades would be regarded as remarkably prolific.

Dickinson and Darwin continued to feel a warm interest in plants. Both were fascinated by their relationships to the natural world and felt love and friendship for many nonhuman beings. This expansive sense of kinship extended to the human side, where they felt as connected to people they encountered on paper as to those they met in person. They were both deeply involved with the written world as writers and readers. Books and ideas extended their sense of relationship to thinkers of the past and to contemporaries they would never meet in person. Their relationships with particular people intertwined with these other networks of literary and biological kinship.

During this period, Darwin and Dickinson did not know each other, but their lives were interconnected. As far as human connections went, their closest link was Thomas Wentworth Higginson. His accounts of visiting each of them at their homes in the 1870s offer concrete evidence that Dickinson and Darwin had at least one friend in common. Their other relationships—botanical, animal, and intellectual—are less concrete, though both figures left extraordinary documentary records of their feelings of affinity for a shared network of plant and animal species and their profound engagement with many of the same books, images, and ideas. By their own expansive understandings of kinship, Darwin and Dickinson had many mutual friends, human and nonhuman, animate and inanimate.

Books provided a source of companionship for both Dickinson and Darwin. In his *Autobiography*, Darwin recalled his childhood: "I was fond of reading various books, and I used to sit for hours reading the historical plays of Shakespeare, generally in an old window in the thick walls of the school."[1] As he grew older, Darwin's enthusiasm for some forms of literature waned. Writing when he was in his late sixties, he said, "Later in life I wholly lost, to my great regret, all pleasure from poetry of any kind, including Shakespeare."[2] Though he lamented the change in his tastes, he felt strongly: "I cannot endure to read a line of poetry: I have tried lately to read Shakespeare, and found it so intolerably dull that it nauseated me."[3] Some historians have pounced on Darwin's statement, using it as evidence that Darwin was so committed to science that he completely rejected aesthetic ways of understanding the world. The critics Gillian Beer, Robert Richards, and George Levine dispute this anti-aesthetic reading of Darwin's *Autobiography*. Darwin expressed great regret over his "curious and lamentable loss of the higher aesthetic tastes." He could not understand the "atrophy of that part of the brain ... on which the higher tastes depend" and blamed himself: "A man with a mind more highly organised or better constituted than mine, would not I suppose have thus suffered; and if I had to live my life again I would have made a rule to read some poetry and listen to some music at least once every week; for perhaps the parts of my brain now atrophied could thus have been kept active through use. The loss of these tastes is a loss of happiness, and may possibly be injurious to the intellect, and more probably to the moral character, by enfeebling the emotional part of our nature."[4]

Darwin's comment "If I had to live my life again" clearly shows that he blamed his own behavior for his loss of interest. His primary concern in this passage seems to be about his aging brain. He feared that without the arts he would lose happiness, intellectual capacity, and moral character because the emotional part of his nature would be enfeebled. These are not the reflections of a person who devalues poetry and music.

It comforted Darwin that at least he still appreciated novels. As he put it, "On the other hand, novels which are works of the imagination, though not of a very high order, have been for years a wonderful relief and pleasure to me, and I often bless all novelists. A surprising number have been read aloud to me, and I like all if moderately good, and if they do not end unhappily—against which a law ought to be passed. A novel, according to my taste, does not come into the first class unless it contains some person whom one can thoroughly love, and if it be a pretty woman all the better."[5]

Darwin held novels in low regard, perhaps because he associated them with women writers and readers. From his stipulations—happy endings, lovable characters, and "pretty woman" protagonists—we can surmise that his favorite novels were centered on the marriage plot. Darwin's vague embarrassment about his taste in literature resonates with twenty-first-century attitudes toward romantic comedies. He poked fun at his own tastes, but he freely admitted his enjoyment.

Many novels were read aloud in the Darwin family circle. In his sixties Darwin adopted a new motto: "It's dogged as does it," from Anthony Trollope's *Last Chronicle of Barset*. He not only enjoyed Trollope and most other popular novels but was also inspired by them. Adam Gopnik and Gillian Beer have traced how Darwin plotted his own written works in novelistic ways. The novelist to whom he was closest was George Eliot (Mary Ann Evans). In the late 1860s, the two became friends. Eliot admired Darwin's work and he admired hers.

As early as 1856, in the pages of the *Westminster Review*, Eliot had argued, "All truth and beauty are to be attained by a humble and faithful study of nature, and not by substituting vague forms, bred by imagination on the mists of feeling, in place of definite, substantial reality."[6] After Darwin published *On the Origin of Species*, Eliot began to plot her novels around Darwin's view of nature. Because Eliot had decades to work, her writings show much more fully worked out responses to Darwin than Barrett Browning or Thoreau had been able to achieve.

Eliot's *Middlemarch* was published in 1871. The novelist Henry James recognized it as profoundly Darwinian. His review described *Middlemarch* as "an echo of Messrs. Darwin and Huxley."[7] More than a century later, in *Darwin's Plots*, Gillian Beer would go further, arguing that Eliot went far beyond being "an echo" of Darwin, instead adapting the language and metaphors of Darwinian science for her realist novels. As Beer has explained, Eliot "creates a sense of inclusiveness and extension. Nothing is end-stopped. Multiplicity is developed through the open relation created between narrator and reader, through participation in the immanent worlds of others and through the unlimited worlds of ideas."[8] Eliot's essays and novels held out hope for a green empiricism. They built literary worlds that emphasized interconnection and kinship.

Around this time George Eliot began to occupy an important place in Dickinson's imagination. Though Dickinson saw no reason to publish, it was not difficult for her to imagine herself in company with Eliot. What Eliot did in four-hundred-page novels, Dickinson would do in four-line poems. In 1883 she would send another copy of "Further in Summer than the Birds" to another editor—Thomas Niles, of Roberts Brothers in Boston—in thanks for his gift of a biography of George Eliot. The poem she had sent to Higginson when Carlo died was an expression of Dickinson's own expansive sense of kinship with the natural world and the literary world.

Emily Dickinson knew she was a great poet. She was secure in her social position and happy at the Homestead. In her later years, her social circle became increasingly female. Some of her poems expressed delight at escaping marriage, while others described being a bride as a form of death. Because so many men died during the Civil War, it was common for women of her generation to form households and communities together. Louisa May Alcott's 1873 novel *Work* described such a group of women as "a loving league of sisters."[9] By the late 1870s, the Homestead had become such a sisterhood. Men were not wholly excluded from her circle—her father was part of the household until his death in 1874; her brother and nephews were just across the yard. She fell in love with Otis Phillips Lord in the late 1870s, though their relationship was mostly conducted by mail. He lived a hundred miles away in Salem, Massachusetts. For Dickinson in this period, the core relationships were with her sister, Lavinia; her sister-in-law, Susan; her housekeeper, Margaret Maher (Maggie); and her mother, Emily Norcross Dickinson.

For Dickinson, privacy was liberating. Her niece Martha Dickinson Bianchi recalled Emily's joyous exclamation, "Matty, child, no one could ever punish a Dickinson by shutting her up alone!"[10] According to Bianchi, Dickinson would often "stand looking down, one hand raised, with thumb and forefinger closed on an imaginary key, and say, with a quick turn of her wrist, 'It's just a turn – and *freedom!*'"[11] Dickinson did not imagine locking herself away as much as she imagined locking everyone else out. She did not want to publish because she did not want her work to be confined or subjected to others.

Meanwhile, Darwin continued to publish prolifically. Most of his subsequent works elaborated on his theories of natural selection. At times he went big (with the *Descent of Man* in 1871, for example), while at other times he stayed small (in 1881 he published *The Formation of Vegetable Mould through the Action of Worms, with Observations on Their Habits*). The Darwins' children were grown, and he and Emma began to travel a little, spending winters in London and hosting large family vacations in the Lake District. As he ventured away from Down House, his circle of acquaintances grew wider. In 1868 Darwin spent time on the Isle of Wight with the poet Tennyson and the photographer Julia

Cameron (Virginia Woolf's great aunt). In 1873 he attended a soirée hosted by George Eliot in London. In 1879, during one of his Lake District vacations, he got to know John Ruskin. By this time Darwin had grown accustomed to being a celebrity. He sat for many portraits, photographic and painted, though the results did not always please him. He thought the Walter William Ouless painting made him look "a very venerable, acute, melancholy old dog."[12]

The Expression of the Emotions, Darwin's last great theoretical book, synthesized the observations of humans and animals that Darwin had recorded over many decades. Just as *On the Origin of Species* would become a foundational text for the new field of ecology, *The Expression of the Emotions* would open the new fields of psychology and behavioral science. William James and Sigmund Freud would both start their own great projects by responding to Darwin. Though *The Expression of the Emotions* was a significant work that took Darwinian science in a new direction, in the popular imagination Darwin remained inextricably tied to evolution.

The popular conception of evolution was moving toward a Spencerian "survival of the fittest" model of social Darwinism that Darwin thought was overly harsh. His own evolutionary thinking emphasized coadaptation, kinship, and connection among various branches of the tree of life and pushed him toward kindness and away from cruelty. Acting on their own understanding of evolutionary kinship, Charles and Emma became active in the political movement against cruelty to animals. Although they tried to avoid controversy, they felt strongly about steel traps and inhumane animal experiments. In the early 1870s, both the Darwins corresponded with Frances Powers Cobbe, the Anglo-Irish writer who campaigned for women's suffrage and against cruelty to animals. Charles liked her essay on "Consciousness of Dogs" and shared many of her concerns about vivisection and fur trapping. Their friendship was strained, however, when Charles agreed to sign a petition to Parliament in favor of women's suffrage and then learned that Cobbe intended to present the petition with Charles Darwin as the sole signer. Though he expressed a "general belief that women are not treated with full justice," Darwin was upset at the idea of being put forward alone, as the sole spokesman for women's political rights.[13] Later, Darwin refused to sign a second petition that argued against all animal experiments. He was against "reckless cruelty," but he could not quite bring himself to argue publicly against experimentation.[14]

Darwin was aware that he had become an iconic figure for freethinkers and feminists as well as scientists. He knew that his ideas had taken on a life of their own and that "Darwinism" was often invoked in politically contentious ways. At times he privately agreed, but he avoided parliamentary questions and public debates and usually dodged large social gatherings. He continued to stay away from most scientific meetings, afraid that controversy might erupt.

Darwin was not a total recluse. At home at Down House, there was always a crowd—at any time, a few of Darwin's adult children were likely to be in residence, along with Charles and Emma and a large number of servants. When Asa and Jane Gray visited from Massachusetts, Jane reported to her sister that serving dinner required a butler and four footmen. Jane Loring Gray's letter paints a picture of a pleasant house party at Down:

> A grand piano at one end, a book-case at the other, two writing-tables, sofas, little tables, étagères, small-book-cases, easy chairs of all shapes & kinds.... The fire-place was opposite the window & the sun shone cheerfully in all day – People sewed, or wrote, or scattered to do as they pleased, or read, or chatted. Generally the walkers started for a long tramp at about 11 or ½ past, to get back to lunch at 1½.[15]

Passing around photographs and stereoscope images was a source of great entertainment for the party. The Grays had brought stereoscope images of Yosemite, which they displayed to everyone in the drawing room. Jane's letter related, "One morng. Mr. Darwin brought in some photographs taken by a Frenchman, galvanizing certain muscles in an old man's face, to see if we read aright the expression that putting such muscles in play should produce.... It came out at dinner, that several of us had been trying to move certain muscles before the glass!" These startling images would soon be reproduced in *The Expression of the Emotions*.

Jane Gray's sketch of Charles Darwin in 1868 is worth reading in full:

> As for Mr. Darwin, he is entirely fascinating – He is tall & thin, though broad framed, & his face shows the marks of suffering & disease, for he has been a very great invalid, & still leads a life of rule & regulation – He never stayed long with us at a time, but as soon as he had talked much, said he must go & rest, especially if he had a good laugh – His hair is grey, & he has a full, grey beard cut square across the upper lip, but the sweetest smile, the sweetest voice, the merriest laugh! And so quick, so keen! He never hears a remark, it seemed to me, but he turns it over, he catches every expression that flits over a face & reads it, he is full of his great theories & sees the smallest things that bear upon them, & laughs more merrily than anyone at any flaw detected, or fun made – Full of warmest feelings & quick sympathy, reads or has read to him novels of every kind, & yet carries on there profound investigations, with the most minute & patient experiments; & the number of topics he has taken up & studied & experimented on.[16]

Gray noticed that despite the crowded social atmosphere of a country house party, Darwin kept to himself most of the time. As she put it, "He never stayed long with us." But despite his frequent withdrawals, she found Darwin very sweet: "The sweetest smile, the sweetest voice, the merriest laugh!" She found him sharp and warm, and she commented on his "quick sympathy," as well as his voracious appetite for novels and experiments.

In the late 1860s, Emily Dickinson's social circle had contracted more than Darwin's, in part because she did not have a helpful wife, a butler, and four footmen to protect her daily routines. She tended to stay away from the house parties that Susan hosted across the garden. In person, she had started to avoid most people. On paper, however, in her poems and letters, Dickinson often expressed quick sympathy and great insight. She might have become socially anxious or increasingly impatient, deciding that most social encounters were a waste of time.

She did not cut herself off completely. Dickinson would meet Thomas Wentworth Higginson face-to-face for the first time in 1870, when he visited her in Amherst. Helen Hunt Jackson also visited her in 1876 and 1878 after Higginson reintroduced them. A few years later in the late 1870s, her small circle of confidants would expand again when she and Otis Phillips Lord, the family friend whom she had known since childhood, formed an attachment. These were the exceptions. Dickinson had very few visitors during these years. Thomas Wentworth Higginson would later describe her as "my eccentric poetess Miss Emily Dickinson who never goes outside her father's grounds & sees only me & a few others."[17]

Since Dickinson's circle was so limited, it is remarkable that one of the only friends who visited her in Amherst in the 1870s also visited Darwin in Kent. During the 1870s, Thomas Wentworth Higginson alternated two visits to Dickinson in Amherst with two visits to Darwin in Downe.

In Darwin's first letter to Higginson, a few months after Higginson's first visit to Down House in 1872, Darwin expressed his "sincere admiration" for Higginson and his appreciation of Higginson's memoir *Army Life in a Black Regiment*.[18] Higginson responded, "I count it an honor to be remembered by you at all; & I am delighted that my book gave any pleasure to one who has given me so much."[19] The letters between Dickinson and Higginson were much less formal. By 1873 they had been corresponding for more than ten years. Higginson would save more than seventy of Dickinson's letters to him. Although few of Higginson's letters to Dickinson survive, one from just after he visited her in December 1873 expressed his deep friendship. "Each time we seem to come together as old & tried friends; and I certainly feel that I have known you long & well, through the beautiful thoughts and words you have sent me. I hope you will not cease to trust me and turn to me; and I will try to speak the truth to you, and with love."[20] Though most of the letters Dickinson received in her lifetime were burned after she died, this was one of the few letters to Dickinson that escaped the flames.

In contrast, Higginson did not burn the letters that he received. He carefully preserved as many as he could. Since Higginson himself provided the closest social link between Darwin and Dickinson, the papers in his writing desk offered their closest material proximity. Darwin and Dickinson would never shake hands with each other, but it is likely their handwritten letters brushed against each other in Higginson's jumble of papers.

Higginson first met Dickinson in August of 1870, in her shadowy front parlor. She entered the room wearing a loose white wrapper with a fine blue shawl arranged around her shoulders and handed him two daylilies.[21] They talked for hours. He left carrying the flowers and a photograph of Mrs. Browning's tomb, intended for his wife. After the visit he spent the evening looking at the slabs of stone imprinted with fossil footprints in the Appleton Cabinet. Over the next day or two, he wrote notes on his conversation. On the train from Amherst, he wrote to his wife. He folded his letter around the picture of the poet's grave and a sheet of his notes from his conversation with Emily Dickinson. Four years after Carlo's death, the dog was still on her mind. She told Higginson that "her great dog 'understood gravitation.'"[22]

Two years later, in 1872, Higginson visited the Darwins at Down House. Though Higginson and Darwin were not such "old & tried friends,"[23] they got along well. Higginson described Charles Darwin as "tall and flexible with the overhanging brow and long features best seen in Mrs Cameron's photograph" and said that "he either lay half-reclined on the sofa or sat on high cushions, obliged continually to guard against the cruel digestive trouble which haunted his whole life."[24] Higginson recalled that they discussed *Alice in Wonderland*. He commented, "It was altogether delightful to see the man who had revolutionized the science of the world giving himself wholly to the enjoyment of Alice and her pretty nonsense."[25] When Higginson returned to Britain in 1878, he was invited for a longer stay.

Before his second visit to Darwin, Higginson paid another call on Dickinson. In December 1873, they met for the second (and last) time. This time, Dickinson "glided in, in white, bearing a *Daphne odora* for me, & said under her breath 'How long are you going to stay.'"[26] She also gave him a poem: "The Wind begun to rock the Grass."[27] Afterward he wrote to thank her. "Your poem about the storm is fine – it gives the sudden transitions."[28] On top of praising her, he encouraged her to read other women poets—the letter quoted a poem by Julia Ward Howe and urged her to read Helen Hunt Jackson's newest book, which he had enclosed. Dickinson treasured her friendship with Higginson, in part because it made her feel connected to the literary world at just one remove.

When Higginson returned to Britain in 1878, he was saddened by Darwin's declining health. He recounted, "He went to bed early that night, I remember, and the next morning I saw him, soon after seven, apparently returning from a walk through the grounds, – an odd figure with white beard, and with a short cape wrapped round his shoulders, striding swiftly with his long legs." Darwin's son commented, "There it is: he pretends not to be at work, but he is always watching some of his little experiments, as he calls them, and gets up in the night to see them."[29] Higginson enjoyed the loving, lighthearted atmosphere at Down House very much. He reported, "Nothing could be more delightful than the home relations of the Darwin family."[30]

As much as he enjoyed staying at the Darwin house, Higginson took an early departure the next day because he had an even more thrilling invitation. He later reported, "On this same day . . . I passed from Darwin to Browning, meeting the latter at the Athenæum Club. It seemed strange to ask a page to find Mr. Browning for me."[31] With this brief encounter, Higginson connected Dickinson to the Brownings, just as he had offered her a sort of association, once removed, with the Darwins. Dickinson would never meet Darwin or the Brownings in person, but through Higginson, she came close.

Another highlight of Higginson's 1878 visit to England was meeting Britain's poet laureate, Alfred Tennyson, and his neighbor, the photographer Julia Cameron. While he was visiting Tennyson at his house in the country, they learned that Cameron's youngest daughter was very ill nearby. Higginson went along with Tennyson to visit the sick child, who died a few days after their visit. Higginson wrote, "I shall never forget the scene when Tennyson bent over the pillow, with his sober Italian look, and laid his hand on the unconscious forehead; it was like a picture by Ribera."[32] Cameron offered him photographic souvenirs of

Left to right: *Thomas Carlyle* by Julia Margaret Cameron (1867); *Alfred Tennyson* by Julia Margaret Cameron (1869); *Charles Darwin* by Julia Margaret Cameron (late 1860s).

this strange bedside encounter. He chose an image that reenacted Tennyson's poetry, an image of two angels beside an open tomb, and "three large photographs, of Darwin, Carlyle, and Tennyson himself."[33] There was something curiously circular about this exchange. Meeting Darwin, Higginson had remarked on the man's resemblance to his famous photograph; he would return to the United States with a copy of the image received directly from the photographer.

In the age before networked computers and phones, physical copies of photographic prints, stereoscope cards, and photographic calling cards on paper or cardboard were necessary for sharing images and building social connection.

Back in Amherst, Dickinson refused to be photographed, but she enthusiastically collected images of figures she admired. She had hung a portrait of Elizabeth Barrett Browning in her bedroom shortly after the poet died. After George Eliot died in 1880, Dickinson added a portrait of Eliot. Dickinson's niece Martha would later recall that in her last years there were three portraits hanging in her bedroom: Barrett Browning, Eliot, and Carlyle. It is easy to understand the connection Dickinson felt to Barrett Browning and Eliot. Both were great writers celebrated for their remarkable intellects and their surprising ability to make modern scientific thought glow with warmth. In contrast, the Carlyle portrait has long puzzled scholars because Dickinson never mentioned him in her writing. She did not share the anti-egalitarian views of his later work (the anti-scientific arguments of "The Mechanical Age" were symptomatic of Carlyle's increasingly reactionary conservative attitude).

Today the Emily Dickinson Museum has restored the bedroom to the moment just before Carlyle was added to her bedroom walls. In his provocative discussion of the bedroom portraits, Páraic Finnerty has proposed that Dickinson might have hung the portrait of Carlyle near Barrett Browning and Eliot in order to protest Carlyle's exclusively male literary canon, or at least to propose that Barrett Browning and Eliot belonged among the Carlylean heroes. It is also possible that there was some triangulation with Higginson at work. Where Higginson treasured images of Tennyson and Darwin, Dickinson substituted Barrett Browning and Eliot, women writers whose empiricist enchantment she hoped to emulate. The image of Carlyle, clipped from *Scribner's*, was an etching of Julia Cameron's photograph, the same image that Thomas Wentworth Higginson had chosen for himself when he visited Cameron in England. Dickinson might have chosen to add Cameron to her personal gallery, rather than the photographer's subject. Perhaps, for Dickinson, the picture evoked Higginson. Perhaps it connected her to Cameron's other subjects: Alfred Tennyson, John Herschel, and Charles Darwin. For the last five years of her life, the images on Dickinson's bedroom walls helped her feel connected to her intellectual and artistic heroes.

In 1877 Emily Dickinson worked on a new poem, "What mystery pervades a well!," in which she imagined looking at well water (perhaps through a microscope) and reflected on the fascinating (and sometimes frightening) strangeness of nature. The last two stanzas read:

> But nature is a stranger yet;
> The ones that cite her most
> Have never passed her haunted house,
> Nor simplified her ghost.
>
> To pity those that know her not
> Is helped by the regret
> That those who know her, know her less
> The nearer her they get.[34]

The last two lines emphasized the mysteries at the heart of nature and natural science: "That those who know her, know her less / the nearer her they get." But although the poem played with gothic "haunted house" metaphors, the overall effect of the poem was not spooky. To the contrary, for Dickinson, as for Darwin, "looking in an abyss's face" prompted thoughts about courage and kinship.

This sense of mysterious relatedness filled Dickinson with awe (and sometimes with dread—in one draft she tried dread as an alternative for awe). As was her practice, she wrote a few drafts, experimenting with different versions. She copied the last two stanzas out and sent them to Susan, replacing the word "Nature" with Susan's name ("But Susan is a stranger yet"). On a separate piece of paper she added a few more lines, including:

> How adequate the Human Heart
> To it's emergency –[35]

For Dickinson, *emergency* had two meanings—a time of dangerous crisis and a time of emerging, being born or reborn. When she imagined connecting her own "emergency" to nature's deep well of mystery—and to the human heart—she showed, once more, that she and Darwin were "Related somehow."[36]

CHAPTER 15

Perfectly Disinterested: Darwin's Last Days

IT TOOK Charles Darwin forty years to complete the grand, connected narrative he had started writing when he set sail in 1831. There were four major books in the series: *The Voyage of the Beagle, On the Origin of Species, The Descent of Man*, and *The Expression of the Emotions in Man and Animals*. The biologist Edward O. Wilson describes them as "four great classics, flowing along one to the next like a well-wrought narrative."[1] In *The Voyage of the Beagle* (1839), Darwin emulated Alexander von Humboldt, whose *Personal Narrative* had inspired him. With quiet confidence, he presented himself as Britain's answer to Humboldt. Like the German explorer, Darwin became a scientific celebrity. His great popularity helped set the stage for his later works. He had established himself as a widely liked public figure long before he challenged conventional attitudes. *On the Origin of Species* (1859) outlined his theoretical argument for natural selection without explaining where humans fit in to the great tree of life. Though Darwin had long been convinced that humans were ordinary biological organisms that had evolved naturally, he gave the public more than a decade to absorb the implications of natural selection before he announced this conviction in *The Descent of Man* (1871). Shortly thereafter, with the fourth and final theoretical book, *The Expression of the Emotions in Man and Animals* (1872), he went a step further, arguing that human hearts and minds were naturally occurring biological features, not markers of separation from other organisms but signs of kinship with other species. Darwin made it clear in the notes of *The Expression of the Emotions* that he had started the book in 1838, though he did not make his ideas public until 1872. With *The Expression of the Emotions*, Darwin finished the project he had begun in the 1830s.

Taken together, Darwin's books slowly and deliberately dismantled human-centered versions of reality. Whether this was depressing or exciting was a matter of perspective. From a negative point of view, his decades of work seemed to humiliate the human since he made the case that humans were not divine or semidivine beings. From a positive angle, his work could be understood to be enlarging the human family and laying the groundwork for repairing the metaphysical division between humans and all their biological kin. In an 1882 essay in the *Atlantic Monthly*, John Fiske summed up the trade-off: "Again has man been rudely unseated from his imaginary throne in the centre of the universe, but only that he may learn to see in the universe and in human life a richer and deeper meaning than he had before suspected."[2] For Fiske and a few others, Darwin's ideas made life more meaningful because they offered humans a thrilling expansion of their networks of kinship and sympathy. This view was relatively uncommon. More often, late nineteenth-century thinkers associated Darwin with meaninglessness. Readers of the *Atlantic Monthly*, however, had the chance to read Asa Gray, Thomas Wentworth Higginson, and John Fiske. Emily Dickinson's view of Darwin was shaped by their accounts.

After *The Expression of the Emotions*, Darwin continued to write books about specific plants and animals (including biographical and autobiographical writings about himself and his family), but there was no need for another theoretical book. Instead, he focused on defending and supporting the theories he had already published.

By the 1870s there was no question that personal popularity was an important tool for establishing scientific credibility. Darwin curated his image carefully, commissioning new photographic calling cards every few years. His allies worked to promote him as a personally heroic scientist, in hopes that his character would help substantiate his theories.

Darwin participated in these efforts, though he needed to tread delicately since the goal was to present him as a disinterested man of science. In his final decade, Darwin had become a scientific celebrity, a symbol of Victorian science for many people around the world. As much as he enjoyed being lionized, he was increasingly amazed by the many, often contradictory interpretations of his work.

Most people who met him agreed that Darwin was very likeable. The archives include hundreds of affectionate letters and reminiscences like Jane Loring Gray's, depicting him as charming and "entirely fascinating."[3] Yet even the warmest accounts show that he was a complicated person—shy and sociable, sharp-witted and warmhearted, anxious and daring, fragile and tough. In one sense Darwin was somewhat unlucky: his health was a constant trouble for many decades. In most other respects, he was remarkably fortunate. Born into one of the wealthiest, most scientifically and technologically advanced families in England at a time when the British Empire was rising to become the greatest power the world had ever seen, his opportunities were incomparable.

At times Darwin's privilege made him blind. His writings are shot through with nineteenth-century assumptions about race, class, gender, and empire that offend many twenty-first-century sensibilities. Darwin was a Victorian gentleman who was perfectly comfortable with his social position and rarely questioned—or even noticed—how social inequities affected others around him. There were paradoxes here too, however. Darwin's works often opened possibilities for undoing the oppressive social hierarchies that structured his world.

In matters of belief, Darwin was determined to remain uncertain. He often said that his judgments about theological topics and political issues changed from day to day. He continued to hope that "a firm belief in the laws of nature will some day reign supreme."[4] However, he never decided whether such a belief could be squared with Christianity or not. He assembled a team of three close allies (Thomas Huxley, Joseph Hooker, and Asa Gray), all of whom he loved and trusted completely. Huxley was a brash agnostic who enjoyed antagonizing conventional Christians. Hooker was a polite and cautious British gentleman who attended church and thought science and religion should be separate from each other. Gray was an American, a devout Christian who insisted that science and religion were mutually reinforcing. Darwin valued—and sometimes fleetingly shared—all three stances. Following Huxley, many atheists and agnostics claimed Darwin as an icon. In response, some conservative Christians anathematized him, imagining Darwinism as a new nineteenth-century heresy. Other Christians followed Hooker's lead and tried to keep Darwinian science and religion separate. (Stephen Jay Gould would later describe this approach in terms of "non-overlapping magisteria.")[5] Finally, devoted Christians of a more liberal stripe—those whose views aligned with Gray—saw Darwin as a prophetic figure who pointed toward a new, modern synthesis of empiricism and transcendence.

When Darwin was attacked, it was easy enough. He stayed out of the conflict while his allies (Huxley, Hooker, Gray, and others) defended him. Though he shrugged off the anti-Darwinians with ease, he was more troubled by the enthusiastic self-styled Darwinians who claimed that his theory supported their widely varying ideologies. When people claimed him as their champion, it was sometimes awkward. He wrote many polite letters begging his supposed followers to desist. In response to the liberal American theologian Francis Abbot's request to reprint an extract from one of his personal letters, he hedged, "I fully believe & hope that I have never written a word, which at the time I did not think; but I think that you will agree with me that anything which is to be given to the public ought to be maturely weighed & cautiously put."[6] Despite Darwin's hesitancy, extracts from his letter to Abbot were widely disseminated. Emily Dickinson may have read the letter in the *Springfield Republican*, where it was published in 1872.[7]

Darwin's exchanges with Abbot show that he was sympathetic toward liberal theology but reluctant to take a stance. Later, the playwright George Bernard Shaw described Darwin as "an honest naturalist working away at his job with so little preoccupation with theological speculation that he never quarreled with the theistic Unitarianism into which he was born, and remained to the end the engagingly simple and socially easy-going soul he had been in his boyhood."[8] Shaw's description of Darwin as engaging is certainly accurate, but many scholars balk at calling Darwin "simple" or "easy-going." Darwin's ideas—his attitudes—were often complex and contradictory.

From the start the implications of Darwinism were double-edged. The historian Beryl Satter makes a useful distinction between more conservative "social Darwinists" and more progressive "reform Darwinists."[9] Charles Darwin had close personal connections to both of these opposing political stances. On the conservative side, he rarely questioned the class system that had afforded him such remarkable privilege or the British Empire that had sponsored the *Beagle*'s naval voyage. At times Darwin tacitly accepted the racist, sexist, and colonialist assumptions of the social Darwinists. Sometimes his ties to social Darwinism were more explicit: In *The Descent of Man*, he frequently cited "the admirable labours" of his cousin Francis Galton, whose theory of eugenics was central to social Darwinism.[10] On the other hand, in his "Introduction" to the biography of his grandfather Erasmus Darwin, he praised his grandfather's egalitarian impulses and quoted lines from *The Loves of the Plants*:

> Hear him, ye Senates! hear this truth sublime,
> He, who allows oppression, shares the crime.[11]

Like his grandfather, Charles Darwin was sympathetic to reformers who tried to fight oppression. His personal correspondence with Frances Powers Cobbe linked him to women's rights while his friendships with Asa Gray and Thomas Wentworth Higginson tied him to antiracist movements. In spite of these links, Darwin avoided making public statements that would tie him too closely to any political faction.

In the realm of economics, social Darwinist capitalists justified their most competitive exploits by invoking Darwin. Walter Bagehot, the editor of the *Economist* and author of *Physics and Politics*, claimed that Darwin's theory of natural selection precisely supported the existing British social order. In contrast, on the reform side, Socialists, Communists, and anarchists used their own versions of Darwin to argue for new social structures based on mutual aid.

In discussions of race, social Darwinists used Darwin's name and Spencer's "survival of the fittest" slogan to justify eugenics and

Anglo-Saxon supremacy. Meanwhile, Darwin's strong arguments for human *monogenesis*—the notion that all human beings were members of the same species and that variations between groups had little biological significance—was a fundamental premise for antiracist reformers.

When it came to women, Darwin's name provided a rationale for social Darwinists who wanted to confine women to the domestic sphere on account of their weakness. In contrast, reformers who supported women's equality often cited Darwin's description of female choice in reproduction to justify their feminism.

Darwin himself tried to remain above the fray, refusing to clarify the ideological implications of natural selection. The ironic consequence was that Darwin himself "was not a Darwinian," George Bernard Shaw observed.[12] In Shaw's account, Darwin's attempt to avoid conflict backfired: "He had the luck to please everybody who had an axe to grind."[13]

Shaw thought of Darwin as a humble, hardworking naturalist who refused to be drawn into political or theological debates. However, some later critics are skeptical of this narrative because Darwin's ideologically disengaged stance was a remarkably effective tool for advancing his arguments. According to Janet Browne, Darwin's work benefited from the fact that he "was associated with rising ideologies of meritocracy, assiduity and respectability and soon afterwards came to stand for the highest principles of pure scientific enquiry."[14] Browne suggests, "Darwin's public persona in itself materially contributed to the crafting of these new visions of respectable, morally upright science."[15] There was an element of strategic self-promotion in Darwin's political and theological disinterestedness.

Of course, Darwin could have been both strategic and sincere. The fact that his ideological indeterminacy helped him build a reputation for scientific purity does not necessarily make him disingenuous. He was entirely consistent for many decades about his refusal to enter into political or religious conflicts. There is no reason to conclude that his reticence was part of a devious, underhanded scheme to advance one ideology over another. To the contrary, the multivalent implications of

Darwinism were characteristic of Darwin's innermost thoughts. On the most concrete level, his theory blurred specific boundaries and celebrated variation. For Darwin, the world was alive and ever-changing. Likewise, his mind was as changeable as it was quick.

By the end of his life, Darwin knew that his own name had become a slippery signifier, almost impossible to pin down. For the most part, he took it in stride. At times he seemed to enjoy the different possibilities. The only cause he supported wholeheartedly was the fight against cruelty to animals—though even here, he hedged when anti-vivisectionists argued against all animal experimentation. He was against cruelty but in favor of experiments. In other arenas he encouraged many (often contradictory) uses of his theory. In *Darwin Loves You*, George Levine comments on the "polysemic nature" of Darwin's arguments and language, arguing that ultimately "the range of possible interpretations seems boundless."[16]

When Darwin was young, "scientist" had not yet been coined and "science" was defined in many contradictory ways. By the end of his life, he had come to personify the ideal scientist. The definition of "science" had hardened and clarified around Darwin's naturalist methods. But for Darwin, science itself was as full of change and possibility as it had ever been. It was perfectly fitting that he had become an icon of uncertainty.

Charles Darwin died on April 19, 1882, at age 73. He had been in decline for a few months, though he continued to tend his experiments in the garden and greenhouse. He was confined to his bed for the last two days. Fifteen minutes before he died, he said to Emma, "I am not the least afraid to die. Remember what a good wife you have been."[17] He did not express any change of heart about his religious or scientific views. In her diary, Emma Darwin noted "3 ½," meaning 3:30 p.m. She did not write any more that day. The next day, April 20, she wrote, "Polly died." Polly was the family dog. Emma and the children believed that Darwin's beloved terrier had "died of heartbreak."[18]

Dozens of obituaries were published in papers around the world. In the *Academy*, the science writer Grant Allen wrote, "The news that Charles Darwin was no more fell upon the world like thunderclap. It is true, his years might have led us to suspect that he had no long span of life yet before him; but his scientific activity was still so ceaseless, his powers were still so fresh and vigorous, his old age was still so green and vital." As Allen put it, "We looked forward to many another of the familiar green-bound volumes, rich with teeming facts and marvellous applications of minute discovery."[19] Allen linked the green bindings of Darwin's books to his "green and vital" attitude.

By all accounts, Darwin had wanted a quiet funeral in the village of Downe, twenty miles outside London. He had grown increasingly reclusive in his late years, and he had probably imagined that he would be buried in "the sweetest place on earth,"[20] the family plot where his much-loved brother and two of his children had already been interred. As he approached death in the spring of 1882, he may have imagined a quiet burial with family and neighbors and a few old friends. But within hours of his death, his scientific friends had clubbed together to petition for his interment in Westminster Abbey, in London. A few days later, April 26, 1882, Darwin's remains were placed near Isaac Newton's monument in the north choir aisle of the elaborate vaulted cathedral. His grave was directly beside John Herschel, the man whose book—and whose busy, happy scientific family in their comfortable country house—had helped inspire his career.

When his death was reported in the *New York Times*, the newspaper attributed nineteenth-century attitudes of "scientific unbelief" to Darwin's influence.[21] But other publications (including the *Popular Science Monthly* and the *Atlantic Monthly*) cited prominent Church of England clergymen who did not equate Darwinism with unbelief, quoting Canon Liddon's sermon at Saint Paul's Cathedral in London, which averred that natural selection was "not necessarily hostile to the fundamental truths of religion."[22] In the *Atlantic Monthly*, John Fiske commented, "As theologians are no longer frightened by the doctrine of gravitation, so they are already outgrowing their dread of the doctrine of natural selection."[23] The editors of the *Popular Science Monthly* explained, "Now that Darwin is dead, there is a universal burst of admiration for the man ... and he is laid in Westminster Abbey alongside of Newton, while the most eminent preachers of London agree in declaring that there has been nothing in his teaching that is not wholly consistent with the soundest Christian belief."[24]

Funeral of Charles Darwin (1882). From *The Graphic*, May 6, 1882, Wellcome Collection.

Darwin's funeral was the opposite of what he and his family had expected. "It gave us all a pang not to have him rest quietly" in the village cemetery, his wife Emma recalled.[25] The great cathedral at the heart of the city did not seem like a place where Darwin would be at peace. He had long refused to attend church and been reluctant to visit London. Yet despite their misgivings, the Darwin family decided to give Darwin's body to England. The rough coffin made by a neighbor was changed for a more elaborate one, glossy and draped with velvet. It was carried through the enormous crowd at the abbey by ten men.[26]

Darwin's funeral was an occasion of state. Representatives from the embassies of France, Germany, Italy, and Spain attended. The *Times* described the guests as "a remarkable and representative crowd of distinguished men, such as only an occasion of deep and general interest would bring together. Leaders of men and leaders of thought; political opponents, scientific co-workers; eminent discoverers, and practitioners of the arts."[27] Huxley and Hooker were both pallbearers. Another pallbearer was U.S. ambassador James Russell Lowell, the poet whose writing had been important to Emily Dickinson even before he became the editor of the *Atlantic Monthly*. As a child she had fiercely insisted that Lowell was as good as any English poet. Years later, when Lowell went to England to serve as ambassador, he became a friend of the Darwin family.

Lowell had a knack for quick verbal sketches. In a letter to the family, he described Darwin as "almost the only perfectly disinterested lover of truth I ever encountered."[28] These words might have served as an epitaph. They summed up the way the family hoped Charles Darwin would be remembered. However, in the end his family decided against any epitaph at all. Darwin's stone provides a striking contrast to Herschel's.

The inscription on the white marble is understated, even minimalist. It says:

> Charles Robert Darwin
> Born 12 February 1809
> Died 19 April 1882

Other than his name and dates, there are only two words, both in English: *Born* and *Died*. The white gravestone lies directly beside Herschel's black one, which is inscribed with Latin and includes a quotation from the Bible, also in Latin. Even Herschel's dates are in Latin, a string of roman numerals chiseled into the stone. Only eleven years had passed between the two burials, but the message was clear: Darwin represented a new, modern sensibility.

It would soon be commonplace to think of Darwin as a harbinger of modernity. When Lytton Strachey was writing his celebrated critique of *Eminent Victorians* (1918), he planned to include Darwin but eventually decided to exclude him because he was "on the side of the moderns."[29] Janet Browne explains that Darwin would have complicated—and probably weakened—Strachey's attack on Victorian attitudes. Although his circumstances may have positioned him to be the quintessential Victorian gentleman, Darwin was not a typical Victorian.

Julia Cameron's portraits of the two scientists also highlight Darwin's modernity. In his velvet cap and shawl, Herschel could be a contemporary of Milton or Galileo. Darwin, on the other hand, wears a loose-fitting coat that would not be out of place in a laboratory today.

Darwin's legacy was both a privilege and something of a burden for his friends and family. He left financial bequests for Hooker and Huxley

Gravestones of John Herschel and Charles Darwin in Westminster Abbey. Wikimedia Commons.

John Herschel by Julia Margaret Cameron (1867).

Charles Darwin by Julia Margaret Cameron (late 1860s).

and a large fortune for his family. All of them would continue to defend his ideas—and his reputation—for the rest of their lives. His son Francis would spend many years editing and publishing Charles's vast collection of letters and papers.

The entire family pitched in to help keep the Darwin industry going. Shortly after his fathers' death, his oldest son asked the Harvard professor William James for help interpreting some of Darwin's correspondence with the progressive theologian Francis Abbot. William James was staying in London, house-sitting for his novelist brother Henry, who was in the United States. The request made sense since Abbot had been a student of William James's. Interestingly, the words that James used to explain Darwin's ideas about morality would later recur in his own writings.[30] James's pragmatic moral philosophy was perfectly aligned with his interpretation of Darwin.

During his lifetime, Darwin and his closest allies had focused on establishing his reputation as disinterested, nonpartial, above the fray. After his death, members of his family (particularly his son Francis and his granddaughter Nora Barlow) put most of their energy into preserving Darwin's reputation as a resolutely neutral scientist who kept himself separate from philosophical, political, and cultural questions. Although contradictory versions of Darwinism would proliferate in the years to come, almost everyone agreed that Darwin was more of a scientist than a Darwinist. The prevailing consensus was that Darwin's version of science was objective rather than subjective. Hence, most versions of Darwinism assumed an impersonal point of view, stripped of emotion and individuality. Whether *Darwin* meant greed or altruism, up/down hierarchy, or horizontal relationships, it almost always denoted the hard-edged, emotionless, impersonal perspective. In this version of Darwin, there was little room for mystery or magic. Thomas Wentworth Higginson summed up the prevailing understanding in his memoir, *Cheerful Yesterdays*, when he expressed regret for his youthful sense of "pre-Darwinian wonder."[31] For Higginson—and for most of his contemporaries—the world after Darwin was a world without wonder.

Equating Darwin and Darwinism with the absence of emotions was unjust to Darwin and his work. Personally, he had never lost his intense emotional responses to the natural world. His writings were threaded through with expressions of fascination and excitement. From his first account of his voyage on the *Beagle* to his final paragraphs about worms, he consistently invited readers to share in his delight. *The Voyage of the Beagle* opened with Darwin's description of his first moments on a tropical island: "The scene, as beheld through the hazy atmosphere of this climate, is one of great interest; if, indeed, a person, fresh from the sea, and who has just walked, for the first time, in a grove of cocoa-nut trees, can be a judge of any thing but his own happiness."[32] More than forty years later, at the end of his last book, Darwin once again invited readers to view the natural world with emotionally engaged appreciation: "When we behold a wide, turf-covered expanse, we should remember that its smoothness, on which so much of its beauty depends, is mainly due to all the inequalities having been slowly levelled by worms. It is a marvellous reflection that the whole of the superficial mould over any such expanse has passed, and will again pass, every few years through the bodies of worms."[33] From the happiness of the young man in the coconut grove to the "marvellous reflection" of the old man who finds beauty in the way that worms slowly level and smooth the earth, Darwin's emotions and his sense of wonder were always central to his narratives.

In a strange twist of cultural history, the attitudes toward the natural world that Darwin expressed did not fit with prevailing understandings of naturalist writing any better than they fit with expectations for Darwinism. By the 1880s Darwin was not only not a Darwinist—he was also not a naturalist. The term "naturalist" had taken on a completely different meaning. No longer a simple description of people who collected wildflowers and beetles, it had become a description of a literary approach. In literature, Darwinian naturalism became associated with figures like Emile Zola and Stephen Crane, whose harshly deterministic narratives made little room for human emotion. The naturalist literary movement explored the idea that nature was not at all beneficent. Instead, as Stephen Crane put it in "The Open Boat," late nineteenth-century naturalists believed that nature was "indifferent, flatly indifferent." "Naturalism" had changed from green to gray.

Though "Darwin" was iconic for latter-day naturalists such as Crane, when they referred to "Darwin," they did not mean the particular historical person Charles Darwin. They certainly were not referring to the young voyager on the *Beagle*. They did not even intend to refer to his specific ideas about nature and natural selection. Instead, "Darwin" became a shorthand way of referring to a version of materialism that saw the word as wholly disenchanted, meaningless, and dispiriting.

This is a tricky point that is worth emphasizing. In late nineteenth-century literary circles, Darwin had become an anti-wonder icon. Writers who viewed the natural world appreciatively were not seen as naturalist writers, much less as sophisticated scientific thinkers. Emotional engagement with the natural world was decidedly not associated with Darwin.

In this context, Dickinson's first readers were unlikely to understand that her sense of nature and the natural world was related to Darwin's thought. Since Higginson had come to equate Darwinian science with emotional disengagement, he would not have seen Dickinson's attitude as Darwinian. Her descriptions of nature are nothing like the "naturalism" of late nineteenth-century literature. Further, since Dickinson never mentioned Darwin directly in any of her letters to Higginson, it is understandable that Higginson could have overlooked Dickinson's interest in Darwin.

Even beyond her correspondence with Higginson, Dickinson rarely mentioned Darwin. There are only two mentions of Charles Darwin's name in Dickinson's surviving letters.[34] She does not use Darwin's name in any of her poems. The same can be said of many of Dickinson's most important influences. None of her poems mention Elizabeth Barrett Browning or George Eliot by name, though portraits of both hung near her writing desk. Even Shakespeare, the writer whom Dickinson cited most frequently, is mentioned by name in only one poem and a handful of letters. Readers who are familiar with Shakespeare's writings find dozens of allusions in Dickinson. Non-Shakespeareans miss the oblique references.[35] Most of Dickinson's allusions to Darwin are similarly subtle.

Since she rarely mentioned her influences by name, it is particularly notable that Dickinson mentioned Charles Darwin in a letter to Otis Phillips Lord eleven days after he died:

Mrs Dr Stearns called to know if we didnt think it very shocking for Butler to "liken himself to his Redeemer," but we thought Darwin had thrown "the Redeemer" away. Please excuse the wandering writing. Sleeplessness makes my Pencil stumble. Affection clogs it – too. Our Life together was long forgiveness on your part toward me. The trespass of my rustic Love upon your Realms of Ermine, only a Sovreign could forgive – I never knelt to other – The Spirit never twice alike, but every time another – that other more divine. Oh, had I found it sooner! Yet Tenderness has not a Date – it comes – and overwhelms.

Dickinson was in her fifties. Lord was about to turn 70. But there is no question that this is a passionate love letter. Dickinson's friendship with Lord had taken the turn toward romantic love around 1880. Around that time, Dickinson described Judge Lord as her "father's closest friend."[36] After Lord was widowed in 1877, his visits to Amherst became more frequent. When he was home in Salem, Lord and Dickinson exchanged voluminous letters, writing to each other every Sunday. Judge Lord was probably a good epistolary match for Dickinson—he was widely celebrated as a wit and wordsmith. His letters to her were destroyed, as were the letters he received from her. A few of Dickinson's drafts remain, though they have been heavily edited by a person who used scissors to excise large portions. In this context it is surprising that the love letter that mentioned Darwin survived. "Tenderness has not a Date," the letter concludes; "it comes – and overwhelms."

This piece of "wandering writing" moves directly from mocking a neighbor's piety by declaring allegiance to Darwin to making an avowal of "rustic Love." In the weeks after Darwin died, Dickinson was highly aware of the impact of his ideas in the United States. As Kim Hamlin explains in *Eve to Evolution*, freethinking feminists of the 1870s and 1880s had turned to Darwin for scientific arguments that bolstered their case for equality. The Darwinian argument for women's rights hinged on Darwin's argument in *The Descent of Man* that in most species of animal, females choose their mates. By extension, American feminists argued, Darwin had proved that "female choice" in humans was natural. Declaring herself a freethinking Darwinian, Dickinson leapt to a declaration of her own desire.

The desires that coursed through the Homestead that year were complicated. In addition to Emily's tenderness for Otis Lord, there was also the scandalous affair between her brother, Austin, and their much younger neighbor, Mabel Loomis Todd. In 1882, Emily and Lavinia would tacitly agree to let Austin use their house as a meeting place for his assignations with his lover. Once, after listening to Todd play the piano at the Homestead, Dickinson sent her a piece of cake and a glass of sherry, but Dickinson would never meet Todd face-to-face. Susan, Emily's dearest love and Austin's wife, was deeply hurt by her husband's affair and the fact that her sisters-in-law accepted it. Upstairs, Emily Norcross Dickinson, the poet's mother, grew weaker.

The intensity of the 1880s inspired Dickinson to write some of her most fascinating poetry. Her late works were passionate, Darwinian, and profoundly magical. She had lost interest in making final copies. Her poems were full of alternative words, while her papers and letters contained many variants of different poems. She seemed to love variation and change even more than she had when she was younger.

In 1882, a few months after Darwin died, Emily Dickinson wrote a note to Susan, the sister-in-law whom she loved passionately but from whom she was sometimes distanced. Three drafts survive. The one she finally sent to Susan was folded around a handful of rose petals. It read, "Excuse Emily and her Atoms – The 'North Star' is of small fabric, but it denotes much." By "Atoms," Dickinson probably meant short poems. She may also have meant to refer to her own materialist leaning—her tendency to focus on concrete and particular natural objects. Perhaps the fading rose petals were "Atoms." In the 1950s the editors of Dickinson's letters mentioned that the dried flowers were still enclosed in the folded paper. But in 1996, Martha Nell Smith sadly reported: "Rose petals that were still enclosed with the note when Johnson was working on the variorum are no longer there."[37] The scattered collection of drafts and the vanished remnants of dried flowers are very different from the neatly bound, carefully preserved *Herbarium* and fascicles of Dickinson's early days.

On the thin strip of paper where she first drafted the North Star note to Susan, Dickinson also drafted a poem: "Cosmopolites without a plea." Given the context, the "cosmopolites" Dickinson referred to might have been rose petals, or perhaps migrating birds or butterflies. The poem moves through many frames of reference:

> Cosmopolites without a plea
> Alight in every Land
> The Compliments of Paradise
> From these within my Hand
> Their dappled Journey – to themselves
> A compensation fair –
> Knock and it shall be opened
> is their Theology   [their] Philosophy[38]

On the scrap of paper, the words that make up the draft of this poem are entangled with the words of the note to Susan. "The North Star is of small fabric yet it implies (achieves) much" is just a few inches from "Theology" and "Philosophy,"[39] the words that Dickinson proposes as possible conclusions to "Cosmopolites without a Plea." In the last line, on the other side of the note to her dear Susan, Dickinson quotes the most open of Bible verses ("Knock and it shall be opened"). She claims this as a "Theology" and then, directly below, she suggests an alternative word: perhaps it is a "Philosophy."

These scattered notes and drafts—with their disappearing traces of rose—materialize Dickinson's sense of Darwinian kinship, the entanglement of human and nonhuman, animal and plant, natural and divine. "Theology" and "Philosophy" are not synonyms here. Yet although they are distinct ways of knowing, Dickinson affirms their interdependence when she makes them alternatives for each other.

Like Darwin, Dickinson was one of the moderns. She, too, loved variation and change. She preferred the open to the defined, and she loved paradox as much as Darwin.

CHAPTER 16

Nature Is a Haunted House: Dickinson Faces Death

"I WOULD like it not to end," Edward Dickinson said to his daughter, pleased that she had spent Sunday afternoon in the parlor with him, reading quietly in the cool shadows as the June sun blazed outside. Emily was "almost embarrassed" by this unusual expression of affection, however indirect.[1] The next day, Edward was back to work. Early Monday morning, he took the train to Boston to represent Amherst in the state legislature. On Tuesday he spent the morning at the State House. Feeling unwell, he walked back to his hotel early in the afternoon. He had started to pack his bags for a return to Amherst when he collapsed and died. The newspapers reported an "apoplectic fit."[2]

Edward Dickinson's death—on June 16, 1874, at age 71—came as a shock. The previous year, he had retired from his position as treasurer of Amherst College, succeeded by his son, Austin Dickinson. No one expected his life to end so soon afterward. In 1873 the *Boston Evening Journal* described him as one of the "River Gods" of the Connecticut valley.[3] That fall he had been elected to the Massachusetts state legislature and started commuting regularly to Boston by train. He still loved the railroad. In fact, his motive for returning to politics was to fight for the expansion of the Amherst railway. It seemed very wrong that the Squire of Amherst had died in Boston, alone in a hotel room.

Emily Norcross Dickinson, the poet's mother, was overcome by grief. On June 15, 1875, a year after Edward's departure, she suffered a stroke that caused partial paralysis and impaired her memory. She would never recover full mobility or recall her husband's death. In the following years, she asked where he was many times every day and felt hurt and worried when he did not come.

After Edward's death, the center of gravity at the Dickinson compound shifted from the Homestead to the Evergreens. Emily's brother, Austin, took over the family finances as well as their father's law practice and his position as treasurer of the college. The Homestead women became dependent on Austin. Their family relationships were a tangle of love and regret. Dickinson wrote to Higginson, "Home is so far from Home, since my Father died."[4] Although the situation was not materially different, the mood was significantly altered. An atmosphere of intense and complicated emotion suffused the Homestead, the Evergreens, and the gardens that connected the two houses. Emily and Lavinia no longer occupied the position of the Squire's daughters—instead, they were consigned to the role of maiden aunts to Austin and Sue's three children, Ned, Mattie, and then Thomas Gilbert Dickinson, born August 1, 1875. Little Gib was a delightful child, a bright spark in an otherwise dark period for the family. Though Emily enjoyed the children very much, she felt isolated. She told Higginson that writing was her only "Playmate."[5] Over the next eight years, there was little time for play. Emily and Lavinia cared for their mother as her health declined.

In 1877 Emily wrote, "Since my Father's dying, everything sacred enlarged."[6] With her heightened sensibilities, everything around her seemed newly mysterious, rich with significance. Her letters after 1874 speak of a new "Intimacy with Mystery."[7] In the shadow of death, Dickinson was often strangely exhilarated. In a private note, she described "the Dead as exhilerants."[8] A poem connected exhilaration to "Enchanted Ground":

Exhiliration is the Breeze
That lifts us from the Ground
And leaves us in another place
Whose statement is not found –

Returns us not, but after time
We soberly descend
A little newer for the term
Upon Enchanted Ground –[9]

On the same leaf, she copied a second poem. She gave the page to Susan:

Best Witchcraft is Geometry
To the magician's mind –
His ordinary acts are feats
To thinking of mankind –[10]

"Exhiliration is the Breeze" and "Best Witchcraft is Geometry" help clarify Dickinson's thinking about the relationship between death and magic. Her invocation of "Enchanted Ground" links magic to the earth, while "Best Witchcraft is Geometry" connects it to mathematical thinking. Taken together, these poems show Dickinson trying to think through—and to express—a sense of natural magic (as opposed to supernatural magic). For her, precise and logical empirical approaches to the natural world did not foreclose enchantment. Thinking about death exhilarated her because it brought mystery to the fore. Dickinson's poems and letters during these years of illness and death frequently invoked magic. She often mentioned ecstasy, awe, and bliss—and sometimes fear and trembling. She danced away from direct expressions of grief, preferring phrases such as "happy Sorrow."[11]

Four years after her father's death, in January 1878, Dickinson's old friend Sam Bowles died. Although his health had been poor for a few years, his death was particularly shocking because he was of her own generation. The editor of the *Springfield Republican* had been one of the first literary people to champion Dickinson and one of her first friends outside the immediate Amherst circle. Dickinson felt his loss keenly. She wrote to Maria Whitney, a professor of modern languages at Smith College who was also grief-stricken at Bowles's death. The two began a correspondence that continued for many years. Though they did not know each other particularly well, their predicament was similar. Neither Whitney nor Dickinson had been members of Bowles's immediate family. Like Dickinson, Whitney was an "unmentioned Mourner" whose deep feelings for Bowles were not acknowledged by many others.[12]

When George Eliot died in 1880, Dickinson mourned from afar, as she had once grieved for Elizabeth Barrett Browning. Then, in October 1881, Josiah Holland died. Holland and Bowles had been coeditors of the *Springfield Republican* in 1852, when Dickinson's first poem was published. Emily and Lavinia's visit to Josiah and Elizabeth Holland's house in 1853 had launched a lifelong friendship. Dickinson exchanged frequent letters with Elizabeth Holland for decades. In the aftermath of Josiah's death, Dickinson wrote, "All grows strangely emphatic."[13] Sympathy intensified every emotion.

Despite—or perhaps because of—the deaths and illnesses, Dickinson's circle of correspondents enlarged a bit during these years. After her father died and her mother became incapacitated, Dickinson deepened her connection to a few others outside her immediate family. She renewed her acquaintance with childhood friend Helen Fiske, now known as Helen Hunt Jackson, the author of the best-selling novel *Ramona*. The two began to exchange frequent letters. Jackson even managed to convince Dickinson to publish one of her poems in *A Masque of Poets*, a book of anonymous verses. It probably delighted Dickinson when "Success is counted sweetest" was attributed to Ralph Waldo Emerson, whom she had admired since she was a child. She opened a regular correspondence with Thomas Niles, the publisher of the anonymous volume to which she had contributed her poem. Niles repeatedly mentioned his eagerness to publish more of her poetry. Here again, Dickinson's feeling of loss was balanced by her sense of expanding possibilities.

In the spring of 1882, three men whom Dickinson had admired from a distance died in quick succession. Charles Wadsworth died April 1. Charles Darwin died April 19. Ralph Waldo Emerson died April 27. Her father, Sam Bowles, and Josiah Holland had died some time before. By the summer of 1882, most of the men whom Dickinson may have imagined as masters were gone. She seems to have felt somewhat liberated. She preferred mystery to mastery anyway. Her grief was mingled with a sense of excitement.

The next group of deaths was much harder. On November 14, 1882, the poet's mother, Emily Norcross Dickinson, died. Emily and Lavinia were on their own in the Homestead, completely dependent on Austin and Sue, who were increasingly at odds with each other.

Nothing could have prepared any of them for the next loss. On October 5, 1883, the youngest member of the Dickinson clan, Thomas Gilbert Dickinson, known as Gib, died a few days after contracting typhoid fever. The poet watched at her 8-year-old nephew's bedside through his last agonizing night. Later, she wrote, "I see him in the Star, and meet his sweet velocity in everything that flies – His life was like the Bugle, which winds itself away, his Elegy an echo – his Requiem ecstasy – Dawn and Meridian in one."[14] In a letter to Susan, she offered a blessing: "Pass to thy Rendezvous of Light, Pangless except for us, who slowly ford the Mystery Which thou hast leaped across!"[15]

On March 13, 1884, six months after her nephew Gib's death, Emily's beloved Otis Phillips Lord died. Dickinson survived this loss as she had survived the others, but she was deeply affected. Death was never far from her thoughts. Her own health was fragile, and many of her letters were notes of condolence or thanks. She exchanged books and flowers with many of her acquaintances. Early one morning, she sent "Carmine Sheaves" of sweet peas to Sam Bowles's son.[16] She sent small gifts to many other friends and neighbors. When they gave her presents, she responded with playful thank-you notes. She wrote to her Norcross cousins in Boston that she was "watching like a vulture" for the next volume of George Eliot's biography.[17] She gave her copy of the first volume to Higginson.

In the winter of 1885, Dickinson found some escape by reading a thriller by Hugh Conway, *Called Back*. Set in London, the novel is about a young couple caught up in a murderous plot by a group of Italian revolutionaries turned anarchists. The hero, Gilbert Vaughan, is temporarily blinded while the heroine, Pauline March, temporarily loses her memory. Eventually, they discover that when they hold hands their psychic link allows him to see her repressed memories of the murder. Gilbert makes an arduous journey across Russia to a Siberian penal colony to confront one of the murderers, and what he learns helps Pauline to recover her sanity. *Called*

Back is sensational and melodramatic, full of fateful coincidences and unlikely but highly entertaining plot twists. Dickinson recommended it to her cousins: "It is a haunting story, and as loved Mr. Bowles used to say, 'greatly impressive to me.'"[18] As ever, she chose her language carefully. Though *Called Back* is not about ghosts, its plot turns on sense impressions transmitted from one person to another when their nervous systems make a mysterious electrical connection. For Dickinson, the tale of inexplicable communication was "haunting."

Dickinson had always been fascinated by the notion that electrical impulses in her brain traveled through her hands to make impressions on paper that could spark thoughts within her readers' brains. In the mid-1880s, she wrote more and more often to bereaved friends, aware that her writing might evoke grief simply by evoking memories. In 1885 she asked Higginson if their mutual friend Helen Hunt Jackson had died. "What a Hazard a Letter is!" she commented, "When I think of the Hearts it has scuttled and sunk, I almost fear to lift my Hand to so much as a Superscription."[19] Dickinson's fears were confirmed. Helen Hunt Jackson, her childhood friend and fellow Amherst writer, had died on August 12, 1885. Jackson was 54, the same age as Dickinson. In one of her last letters to Higginson, Dickinson enclosed her last poem, an elegy for Helen:

> The Immortality she gave,
> We borrowed at her Grave –
> For just one Plaudit famishing,
> The might of Human Love –[20]

Dickinson believed that Helen's death had inspired Higginson—as it inspired her—to write immortal lines of poetry. Her poem—her last poem—commemorated a community of women writers who encouraged each other. Though the poet feared that she would forever hunger for another word of praise from her childhood friend, the final line celebrated both the power and possibility—the "might"—of "Human Love."

A few months later, in November 1885, Dickinson fell ill. She spent most of the winter confined to her bed. At the time she was diagnosed with Bright's disease, a nineteenth-century term for illness associated with kidney problems. Her doctor urged her to rest—she described herself as "Bereft of Book and Thought, by the Doctor's reproof."[21] When spring came, she recovered briefly. In a letter to her Norcross cousins, she compared herself to a spring flower, "stirring, as the arbutus does, a pink and russet hope."[22] In a letter to Mrs. George S. Dickerman, wife of the pastor of the First Church, she asked, "If we love Flowers, are we not 'born again' every Day?"[23] To the end she put more faith in flowers than in Christian traditions. She never stopped questioning God. Her last letter to Thomas Wentworth Higginson was just two lines:

> Deity – does he live now?
> My friend – does he breathe?[24]

Though she seems to have been more worried about the prospect of Higginson's death than her own, she knew that her life was ending. She wrote to her Norcross cousins one last time:

> Little Cousins,
>   Called Back.
>
>   Emily.[25]

This last note alluded to the Conway novel that had made such an impression the previous winter. Dickinson was jocose, strangely cheerful in the face of dying.

Emily Dickinson died on May 15, 1886, at age 55. She was buried four days later. The poet was radiant in death. She wore a fresh white dress. Her hair was arranged to show its bright auburn gleam. Thomas Wentworth Higginson, who had come one hundred miles across Massachusetts for her funeral, remarked on the "wondrous restoration of youth" in her face and the "perfect peace on the beautiful brow."[26] Her sister Lavinia put two heliotropes in her hand "to take to Judge Lord."[27] A bunch of violets and a single wild orchid (a pink lady's slipper, *Cypripedium acaule*) rested above her heart.

At the funeral, Higginson read a poem by Emily Brontë, as Dickinson had requested. Men in formal clothes—including the president of

Amherst College and a few professors—carried the small white coffin out of the back door of the house where she had been born in 1830. Once outdoors, a group of Irish laborers who had worked for the Dickinson family for many years stepped in to bear her across the fields. As she had planned, they carried her around her flower garden and through the wide barn, then along a grassy footpath to the cemetery. By all accounts, it was a glorious spring day. The air was filled with birdsong and the fresh green scents of May.

Susan lined the grave with evergreen boughs. After her burial a flat granite stone with her initials—EED—marked the site. Later her niece Martha put up a white marble tombstone. The full inscription reads:

> Emily Dickinson
> Born
> December 10, 1830
> Called Back
> May 15, 1886

Dickinson's epitaph is almost as laconic as Darwin's. His stone is inscribed with his name and dates and the words "born" and "died." Hers is very similar, but instead of "died" her family chose "Called Back." They were quoting her last letter to her cousins, which itself had referenced the "haunting story" that had diverted her during her last long illness.

"Called Back" is a striking and somewhat mystifying epitaph. It could be interpreted as gesturing toward some type of transcendent religious belief—perhaps Dickinson's niece thought the poet's immortal soul had returned to immortality. At the same time, it could also imply the opposite—that Dickinson had returned to the earth. Ashes to ashes, dust to dust. The inscription is quite fitting for Dickinson because it works both ways, as both transcendent and materialist. For some the phrase might evoke morbid, somewhat spooky emotions. For others it might seem quirky, a little bit humorous. Chiseling the title of a thrilling popular novel about anarchist murderers and psychic lovers onto a tombstone is a bold choice.

What would it mean to be "Called Back" to the earth? In 1885, the year before she died, Dickinson sent a poem to Helen Hunt Jackson in

Emily Dickinson's gravestone. Photograph by Jerome Liebling, Getty Images.

which she described human existence as being "Immured the whole of life / Within a Magic Prison."²⁸ If the Earth was a "Magic Prison" for Dickinson, she was a happy prisoner. In an earlier poem, she had speculated,

> Somewhere upon the general Earth
> Itself exist Today –
> The Magic passive but extant
> That consecrated me –²⁹

Dickinson thought of magic as earthly—she located it "upon the general Earth," implying that it was not confined to particular sacred places but widespread. The magic that interested her—even consecrated her—was not showy or even extraordinary. She described it as "passive but extant." Material existence was enchanting in itself. In the next stanza, the poet declared that she "for right to be – / Would pay each Atom that I am." In the last two lines of the poem, she turned to God—but not to the narrow God of conventional piety. Instead, she prayed:

Oh God of Width, do not for us
Curtail Eternity!³⁰

In this poem, the everyday magic of existence makes the earth—and the poet—sacred. Dickinson suggests that defined religious belief diminishes existence, curtailing eternity as it sets limits on the infinite.

Shortly after Dickinson died, Susan published an obituary that compared her to a magician. In Susan's account, the poet's magic was her ability to capture and communicate details of the natural world. She compared Dickinson's "swift poetic rapture" to "the long glistening note of a bird one hears in the June woods at high noon, but can never see." Susan's images recalled the Romantic naturalists. She mentioned botany and chemistry and described Dickinson's conservatory blooming with "rare flowers" through the winter because "she knew her subtle chemistries." She described the poems as works of a naturalist: "So intimate and passionate a part of the high March sky, the summer day and birdcall." Significantly, Susan did not describe Dickinson as a scientist or even a poet. Instead, she called her a "magician."[31]

Susan had always paid attention to Emily's ideas—her description aligned very well with Emily's statements. The closest Emily had ever come to describing herself as a poet had been in her tear-stained love letter to Susan many years before, when she wrote that she and Susan "please ourselves with the fancy that we are the only poets, and everyone else is *prose*."[32] Susan kept that memory private. She compared Emily to a magician because "she caught the shadowy apparitions of her brain and tossed them in startling picturesqueness to her friends."[33] The magic that she attributed to Emily was biological. Emily's brain contained "shadowy apparitions" that she was able to transform into words.

Before she died, Dickinson had written about the topic of death with unusual courage and persistence. Few poets have described their own burials as often (or as well). Dying—the idea of dying—was a source of great creative energy for her poetry. In one poem, she describes dying as "a height," a high point from which the past and future can be seen clearly. In another, death is a seducer. In Dickinson's poetry, the dead often speak. In some of her best poems, newly buried people talk about their experiences.[34]

Dickinson had long known that she would be buried in Amherst, in the small family plot enclosed by a wrought-iron fence where her parents rested. The graves were in sight of the white clapboard house where the family had lived during the years of exile from the Homestead.

More than twenty years before, Dickinson had written a poem that imagined the perfect conversation among the dead:

> I died for Beauty – but was scarce
> Adjusted in the Tomb
> When One who died for Truth, was lain
> In an adjoining Room –[35]

When Dickinson envisioned her own life after death, whom did she hope to meet? With whom would she have wanted to talk through the endless night? If she thought of herself as dying for beauty, what sort of person would she have imagined dying for truth?

Readers have long associated "I died for Beauty" with the Romantic poets who were fascinated by the correlations between truth and beauty. The poem seems to respond directly to John Keats's "Ode on a Grecian Urn" and Elizabeth Barrett Browning's "Vision of Poets." However, it does not really describe an eternal conversation between two dead poets. Dickinson seems to have dreamed of talking to someone whose life and work had been different from hers. Though beauty and truth had been inextricable for Keats half a century before, by the time Dickinson wrote the two were often separated, even opposed. Poets could still claim beauty as their organizing principle, but truth was increasingly determined by scientists. If the speaker in this poem is an unidentified poet, her companion in death is most likely an unidentified scientist.

In Dickinson's poem, the conversation continues for centuries, but the speakers are never named:

And so, as Kinsmen, met a Night –
We talked between the Rooms –
Until the Moss had reached our lips –
And covered up – Our names –[36]

Dickinson had been as reluctant to describe herself as a poet as Darwin had been to describe himself as a scientist. Nonetheless, if we identify the representative poet as Dickinson, we might also consider Darwin as the nineteenth-century scientist. Of course, Dickinson was interested in many other scientific figures (including the Herschels, Edward Hitchcock, Mary Lyon, and Maria Mitchell, to name a few). The poem keeps it open, allowing us endless alternatives. Even so, it works well to imagine Darwin, that "perfectly disinterested lover of truth,"[37] as Dickinson's deathly kinsman. He was as ambitious—and as industrious—as she was, as fascinated by the natural world, as troubled by late nineteenth-century tendencies toward dichotomies and absolutes, and as determined to resist easy certainties. If Dickinson was willing to die for beauty, Darwin was equally willing to give his life for truth. Like her, he continued to see truth and beauty as related to each other, long after it had become fashionable to imagine them as enemies.

Despite the great distance between Britain and the United States and the different circumstances of their lives, Emily Dickinson and Charles Darwin were intellectual "Kinsmen." Both felt the loss of Romantic ways of thinking and mourned the growing separation of science from poetry. They worried about the disenchantment of the natural world that seemed to go along with modern thought, even as they both saw themselves as champions of unflinching truthfulness.

Of course, for both Darwin and Dickinson the idea of a posthumous conversation was fanciful but not plausible. Neither of them really believed that dead people could communicate from within their graves. They both thought that ghosts—and other supernatural phenomena—were generally metaphorical. Both were amused by ghost stories, and both used common metaphors of haunting. They were both very interested in death. At the conclusion of *On the Origin of Species*, Darwin had remarked that the most exalted forms of life rose from war, famine, and extinction. Shaped by death, he believed, "endless forms most beautiful and most wonderful have been, and are being, evolved."[38] Darwin's vision of death as a generative force was not supernatural. His perspective was profoundly naturalistic.

Dickinson's views were similar. She was interested in death as a natural phenomenon, and she understood haunting as a metaphorical way of discussing large questions about life and death that almost eluded language. In 1876 she had written to Higginson, "Nature is a Haunted House – but Art – a House that tries to be haunted."[39] When she tried to make her own work numinous with the shadows of death, she did so in order to express her perception that the natural world was a "Haunted House."

There is no question that Dickinson valued enchantment. Decades before, in "I think I was enchanted," she had praised Elizabeth Barrett Browning's poetry for being "Magic." She had asked Higginson if he could help her to learn poetry's witchcraft. As she quietly wrote hundreds of poems, she honed her own sense of witchcraft—her ability to express natural magic.

She imagined disenchantment as a terrible loss—or worse, as an act of perjury:

Whoever disenchants
A single Human soul
By failure or irreverence
Is guilty of the whole –

As guileless as a Bird
As graphic as a Star
Till Perjury – insinuates
Things are not what they are –[40]

She imagined the natural world—birds and stars, insects and grasses—as saturated with magic. To "disenchant" was to lie about the nature of existence, to insinuate that "Things are not what they are –."

Dickinson's description of disenchantment is a direct reversal of the way that late nineteenth-century literary naturalists would have explained it. The prevailing assumption at the time was that enchantment obscured the material world with a cloud of superstition and sentiment that made it impossible to see things as they really were.

Dickinson disagreed. She condemned disenchantment. She loved to feel enchanted, and she made every effort to write poems and letters that would enchant her readers. At the same time, however, she was deeply skeptical of the supernatural. She preferred the material magic of "Things"—birds, stars, plots of earth—to idealized, dematerialized concepts. She could not accede to religious or spiritual schemes that denied natural "Things" any more than she acceded to disenchantment. She refused to "give up the world."[41]

Neither Darwin nor Dickinson believed that supernatural forces were necessary for the creation of the "forms most beautiful and most wonderful" that Darwin had celebrated in *On the Origin of Species*. Dickinson's least supernatural, most Darwinian account of haunting occurs in a poem about a Brazilian moth:

> A Moth the hue of this
> Haunts Candles in Brazil –
> Nature's Experience would make
> Our Reddest Second pale –[42]

The first two lines bring Darwin to mind because they discuss South American entomology. The moth haunts the candle in a naturalistic way, regularly returning to the flame. The next lines twist the haunting metaphor away from a straightforward naturalist's description, implying something dramatic about the moth. Perhaps it is very brightly colored, or perhaps it is highly toxic—the possibilities are wide open. All the reader knows is that there is something about the moth that is much more intense and vivid than human experience. "Nature's Experience would make / Our Reddest Second pale –." The way that human beings haunt— or are haunted—pales in comparison to the way the moth haunts the candle. Humans may be appalled by the evolutionists' vision of "Nature – red in tooth and claw," but the "Reddest Second" of human experience dims in comparison to "Nature's Experience."

Dickinson's entomological account of haunting was vivid and vital. It was nothing like the disenchanted gray palette of nineteenth-century literary naturalists, but it was also completely unlike the spiritualist accounts of haunting that were becoming increasingly popular.

During the nineteenth century, performances by spiritualists and mediums were very common. Many members of Darwin's and Dickinson's social circles were interested in communicating with the spirits of the dead. At the time, many serious scientists and philosophers, including Alfred Russel Wallace, the codiscoverer of natural selection, and William James, the great pragmatist, were fascinated by psychic phenomena. When Darwin's son hosted a famous medium in London, Darwin excused himself from the ritual despite the fact that George Henry Lewes and Mary Ann Evans (the novelist known as George Eliot) were in attendance. "The Lord have mercy on us all, if we have to believe in such rubbish," he wrote.[43] As the tables flew across the drawing room downstairs, Darwin held himself apart.

Although he balked at attending séances, Darwin also hesitated about completely dismissing the possibility of some sort of afterlife. Early in their marriage, his wife, Emma, wrote of her belief that they would "belong to each other forever."[44] Darwin saved the letter for decades and left a brief note in the margin: "When I am dead, know that many times, I have kissed and cryed over this."[45] If he could not speak to Emma after death, he hoped that his written words would speak for him. He trusted paper and ink.

Darwin was right. His words would last long after his death. On the day that he was buried, Grant Allen's obituary declared, "All that remains of Charles Darwin is being consigned with fitting solemnity to its last home in Westminster Abbey; and yet not all that remains, for his voice is with us still, and will be as long as the English speech is yet spoken upon earth – ay, and longer."[46] The conviction that Darwin's voice would resonate through the ages was not necessarily connected to spiritualism, though the persistence of memory was certainly at the crux of the intersection between materialism and spiritualism. For many of his more scientifically inclined contemporaries, the concept of Darwin's voice and Darwin's ghost differed—if they did—"As Syllable from Sound."[47]

Considering Darwin's impatience with the supernatural, it seems a bit perverse that many scholars use Darwin's ghost as an organizing metaphor. Michael R. Rose's book on evolutionary biology in the modern word is titled *Darwin's Spectre* (1998). It announces, "A spectre is haunting the modern world, Darwin's Spectre, Darwinism."[48] In 2012 Rebecca Stott published *Darwin's Ghosts: In Search of the First Evolutionists*. In 2014 Banu Subramaniam issued *Ghost Stories for Darwin: The Science of Variation and the Politics of Diversity*. In 2021 Jude Piesse released *The Ghost in the Garden: In Search of Darwin's Lost Garden*. These accounts offer judicious and useful readings of Darwin, and for the most part they pick up on Darwin's own language. In several letters he described himself as a ghost wandering familiar haunts. Darwin's use of ghost metaphors was more concerned with memory and imagination than anything supernatural. The critics who invoke Darwin's ghost generally avoid the supernatural as well. Nonetheless, the metaphor does not align with dry rationalist versions of Darwin. In various ways these works push back against imagining Darwin as the great disenchanter.

Darwin took the middle path between supernaturalism and disenchantment. It would be as wrongheaded to describe him as a supernaturalist as it is to describe him as a disenchanter. He was neither. His attitude toward the natural world made space for enchantment, but the magic and mystery that he cared about were wholly natural. The critic George Levine explains, "Darwinian enchantment entails a direct and equal confrontation with the myriad otherness that constitutes the evolved world. Out of barnacles, sea-slugs, ants, worms and vegetable mould, and climbing plants, he created a sublime."[49] Yet Darwin's sense of enchantment was often hard for late nineteenth-century readers to grasp because the concept of natural magic had been lost in the great debates between science and religion. The middle path was difficult to imagine, much less to walk.

After Darwin's death, few people were able to fully comprehend his work. His enchanted view of the natural world was frequently overlooked. Dickinson was one of the few writers whose works captured the full nuance and complexity of Darwinian enchantment. However,

Dickinson's work was even harder to understand. Her poems and her prose—even her casual notes and letters—were often multivalent, open to many different interpretations. At times she couched her most trenchant observations as riddles or logic puzzles. Most of Dickinson's first readers overlooked her discussions of natural science. Those who noticed her scientific interests tended to link science to religious skepticism and overlook her sense of natural magic.

Gender surely played a role in many of the misreadings of Dickinson. The poet was often pathologized. Even Higginson, one of her best friends, sometimes went along with characterizations of Dickinson as mentally ill. In 1873, after he visited Dickinson in Amherst, he wrote to his sisters, reporting his wife's question: "Why do the insane so cling to you?"[50] In 1876, in a letter to his sister Anna, he described attending a party where he received a humorous letter purporting to be from "my partially cracked poetess."[51] These lines of Higginson's are often quoted without their full context—in both cases, he was telling his sisters about other people's responses to Dickinson. His own reactions were more complex. He certainly thought Dickinson was unconventional. In 1873 he described her as "eccentric."[52] By 1891 he remembered her as an "enigmatical" being who lived an "abnormal" life.[53] He knew that she refused to go along with many social expectations, but he may have attributed her behavior to artistic genius rather than illness.

Over the years, Dickinson's readers have diagnosed her with many hypothetical illnesses and other conditions. She has been represented as consumptive, epileptic, hypertensive, neurotic, hysterical, agoraphobic, and traumatized; as an extremely timid suffering martyr and a shockingly saucy dramatic trickster; as a vestal virgin, a polyamorous lover, and a tragic spinster; as a nun, a witch, a fierce atheist, a mild skeptic, and a profound mystic; as a girlish victim, a lonely woman, and a feminist goddess.

James Russell Lowell's great niece, the poet Amy Lowell, imagined Dickinson as a "frail little elf." In "The Sisters," Lowell described Dickinson as a self-denying obsessive:

But Emily hoarded – hoarded – only giving
Herself to cold, white paper. Starved and tortured,
She cheated her despair with games of patience
And fooled herself by winning. Frail little elf,
The lonely brain-child of a gaunt maturity,
She hung her womanhood upon a bough
And played ball with the stars – too long – too long –
The garment of herself hung on a tree
Until at last she lost even the desire
To take it down.[54]

Published in 1922, Lowell's lines gestured toward Dickinson's interests in botany, astronomy, and metaphysics. For Lowell, Dickinson's engagement with scientific ideas was a renunciation of womanhood. As Lowell phrased it, "She hung her womanhood upon a bough / and played ball with the stars." The assumption is clear: in order to "play ball" with scientists and philosophers, Dickinson needed to deny her female identity. By the early twentieth century, it had become hard to imagine women as scientists.

A century before, when Dickinson was in school, the reverse had been true: it had been hard to imagine intelligent women studying anything other than science. During Dickinson's lifetime, everything changed. When she was young, girls and women were encouraged to study the natural world. The arts and sciences worked together to elucidate nature. Scientific thinking was presumed to be not only philosophical but also fundamentally linked to existential mysteries.

By the time she died, these associations had been pried apart. Natural science had been reimagined as inherently masculine. Girls and women were increasingly discouraged or excluded from scientific work. Science was imagined as completely separate from the arts, and science was positioned outside of—or even against—the arts and the humanities. As the rhetoric of disenchantment took hold, human beings' emotional connections to the nonhuman natural world were often dismissed as superstitious or sentimental.

These changes reinforced each other. As the figure of the twentieth-century scientist emerged, he was imagined as masculine, emotionless, and disconnected from nature.

Emily Dickinson, on the other hand, was a female poet who cared passionately about the natural world. She had studied science, and she observed the natural world with exact precision, but few of her readers thought of her as a scientific thinker. Even fewer readers caught on to her interest in Darwin. Darwin's sense of enchantment meshed with Dickinson's vocation. She had worked urgently, writing poetry that gave voice to natural magic. When she died, however, no one knew the full extent of her work, much less its implications for thinking about the natural world.

It took about five years for Dickinson's poetry to catch wide public attention. At the time of her death, Dickinson had published one poem in a book and nine other poems in newspapers or magazines. Her correspondents knew that she liked to write poetry because she often included small poems with her letters. Even so, her sister, Lavinia, was shocked to discover almost two thousand manuscript poems stored in Emily's bedroom drawers. She followed her sister's request and burned almost all of Dickinson's personal papers except for the three passionate and mysterious texts known as "Master" documents. She could not burn any of the poetry.

Susan Dickinson was not as surprised as Lavinia. She had a better idea of the scale of Dickinson's work since the poet's most prolific years had been at the height of their intimacy.

Yet although Susan understood Dickinson's poetry better than anybody, she also understood Dickinson's reticence. If Lavinia had given the manuscripts to her brother's wife, Susan, would we know about them today? Or would Susan have kept them for herself, as private treasures? There is no way to know. What we do know is that Lavinia chose to give most of the papers to the other woman in Austin's life, Mabel Loomis Todd. Todd reached out to Thomas Wentworth Higginson, and with his help, she started editing. The first edition of *Poems of Emily Dickinson* edited by Todd was published in 1890. A small book with a grass-green cover, it included fewer than two hundred poems. Because the book was very well received, Todd released two more volumes of Dickinson poems and a two-volume collection of her letters. Susan was furious. The controversy flared into the public with a dispute over a piece of land that Austin had bequeathed to Mabel. Battles over the manuscripts would continue for decades. Susan was hesitant about publishing the private papers the poet had sent to her. After Susan's death, her daughter published an edition of poems from Susan's trove of papers (*A Single Hound: Poems of a Lifetime*, published in 1914). Poems and letters would be published piecemeal until the first complete edition in 1955.

Despite their bitter disagreements, all the members of Dickinson's not-always-entirely-loving league of sisters agreed that Dickinson's poetry was extraordinarily important. Mabel Loomis Todd, her daughter Millicent Todd Bingham, and Susan's daughter Martha Dickinson Bianchi all built their careers around editing and promoting Dickinson's work. They believed in it. Readers were interested in her poetry right from the start, but it would be decades before she came to be regarded as a major poet.

Although many of her earliest readers recognized that Dickinson associated poetry with magic, Susan may have been the only one who understood that Dickinson also associated nature—and the careful study of nature—with magic. Though natural science had often been associated with natural magic when Dickinson was born, the concept of natural magic had all but vanished by the time she died. By the late nineteenth century, science had come to be defined in opposition to the supernatural, and even to the emotional. Prevailing cultural assumptions had made it hard for late nineteenth-century thinkers to recognize that Darwin responded to the natural world with emotions of wonder that recalled older conceptions of natural magic. These assumptions also made readers blind to Dickinson's scientific acuity. Readers could not deny that Dickinson saw the earth as "Enchanted Ground," but they were unable to reconcile her responses to the natural world with their concept of science or scientists.

Darwin and Dickinson shared a passion for the deep mysteries that give meaning to life on our planet. Neither of them was a nihilist. Nor were they satisfied by pious dogmatisms of any sort. Instead they were both relentless in their pursuit of meaning and endlessly fascinated by the complex and interconnected ecologies that sustain the earth. They saw the natural world as the source of truth and beauty.

Years before, Dickinson had imagined two people who had dedicated their lives to beauty and truth engaged in a posthumous dialogue:

> And so, as Kinsmen, met a Night –
> We talked between the Rooms –
> Until the Moss had reached our lips –
> And covered up – Our names –[55]

In a time when human activity has wreaked frightening changes on our planet, Darwin's and Dickinson's voices can help us recover our sense of ecological meaning. In this precarious moment, their lives and works whisper to each other and to us, telling of the natural magic at the roots of our green world.

AFTERWORD

Hope Is a Strange Invention

Darwin and Dickinson in the Twenty-First Century

MORE THAN a century after their lives ended, the world of Darwin and Dickinson can be hard for us to imagine. Few people today live in close kinship with the natural world. We do not have as many plants or animals in our lives as they did. Our households may include a family pet or two, but compared to our ancestors, many humans have limited contact with other species. Outside our homes, beyond our view, the extinction of animals and plants has been dramatic.

The entangled banks of the twenty-first century host fewer birds, fewer insects, fewer worms, fewer plants. Everything is out of balance. Our oceans have sickened. The sky is full of invisible smoke, and the rain is tinged with poison. The earth's ice caps melt. Across the planet, the forests burn. Dangerous viruses leap from one species to another and proliferate on a new, global scale. A worldwide pandemic has forced everyone to adjust their expectations. Many of us are overwhelmed by a new and growing awareness that our planet is fragile. We are living through a great extinction.

The trouble started before Darwin and Dickinson were born. In their lifetimes, coal-fired steam engines drove the Wedgwood industrial pottery and the hat factories of Amherst while railways crisscrossed Britain and the United States, bellowing smoke into field and forest. Yet although Darwin and Dickinson experienced significant environmental changes caused by industrialization, it was not clear to either of them that human activity threatened the planet. Nonetheless, both were fascinated and troubled by the idea of extinction. The word *Anthropocene* had not come into use, but Dickinson and Darwin knew that the human age was just one geologic age among many.

Edward Hitchcock's fossil footprints and Charles Darwin's *Megatherium* bones were evidence of a deep geological past. The stunning changes in geological thinking in the middle of the nineteenth century expanded the time frame of the planetary story into whole ages before—and after—humans. The central tenet of nineteenth-century geology was that time extended much longer than any life-form. It was "mortifying," according to one historian writing in the *Atlantic Monthly* in 1858.[1] In a later article for the same magazine, John Fiske put on a brave face, declaring, "We care very little what becomes of the black ball of the earth, after all life has vanished from its surface."[2] Despite his bravado, Fiske's extinctionary grief was palpable.

In 1865 Joseph Hooker wrote a mournful letter to Darwin. Their friend, the geologist Hugh Falconer, had died at 56, among the first of their peers to die of ordinary causes (rheumatic disease, not an accident or infection). Hooker was deeply shocked. He railed against the paltriness of human existence, saying that humans were "contemptible," "miserable" beings whose "griefs & pains" "alone are unalloyed."[3]

Darwin responded to Hooker with an even bleaker vision: "Personal annihilation sinks in my mind into insignificance compared with the idea, or rather I presume certainty, of the sun some day cooling & we all freezing. To think of the progress of millions of years, with every continent swarming with good & enlightened men all ending in this; & with probably no fresh start until this our own planetary system has been again converted into red-hot gas. – Sic transit gloria mundi, with a vengeance."[4] His Latin phrase, *sic transit gloria mundi* (Thus the world's glory passes), expresses his intense grief at the prospect of extinction—not merely human extinction but the perishing of the earth and the "entire planetary system," which he imagines being "converted into red-hot gas."

Darwin's leap from personal annihilation to planetary annihilation offers small comfort. In this vision our friends will die, our species will go extinct, all forms of earthly life will perish, and our planet will explode and vanish. *Sic transit gloria mundi*, with a vengeance.

One possible response to the prospect of planetary extinction is bleak nihilism. As I have outlined, many of Darwin's contemporaries wrongly attributed such despair to him. But Darwin's view was longer. Even in his letter to Hooker, he imagined the explosion of the earth as a "fresh start." No extinction was final. Every change was part of an infinite wheel of changes.

Emily Dickinson understood geology and extinction the same way as Darwin. Though she had been surprised when her Valentine poem was published in 1852, its first line, "sic transit gloria mundi," would turn out to be a startlingly apt expression of her poetic purpose. Dickinson had grown up next door to Amherst College, with its astonishing collections of fossil footprints gathered from the banks of the nearby Connecticut River. Beside the Amherst Octagon lay the giant glacial erratic boulder that a merry band of geology students had pried from in front of her home. Surrounded by evidence of the vast expanse of the geological past, she believed that poetry needed to address extinction.

Though Dickinson hesitated to describe herself as a poet in much the same way as Darwin had hesitated to describe himself as a scientist, she had no such reluctance about describing her ideal poet:

This was a Poet –
It is That
Distills amazing sense
From Ordinary Meanings –
And Attar so immense

From the familiar species
That perished by the Door –[5]

Here she described a poet as one who "Distills amazing sense / From Ordinary Meanings –" and also distills "Attar so immense / From the familiar species / That perished by the Door –."[6]

Dickinson probably wrote these lines partly in response to Higginson's "Letter to a Young Contributor," published in the *Atlantic Monthly* in April 1862. He had used a similar botanical metaphor in his essay: "Literature is attar of roses, one distilled drop from a million blossoms."[7] Although Dickinson may have been inspired by Higginson, her twist of the metaphor is notable. The perfume he imagined was attar of roses, distilled from flowers. Dickinson conjured an attar of something else entirely—an attar of extinction, distilled from "the familiar species / That perished" into the fossilized footsteps that marked the rough stone doorsills of a few of her Amherst neighbors. In two astonishingly compressed lines, her poem juxtaposed the immense geological scope of perishing species with the human scale of a doorway—and called it a perfume.

Dickinson probably associated Darwin with the idea that extinction was not a far-off prospect but familiar and close to home. Indeed, it is likely that "This was a Poet –" was prompted as much by her reading about Darwin as by reading Higginson. In July, August, and October 1860, the *Atlantic Monthly* had featured Asa Gray's first three essays on "Darwin on the Origin of Species," which stressed Darwin's rhetoric of familiarity. As Gray explained, at the beginning of *On the Origin of Species* "The author takes us directly to the barn-yard and the kitchen-garden." Gray summarized, "'Variation under Domestication' dealt with familiar subjects in a natural way, and gently introduced 'Variation under Nature,' which seemed likely enough. Then follows 'Struggle for Existence,'—a principle which we experimentally know to be true and cogent, ... then follows, as naturally as one sheep follows another, the chapter on 'Natural Selection.'"[8] Gray's reviews were very likely Dickinson's first introduction to Darwin's theory of natural selection. His description accurately pointed out how *On the Origin of Species* presented evolution as unthreatening (he made it seem as soothing as counting sheep). At the same time, Gray acknowledged that natural selection was terrifying to many.

Dickinson was fascinated by the tension between the terrors of extinction and evolutionary change on the one hand and the familiarity of the animals in the barn and the plants in the garden on the other.

The fact that there were fossil footprints and glacial erratics on her doorstep and that the hills around Amherst were remnants of ancient volcanoes brought it all home to her. "This was a Poet –" declared her intention to write a new kind of poetry. She hoped to capture the piercing beauty of the ordinary, everyday, endless apocalypse in which living beings dwell.

Dickinson's description of a poet was new, in part because she faced a new historical situation. It had been easy for Erasmus Darwin and William Wordsworth to imagine that a single person could write poetry and study the natural sciences. But as the identity of "scientist" was defined, the way people imagined a "poet" changed. Increasingly, poetry was imagined as emotional and ideal, while science was imagined as detached and material. This dichotomy never really worked. In "The Poet," Emerson pushed back against it, arguing that "the poet alone knows astronomy, chemistry, vegetation and animation, for he does not stop at these facts, but employs them as signs."[9] Dickinson built on Emerson, but she imagined "a Poet" differently. Where Emerson thought that poetry should elevate the material world up toward the transcendent symbolic world of ideals, she thought that poetry should press things together, wringing a strange new perfume from the dust of perished species.

Throughout her career, Dickinson questioned the growing separation between scientific and spiritual views of the natural world. In 1861, in a letter to Sam Bowles, she wrote:

"Faith" is a fine invention
For Gentlemen who *see*!
But Microscopes are prudent
In an Emergency!derbysta[10]

Many readers have interpreted this short poem as Dickinson's declaration that she preferred science over religion. However, the poem is a little trickier than it first appears. Dickinson never liked to choose. She enjoyed mixed messages. "'Faith' is a fine invention" gently pushes against religious belief by describing it as invented, but it does not reject faith altogether. Instead, it suggests that gentlemen would be wise to use tools of science—microscopes—alongside tools of religion—such as faith—when they are faced with emergencies.

For Dickinson, "emergency" implied urgent and surprising danger. It may also have contained hints of birth, the emergence of new forms of life. The threat of sudden death and the possibility of new growth are bound together in the idea of "emergency."

As we face the great extinction, it may help us to see that this is also a moment when something new is beginning to emerge. In the nineteenth century, Dickinson did not intend to invoke the climate emergency that envelops us now. Nonetheless, her gently ironic reconciliation of religion and science may be helpful in our current crisis. There is no reason to discard either of them, but we must also acknowledge that even together, science and religion are not enough.

More than fifteen years after she wrote "'Faith' is a fine invention," in 1877 Dickinson wrote a second poem that echoed and expanded it. She sent this new alternative to Susan Dickinson and Thomas Wentworth Higginson. In place of "Faith," Dickinson's later poem proposed "Hope."

> Hope is a strange invention –
> A Patent of the Heart –
> In unremitting action
> Yet never wearing out –
>
> Of this electric adjunct
> Not anything is known
> But its unique momentum
> Embellish all we own –[11]

Dickinson's "Hope" resembles the "Faith" in her earlier poem in that it is an "invention." But in the later variation, she does not even hint at an opposition between hope and science. Instead, Dickinson invites us into the mystery of a living heartbeat "in unremitting action / Yet never wearing out –." Individual hearts may stop beating, but the mysterious electricity that gives momentum to each one of us also connects us—both to other humans and to the world of nonhuman beings.

The energy that drives Dickinson's poetry, like that of Darwin's thought, is both scientific and profoundly mysterious. Dickinson's work connects us to the mystery and the hope—the magic—at the heart of Darwinian science. Together, these unlikely kinsmen insist that the living world can be a source of deep meaning in part because it has grown from an extinctionary past.

Emily Dickinson and Charles Darwin invite us to invent a new kind of hope for our strange moment of planetary despair. Our task is to allow the mysterious electric spark at the heart of all beings to rekindle our sense of kinship and interconnection. Let us hope for a new birth of natural magic.

ACKNOWLEDGMENTS

BOOKMAKING IS a magical art. If there is a spark of magic in these pages, it has been kindled and kept alive by many collaborators.

My superstar agent Lucy Cleland is the reader everyone dreams of. She finds the molten heart of the crustiest prose and explains how to make it glow. At Princeton University Press, editor Anne Savarese is another bright star. I am grateful to her for her careful reading, her fantastic suggestions, and her faith in the book. Without Anne, *Natural Magic* would never have made it beyond the first awkward draft. James Collier, assistant editor, stepped in to help wrangle permissions and illustrations—daunting tasks that he approached with grace and good humor. I am also grateful to the rest of the team at Princeton University Press, including Elizabeth Byrd (production editor), Heather Hansen (designer), David Campbell (copywriter), Erin Suydam (production manager), Alyssa Sanford (U.S. publicist), and Carmen Jimenez (U.K. publicist). At Westchester Publishing Services, John Donohue handled the details of production along with Wendy Lawrence (copyeditor), Erica Van Varick (proofreader), and Ælfwine Mischler (indexer). My botanical ally and fact-checker, Peter Grima of the Massachusetts Department of Conservation, prevented many scientific gaffes. It has been an honor to work with this all-star team. It has also been surprisingly fun. Even after they are printed, these pages will emit a faint happy glow.

My students and colleagues at Simmons University have contributed to my work and my well-being in an infinite variety of ways. I am particularly grateful to my colleagues in literature and writing and the humanities over the years. I have too many friends at Simmons to name.

I continue to treasure the ones who got away, leaving the university during or shortly after the pandemic—Pamela Bromberg, Lydia Fash, Audrey Golden, and Richard Wollman. This project was supported by a grant from the Simmons Fund for Research and a number of Faculty Development Grants. Without support from Simmons, I could never have dreamed of *Natural Magic*.

Dartmouth College has been my research home during these years. Dartmouth reference librarians and their amazing collections have been an invaluable resource. During the darkest days of the pandemic, the masked librarians at Baker were a source of hope and courage along with books.

I am also grateful to the Emily Dickinson International Society (EDIS). Marianne Noble and Dan Manheim welcomed me to EDIS, and it has turned out to be a constant source of scholarly inspiration and personal connection. A "Scholar in Amherst" award from EDIS helped me get started on Dickinson. Martha Nell Smith offered endless encouragement, and Vivian Pollak offered much-needed practical advice. At the Emily Dickinson Museum, Jane Wald and Brooke Steinhauser have been consistently welcoming and supportive. Back in 2019, I had the chance to spend a morning working in Dickinson's room at the Homestead. In just a few hours, the poet came clear to me. At Amherst College in 2022, Michael Elliott showed me around the (possibly haunted) house where Edward and Orra Hitchcock preceded him. Also at Amherst, Karen Sanchez-Eppler and Benigno Sanchez-Eppler have offered hospitality and comradeship for many years. In the summer of 2023, Benigno stood guard in the moonlight while Karen led me and Li-hsin Hsu into the silver shadows deep within a field of flowers. I hope these pages retain a little moondust from that enchanted evening.

In my bibliographic note, I mention many scholars whose works are important to this project. Though most of them are strangers to me, I am so familiar with their ideas that some feel like friends. I am particularly grateful to Richard Holmes for rediscovering the wonder at the heart of Romantic science, Jane Bennett for revitalizing enchantment,

George Levine for reimagining Darwin, and Laura Dassow Walls for making the nineteenth century green again.

Of all my scholarly friends, a few were vital collaborators on other projects while I worked on this book. I am grateful to Betsy Klimasmith and Len von Morzé; Ivy Schweitzer and Tom Luxon; Christian Haines and Mark Noble; and Cristanne Miller and Karen Sanchez-Eppler. In another context, Karen Sanchez-Eppler, Marianne Noble, Caroline Levander, and Christopher Breu read and responded to a different manuscript. Their ideas helped me to structure this book.

I wrote this book in Hanover, New Hampshire, and I am grateful to my friends here. At Saint Thomas Episcopal Church, I have found a community of fellow doubters and a deep well of hope. I am particularly grateful to Guy Collins, Alison Chisholm, and Jack Sammons. I would not survive without the comradeship and nourishment of the Hanover Co-op. My neighbors on Barrett Road have been patient and supportive as I ripped up the lawn and attempted to grow some of the plants and flowers that Dickinson and Darwin wrote about. I feel incredibly lucky to have such great neighbors—and to have cardinal flowers out back, sweet peas in the side yard, and a tangle of tall grass and clover out front by the roses. The birds and the bees are awesome, too.

It is hard to imagine Dickinson and Darwin without their dogs. Similarly, my canine pack has been an integral part of my writing process. Bamse was our pandemic puppy; now he is our boon companion. His doggy relations—Olaf, Nonconnah, Rosie, Pippa, and Popoki—bring chaos and joy in equal measure.

Finally, I am grateful to my human pack. Betsy and David Klimasmith are a constant source of fun and fellowship. Their home is always welcoming and their breakfast conversation is unfailingly delightful. My aunt Martha Lamar has inspired me for my whole life with her intelligence, her good taste, and her warm kindness. My sister Sandy Bergland is a badass aviator and a truly generous person. My brother Chris Bergland is a fellow writer with a remarkable gift of empathy. My mother Mary Jo Litchard is an artist and a storyteller whose spirit of fun infuses our family life.

My biggest thanks and my deepest love go to Kim Brinck-Johnsen, my husband, and Annelise Bergland Brinck-Johnsen, our daughter. The two of them have read drafts, chased down references, and spent hours helping with permissions. They have given me courage and practical help when I needed them. My love and gratitude go far beyond the pages of this or any book. They are the ones who make me believe in magic.

NOTES

Preface

1. Dickinson, "The Gentian weaves her fringes –" (Emily Dickinson Archive: An Open-Access Website for the Manuscripts of Emily Dickinson, edickinson.org; hereafter cited as Emily Dickinson Archive).
2. Dickinson, "A sepal – petal – and a thorn –" (Emily Dickinson Archive).
3. Dickinson, "The Gentian weaves her fringes –" (Emily Dickinson Archive).
4. Dickinson, "In the name of the Bee –" (Emily Dickinson Archive).
5. Dickinson, "To him who keeps an Orchis' heart –" (Emily Dickinson Archive).
6. Alcock, *Enthusiasm for Orchids*, vii; Browne, *Charles Darwin: Power of Place*, 165.
7. Charles Darwin to A. G. More, June 2, 1861, Darwin Correspondence Project, "Letter no. 3174."
8. Charles Darwin to Joseph Dalton Hooker, March 26, 1863, Darwin Correspondence Project, "Letter no. 4061."
9. Charles Darwin to Charles Lyell, June 18, 1858, Darwin Correspondence Project, "Letter no. 2285."
10. Dickinson, "To him who keeps an Orchis' heart –" (Emily Dickinson Archive).

Introduction

1. Shelley, *Frankenstein*, 47.
2. Weber, *Charisma and Disenchantment*, 18. In German, Weber's term for disenchantment was *Entzauberung*. The root word, *zauber*, is usually translated "magic." See 131n23.
3. Weber, *Charisma and Disenchantment*, 18.
4. Weber, *Charisma and Disenchantment*, 18.
5. Levine, *Darwin Loves You*, 238.
6. Darwin, *Autobiography*, 138.
7. Darwin, *Autobiography*, 139.
8. Charles Darwin to Joseph Dalton Hooker, June 17, 1868, Darwin Correspondence Project, "Letter no. 6248."

9. Darwin, *Origin of Species* (1859), 490.

10. Darwin, *Descent of Man*, 2:402.

11. Browne, "Botany for Gentlemen," 594.

12. Emily Dickinson to Thomas Wentworth Higginson, July 1862, Johnson Letter 268, in *The Letters of Emily Dickinson*, edited by Johnson (hereafter cited as JL, followed by the letter number).

13. Fuller, *Book That Changed America*, 56–58.

14. Higginson, "Life of Birds," *Atlantic Monthly*, September 1862, in *Out-door Papers*, 295–296.

15. Dickinson, "By my Window have I for Scenery" (Emily Dickinson Archive).

16. Darwin, *Autobiography*, 139.

17. Tolley, *Science Education of American Girls*, 8.

18. Susan Dickinson, "Obituary of Emily Dickinson."

19. Susan Dickinson, "Obituary of Emily Dickinson."

20. Josephson Storm, *Myth of Disenchantment*, 16.

21. James, "Is Life Worth Living?," 10.

22. See Knapp, *William James*.

23. Copenhaver, *Book of Magic*, 372.

24. Copenhaver, *Book of Magic*, 515.

25. Porta, *Natural Magick*, title page.

26. Brewster, *Letters on Natural Magic*, 2.

27. Brewster, *Letters on Natural Magic*, 2.

28. Brewster, *Letters on Natural Magic*, 297.

29. The Library of Congress includes a copy of *Ventriloquism Explained* in its Harry Houdini Collection. Amherst College Library has multiple editions, including one inscribed to Edward Hitchcock (implying that he is not the anonymous author but merely the "instructor" to whom the book is addressed).

30. *Ventriloquism Explained*, x.

31. *Ventriloquism Explained*, 146.

32. Browne, *Charles Darwin: Voyaging*, 206.

33. *Ventriloquism Explained*, 13.

34. *Ventriloquism Explained*, 13.

35. *Ventriloquism Explained*, 10.

36. Josephson Storm, *Myth of Disenchantment*, 87. Josephson Storm borrows the notion of the "Romantic Spiral" from M. H. Abrams, *Natural Supernaturalism*.

37. See, for example, Bartlett, *Natural and the Supernatural in the Middle Ages*; Copenhaver, *Magic in Western Culture*; Cowles, *Scientific Method*; Gosden, *Magic*; Hunter, *Decline of Magic*; and Wootton, *Invention of Science*.

38. Gosden, *Magic*, 13.

39. Gosden, *Magic*, 13.

40. Bennett, *Enchantment of Modern Life*, 4, 5. See also Levine, *Reading Thomas Hardy*, 32–33.

Chapter 1

1. In the historical record, the spelling of Darwin's mother's first name varies. I follow Janet Browne's example in using "Susanna" rather than "Susannah." In *The Radical Potter*, Tristram Hunt uses "Sukey," the affectionate nickname that Josiah Wedgwood used in his letters to his daughter.
2. Temple, "Portrait of a Boy and His Plant."
3. Temple, "Portrait of a Boy and His Plant."
4. Leyda, *Years and Hours of Emily Dickinson*, 1:58.
5. Leyda, *Years and Hours of Emily Dickinson*, 1:59.
6. Leyda, *Years and Hours of Emily Dickinson*, 1:61.
7. Wingate, "Otis Bullard." See also West, "City in Frames."
8. Massachusetts forester Peter Grima explains, "The fig tree cultivation was likely also aided by hired help, since a phenomenal amount of digging is involved to successfully overwinter a fig tree in New England. My own grandfather and great-grandfather used to do so, in southeastern Massachusetts, which involved digging a huge hole undercutting half of the tree and laying the whole tree into the hole, where it was buried in soil to overwinter. Then in spring the whole process was reversed. While half of the root system was thus traumatized, the invaluable half that remained anchored in the soil was the secret to yielding figs in what was then a cooler climate than today, even in living memory." Grima, personal correspondence, August 16, 2023.
9. Quoted in Hunt, *Radical Potter*, 190.
10. In fact, the ancient pottery they thought of as "Etruscan" had been imported to the Roman provinces from Greece.
11. Uglow, *Lunar Men*, xx.
12. Hunt, *Radical Potter*, 189.
13. Hunt, *Radical Potter*, xix.
14. Uglow, *Lunar Men*, xiv.
15. Browne, "Botany for Gentlemen," 597. See also Linné, *Families of Plants*.
16. In 1778, Erasmus Darwin's neighbor, the celebrated poet Anna Seward, gave him a manuscript of her poem "Verses Written in Dr. Darwin's Botanic Garden." Taking her poem as a starting point, Erasmus framed himself as a "Poetical Genius" inspired by a "Botanic Muse" to explain his view of history. Seward did not see herself as a Botanic Muse. She was offended by Darwin's appropriation of her work.
17. Darwin, *Botanic Garden*, 2:6.
18. Darwin, *Botanic Garden*, 2:13.
19. Darwin, *Botanic Garden*, 2:201.
20. Darwin, *Botanic Garden*, 1:94.
21. Richards, *Romantic Conception*, 516.
22. Wordsworth, "Tables Turned."
23. Wordsworth, "I Wandered Lonely as a Cloud."
24. See Kirkby, "'We Thought Darwin,'" 5.

25. "Darwin possesses the *epidermis* of poetry but not the *cutis*; the *cortex* without the *liber, alburnum, lignum,* or *medulla*." Coleridge, *Anima Poetae*, 280, cited and translated by Stott, *Darwin and the Barnacle*, 135.

26. Keats, "Ode on a Grecian Urn."

27. Erasmus Darwin, *Zoonomia*, 2:247. The *Oxford English Dictionary* entry on "evolution" uses this passage to illustrate the very different implications of the term before the mid-nineteenth century.

28. Shelley, *Frankenstein*, 8–9.

29. Leighton, [Recollections of Charles Darwin].

30. Dickinson, "'Arcturus' is his other name –" (Emily Dickinson Archive).

31. Peter Grima reports, "There are nearly identical [roses] still growing at the Homestead, an heirloom variety known as Blush Noisette, which is believed to be persisting from Emily's lifetime." Grima, personal correspondence, August 16, 2023.

32. Emily Dickinson to Thomas Wentworth Higginson, April 25, 1862, JL 261.

33. Phelps, *Familiar Lectures on Botany*, 12.

34. Samuel Fowler Dickinson, address given to the Hampshire, Hampden, and Franklin Agricultural Society on October 27, 1831, in Northampton, Massachusetts, cited in Leyda, *Years and Hours of Emily Dickinson*, 1:17–18.

35. Phelps, *Familiar Lectures on Botany*, 12.

36. Wainwright, "Jonathan Edwards."

37. Edwards and Miller, *Images or Shadows*. Edwards's manuscript essays date from the mid-eighteenth century. They were edited by Miller and first published in 1948.

38. Dickinson, "Flowers – Well – if anybody" (Emily Dickinson Archive).

39. Darwin, *Autobiography*, 139.

Chapter 2

1. Charles Darwin to Caroline Darwin, September 20, 1833, Darwin Correspondence Project, "Letter no. 215."

2. Darwin, *Autobiography*, 28–29.

3. Browne, *Charles Darwin: Voyaging*, 10.

4. Samuel Taylor Coleridge to Samuel Purkis, February 1, 1803, in Coleridge, *Collected Letters*, 2: 919.

5. Samuel Taylor Coleridge to John Thewall, May 13, 1796, in Coleridge, *Collected Letters*, 1:216.

6. Darwin, *Autobiography*, 23.

7. Darwin, *Life of Erasmus Darwin*, 14.

8. Darwin, *Life of Erasmus Darwin*, 14.

9. Darwin, *Autobiography*, 22.

10. Piesse, *Ghost in the Garden*, 41–77.

11. Browne, *Charles Darwin: Voyaging*, 14.

12. Cowburn, *Salopian Anthology*, quoted in Browne, *Charles Darwin: Voyaging*, 24.

13. Darwin, *Autobiography*, 27.

14. Charles Darwin to "Friend," January 4, 1822, Darwin Correspondence Project, "Letter no. 1K."

NOTES TO CHAPTER 2   367

15. Charles Darwin to "Friend," January 4, 1822, Darwin Correspondence Project, "Letter no. 1K."

16. Darwin, *Autobiography*, 43.

17. "History of the School," School of Clinical Medicine, University of Cambridge, December 6, 2021, https://www.medschl.cam.ac.uk/about/history-of-the-school/.

18. Charles Darwin to Robert Waring Darwin, October 23, 1825, Darwin Correspondence Project, "Letter no. 16."

19. Darwin, *Autobiography*, 47.

20. Thomson, *Young Charles Darwin*, 35.

21. Darwin, *Autobiography*, 48.

22. Darwin, *Life of Erasmus Darwin*, 11–12.

23. Wordsworth, "Tables Turned."

24. Erasmus Darwin, *Zoonomia*, 482–537.

25. Seward, quoted in King-Hele, "Introduction," in Darwin, *Life of Erasmus Darwin*, xiii.

26. Erasmus Darwin, *Zoonomia*, 502.

27. Erasmus Darwin, *Temple of Nature*.

28. Coleridge, *Collected Works*, 9:353.

29. Charles Darwin to Caroline Darwin, April 8, 1826, Darwin Correspondence Project, "Letter no. 30."

30. Thomson, *Young Charles Darwin*, 61; Stott, *Darwin and the Barnacle*, 35.

31. Darwin, "Notebook DAR118," quoted in Thomson, *Young Charles Darwin*, 66.

32. Darwin, "Notebook DAR118," quoted in Stott, *Darwin and the Barnacle*, 36.

33. Darwin, *Autobiography*, 52.

34. Darwin, *Autobiography*, 51.

35. Darwin, *Autobiography*, 28.

36. Darwin, *Autobiography*, 68.

37. Darwin, *Autobiography*, 58.

38. Beer, *Darwin's Plots*, 31–33; Richards, *Romantic Conception*, 537–538.

39. Thomson, *Young Charles Darwin*, 63.

40. Darwin, *Autobiography*, 68.

41. Charles Darwin to Caroline Darwin, April 28, 1831. Darwin Correspondence Project, "Letter no. 98."

42. Charles Darwin to John Stevens Henslow, July 11, 1831, Darwin Correspondence Project, "Letter no. 102."

43. Charles Darwin to John Stevens Henslow, July 11, 1831, Darwin Correspondence Project, "Letter no. 102."

44. Charles Darwin to Susan Darwin, September 9, 1831, Darwin Correspondence Project, "Letter no. 122."

45. Darwin, *Charles Darwin's Beagle Diary*, 19.

46. Charles Darwin to Robert Waring Darwin, March 1, 1832, Darwin Correspondence Project, "Letter no. 158." See Dickinson, "Ah Teneriffe – Receding Mountain –" (Emily Dickinson Archive).

47. Charles Darwin to John Stevens Henslow, August 15, 1832, Darwin Correspondence Project, "Letter no. 178."

Chapter 3

1. Emily Dickinson, "A narrow Fellow in the Grass" (Emily Dickinson Archive).
2. Emily Dickinson to Thomas Wentworth Higginson, August 1862, JL 271.
3. Darwin, *Autobiography*, 23.
4. Edward Dickinson to his daughter Emily Dickinson, January 5, 1838, in Leyda, *Years and Hours of Emily Dickinson*, 1:39.
5. Martineau, *Retrospect of Western Travel*, 84–85.
6. Gallenga, *Episodes of My Second Life*, 123.
7. Dickinson, "A narrow Fellow in the Grass" (Emily Dickinson Archive).
8. Habegger, *My Wars Are Laid Away in Books*, 160. Later, in a letter to her nephew, Dickinson sent a quotation from Shakespeare that used masculine pronouns in reference to herself. She remarked, "Excuse the bearded Pronoun." Emily Dickinson to Ned Dickinson, late 1885, JL 1026.
9. Emily Dickinson to Thomas Wentworth Higginson, July 1862, JL 268.
10. Dickinson, "Nature and God – I neither knew" (Emily Dickinson Archive). Enclosed in a letter to Samuel Bowles, 1864.
11. Herschel, *Preliminary Discourse*, 5.
12. Herschel, *Preliminary Discourse*, 6.
13. Charles Darwin to William Darwin Fox, February 15, 1831, Darwin Correspondence Project, "Letter no. 94."
14. Darwin, *Autobiography*, 68.
15. Herschel, *Preliminary Discourse*, 12.
16. Herschel, *Preliminary Discourse*, 14.
17. See Walls, *Emerson's Life in Science*, 54–62.
18. Richardson, *Emerson*, 124.
19. Quoted in Richardson, *Emerson*, 124. See my discussion of Somerville in chapter 4.
20. Quoted in Walls, *Emerson's Life in Science*, 84.
21. Emerson, *English Traits*, 46.
22. Buell, *Emerson*, 8.
23. Buell, *Emerson*, 8.
24. Hitchcock, *Geology of Massachusetts*, 236.
25. Emerson, "Nature."
26. Quoted in Walls, *Emerson's Life in Science*, 98.
27. Emerson, "Nature."
28. Hitchcock, *Geology of Massachusetts*, 113.
29. Hitchcock, *Geology of Massachusetts*, 113.
30. Marché, "Edward Hitchcock's Poem," 5.
31. Marché, "Edward Hitchcock's Poem," 6–7.
32. Marché, "Edward Hitchcock's Poem," 7.
33. Todd, *Letters of Emily Dickinson*, 127–128, quoted in Habegger, *My Wars Are Laid Away in Books*, 163.
34. Herschel, *Preliminary Discourse*, 13.
35. Herschel, *Preliminary Discourse*, 11.

36. Dickinson, "Wonder – is not precisely knowing" (Emily Dickinson Archive).
37. Herschel, *Preliminary Discourse*, 12.
38. Dickinson, "Eden is that old fashioned House" (Emily Dickinson Archive).
39. *History of the Town of Amherst, Massachusetts*, 187.
40. *Reminiscences of Amherst College*, 5.
41. Quoted in Walls, *Emerson's Life in Science*, 98.

Chapter 4

1. Charles Darwin to William Darwin Fox, February 15, 1836, Darwin Correspondence Project, "Letter no. 299."
2. Darwin, *Autobiography*, 101.
3. Browne, *Charles Darwin: Voyaging*, 194.
4. Lyell, *Principles of Geology*, 1.
5. Lyell, *Principles of Geology*, 3.
6. Darwin, *Autobiography*, 101.
7. Pearson and Nicholas, "Charles Darwin's Geological Observations."
8. Darwin, *Charles Darwin's Beagle Diary*, January 23, 1832, Saint Jago, Darwin Online.
9. Alexander von Humboldt to Varnhagen von Ense, Berlin, October 27, 1834, in Humboldt, *Letters of Alexander von Humboldt*, 15–16.
10. Emerson, "Humboldt," in *Complete Works*, 11:457.
11. Darwin, *Charles Darwin's Beagle Diary*, January 23, 1832, Saint Jago.
12. Charles Darwin to John Stevens Henslow, May 18 to June 16, 1832, Darwin Correspondence Project, "Letter no. 171."
13. Charles Darwin to Robert Waring Darwin, February 8 to March 1, 1832, Darwin Correspondence Project, "Letter no. 158."
14. Darwin, *Charles Darwin's Beagle Diary*, February 6, 1832, Saint Jago, Darwin Online.
15. Charles Darwin to Willian Darwin Fox, May 1832, Darwin Correspondence Project, "Letter no. 168."
16. Darwin, *Charles Darwin's Beagle Diary*, February 28, 1832, Bahia, Darwin Online.
17. Wordsworth, "Observations Prefixed to *Lyrical Ballads*."
18. Quoted in Browne, *Charles Darwin: Voyaging*, 189.
19. Charles Darwin to John Stevens Henslow, May 18 to June 16, 1832, Darwin Correspondence Project, "Letter no. 171."
20. Charles Darwin to William Darwin Fox, August 9–12, 1835, Darwin Correspondence Project, "Letter no. 282."
21. Darwin, *Autobiography*, 81.
22. Charles Darwin to Catherine Darwin, May 22 to July 14, 1833, Darwin Correspondence Project, "Letter no. 206."
23. Darwin, *Voyage of the Beagle*, 47.
24. Darwin, *Voyage of the Beagle*, 47.
25. Darwin, *Voyage of the Beagle*, 40.

26. Charles Darwin to Catherine Darwin, May 22 to July 14, 1833, Darwin Correspondence Project, "Letter no. 206."
27. Lister, *Darwin's Fossils*, 40.
28. Darwin, *Voyage of the Beagle*, 46.
29. Charles Darwin to Catherine Darwin, May 22 to July 14, 1833, Darwin Correspondence Project, "Letter no. 206."
30. Darwin, *Autobiography*, 28.
31. Charles Darwin to Catherine Darwin, May 22 to July 14, 1833, Darwin Correspondence Project, "Letter no. 206."
32. Charles Darwin to Catherine Darwin, May 22 to July 14, 1833, Darwin Correspondence Project, "Letter no. 206."
33. Herschel, *Preliminary Discourse*, 115.
34. Herschel, *Preliminary Discourse*, 263.
35. John Stevens Henslow to Charles Darwin, August 31, 1833, Darwin Correspondence Project, "Letter no. 213."
36. Lister, *Darwin's Fossils*, 40.
37. Carlyle, "Signs of the Times," 37.
38. Snyder, *Philosophical Breakfast Club*, 1–7.
39. Whewell, quoted in Costa, *Darwin's Backyard*, 12.
40. Charles Darwin to Catherine Darwin, May 22 to July 14, 1833, Darwin Correspondence Project, "Letter no. 206."
41. Seward, quoted in King-Hele, "Introduction," in Darwin, *Life of Erasmus Darwin*, xiii.
42. Charles Darwin to Catherine Darwin, May 22 to July 14, 1833, Darwin Correspondence Project, "Letter no. 206."
43. Darwin, *Autobiography*, 78.
44. Charles Darwin to Caroline Darwin, April 29, 1836, Darwin Correspondence Project, "Letter no. 301."
45. Charles Darwin to Caroline Darwin, April 29, 1836, Darwin Correspondence Project, "Letter no. 301."
46. Charles Darwin to John Stevens Henslow, July 9, 1836, Darwin Correspondence Project, "Letter no. 304."
47. Charles Darwin to Catherine Darwin, June 3, 1836, Darwin Correspondence Project, "Letter no. 302."
48. Charles Darwin to John Stevens Henslow, July 9, 1836, Darwin Correspondence Project, "Letter no. 304."
49. Holmes, *Age of Wonder*, 462.
50. Holmes, *Age of Wonder*, 462.
51. Darwin, *Autobiography*, 85.
52. Charles Darwin to Catherine Darwin, May 22 to July 14, 1833, Darwin Correspondence Project, "Letter no. 206."

Chapter 5

1. Emily Dickinson to Thomas Wentworth Higginson, Spring 1876 (commenting on the experience of reading *Out-door Papers*, the volume that contained "A Procession of the Flowers"), JL 458.
2. Dickinson, "Prose Fragment 117," *Letters*, Johnson edition, 3:928.
3. Dickinson, "Of Death I try to think like this," (Emily Dickinson Archive). Dickinson often preserved copies of her poems that included variant words and lines. Because the poet indicated that both variants were valid, some twenty-first-century scholars include the variants when they transcribe the poems.
4. Emily Dickinson to Abiah Root, May 7, 1845, JL 6.
5. Edward Dickinson to Emily Dickinson, January 5, 1838 (enclosed in a letter to the poet's mother, Emily Norcross Dickinson), in Leyda, *Years and Hours of Emily Dickinson*, 1:39.
6. Dickinson, "It was given to me by the Gods –" (Emily Dickinson Archive).
7. Dickinson, "I taste a liquor never brewed –" (Emily Dickinson Archive).
8. Emily Dickinson to Abiah Root, May 7, 1845, JL 6.
9. Emily Fowler Ford to Mabel Loomis Todd, *Letters of Emily Dickinson*, Todd edition, 125.
10. Habegger, *My Wars Are Laid Away in Books*, 166.
11. Ackmann, *These Fevered Days*, 25.
12. Dickinson, "Ah, Teneriffe – Receding Mountain –" (Emily Dickinson Archive). In the first chapter of *The Voyage of the Beagle*, Darwin writes: "On the 6th of January we reached Teneriffe, but were prevented landing, by fears of our bringing the cholera: the next morning we saw the sun rise behind the rugged outline of the Grand Canary island, and suddenly illuminate the Peak of Teneriffe, whilst the lower parts were veiled in fleecy clouds. This was the first of many delightful days never to be forgotten." Darwin, *Voyage of the Beagle*, 1.
13. Darwin, *Voyage of the Beagle*, 1.
14. Emily Dickinson to Abiah Root, August 3, 1845, JL 7.
15. Darwin, *Autobiography*, 46.
16. Habegger, *My Wars Are Laid Away in Books*, 142.
17. Habegger, *My Wars Are Laid Away in Books*, 141.
18. Habegger, *My Wars Are Laid Away in Books*, 152.
19. Emily Dickinson to Abiah Root, March 14, 1847, JL 15.
20. Bernhard, "Portrait of a Family," 371.
21. Edward Dickinson to Emily Norcross, June 4, 1826, quoted in Bernhard, "Portrait of a Family," 374.
22. Bernhard, "Portrait of a Family," 374–375.
23. Habegger, *My Wars Are Laid Away in Books*, 33.
24. Emily Dickinson to Abiah Root, May 7, 1845, JL 6.
25. Dickinson, *Herbarium*.
26. Lincoln, *Familiar Lectures on Botany*, 28, 30. Subsequent editions of *Familiar Lectures* were published under the author's married name, Almira Hart Lincoln Phelps.
27. Erasmus Darwin, *Loves of the Plants*, cited in Browne, "Botany for Gentlemen," 612.
28. Browne, "Botany for Gentlemen," 614.

29. Emily Dickinson to Abiah Root, May 7, 1845, JL 6.
30. Ackmann, *These Fevered Days*, 18.
31. Emily Dickinson to Abiah Root, May 7, 1845, JL 6.
32. Emily Dickinson to Abiah Root, May 7, 1845, JL 6.
33. Emily Dickinson to Abiah Root, May 7, 1845, JL 6.
34. Emily Dickinson to Thomas Wentworth Higginson, April 25, 1862, JL 261.
35. Edward Dickinson to Emily Norcross (Dickinson), 1826, in Pollak, *Poet's Parents*, 35.
36. Emily Dickinson to Abiah Root, February 23, 1845, JL 5.
37. [Edward Hitchcock], *Power of Christian Benevolence*, 344.

Chapter 6

1. Darwin, *Autobiography*, 81.
2. Charles Lyell to Adam Sedgwick, December 6, 1835, quoted in Browne, *Charles Darwin: Voyaging*, 336.
3. Browne, *Charles Darwin: Voyaging*, 336.
4. Adam Sedgwick to Samuel Butler, November 7, 1835, in Hughes, *Life and Letters of the Reverend Adam Sedgwick*, 360, quoted in Browne, *Charles Darwin: Voyaging*, 336.
5. Darwin, *Autobiography*, 82.
6. Holmes "In Retrospect," 432.
7. Charles Darwin to John Stevens Henslow, July 9, 1836, Darwin Correspondence Project, "Letter no. 304."
8. Darwin, *Autobiography*, 84.
9. Darwin, *Autobiography*, 82.
10. Darwin, *Charles Darwin's Beagle Diary*, 418.
11. Charles Lyell to John Herschel, May 24, 1837, quoted in Stoddart, "Darwin, Lyell," 206.
12. Charles Lyell to Adam Sedgwick, April 21, 1837, in Hughes, *Life and Letters*.
13. Darwin, *Autobiography*, 82.
14. Darwin, *Autobiography*, 83.
15. Darwin, *Autobiography*, 119.
16. Caroline Darwin to Charles Darwin, October 28, 1833, Darwin Correspondence Project, "Letter no. 224."
17. Charles Darwin to Susan Darwin, May 15, 1838, Darwin Correspondence Project, "Letter no. 413."
18. Charles Darwin to Susan Darwin, May 15, 1838, Darwin Correspondence Project, "Letter no. 413."
19. Charles Darwin to Susan Darwin, April 1, 1838, Darwin Correspondence Project, "Letter no. 407."
20. Charles Darwin to Susan Darwin, April 1, 1838, Darwin Correspondence Project, "Letter no. 407."
21. Charles Darwin to Caroline Darwin, November 9, 1836, Darwin Correspondence Project, "Letter no. 321."
22. Charles Darwin to Caroline Darwin, December 7, 1836, Darwin Correspondence Project, "Letter no. 325."

23. Charles Darwin to Susan Darwin, April 1, 1838, Darwin Correspondence Project, "Letter no. 407."
24. Charles Darwin to Susan Darwin, April 1, 1838, Darwin Correspondence Project, "Letter no. 407."
25. Charles Darwin to Susan Darwin, April 1, 1838, Darwin Correspondence Project, "Letter no. 407."
26. Darwin, *Autobiography*, 120.
27. Charles Darwin to Susan Darwin, May 15, 1838, Darwin Correspondence Project, "Letter no. 413."
28. Emma Wedgwood to Charles Darwin, December 29, 1838, Darwin Correspondence Project, "Letter no. 464."
29. Martineau, *Retrospect of Western Travel*, 85.
30. Martineau, *Retrospect of Western Travel*, 84.
31. Emma Wedgwood to Jessie Sismondi, November 15, 1838, in Litchfield, *Century of Family Letters*, 5–6.
32. Darwin, *Autobiography*, 99–100.
33. Emma Wedgwood Darwin to Elizabeth Wedgwood, April 2, 1839, in Litchfield, *Century of Family Letters*, 40–41.
34. Lyell, *Travels in North America*, 200.
35. Charles Darwin to Edward Hitchcock, November 6, 1845, Darwin Correspondence Project, "Letter no. 925."
36. Dickinson, "A science – so the Savans Say" (Emily Dickinson Archive).

Chapter 7

1. Emily Dickinson to Abiah Root, November 6, 1847, JL 18.
2. Lucy Goodale to her parents, 1838, quoted in Shmurak, "Castle of Science," 315.
3. Bingham, *Emily Dickinson's Home*, 74.
4. Emily Dickinson to Abiah Root, May 16, 1848, JL 23.
5. Kuhn, "Dickinson and the Politics of Plant Sensibility," 153.
6. Kuhn, "Dickinson and the Politics of Plant Sensibility," 153.
7. Emily Dickinson to Austin Dickinson, February 17, 1848, JL 22.
8. Leyda, *Years and Hours of Emily Dickinson*, 123.
9. Emily Dickinson to Austin Dickinson, February 17, 1848, JL 22.
10. Bingham, *Emily Dickinson's Home*, 75.
11. Habegger, *My Wars Are Laid Away in Books*, 198.
12. Habegger, *My Wars Are Laid Away in Books*, 202.
13. Emily Norcross to Hannah Porter, January 11, 1848. quoted in Habegger, *My Wars Are Laid Away in Books*, 203.
14. Emily Dickinson to Abiah Root, March 28, 1846, JL 11.
15. Emily Dickinson to Abiah Root, March 28, 1846, JL 11.
16. Emily Dickinson to Abiah Root, September 8, 1846, JL 13.
17. Emily Dickinson to Abiah Root, May 16, 1848, JL 23 (in editor's note).
18. Leyda, *Years and Hours of Emily Dickinson*, 136.

19. Emily Dickinson to Austin Dickinson, October 21, 1847, JL 16.
20. Thoreau, "Resistance to Civil Government," 66–68.
21. Walls, *Thoreau*, 232–272.
22. Thoreau, "28 October 1853," 123.
23. "I have chosen letters for my profession," he declared to the president of Harvard College, in a request for library privileges. Walls, *Thoreau*, 280.
24. Thoreau "7 June 1851," 245. See also Walls, *Thoreau*, 308–309.
25. Herschel, *Preliminary Discourse*, 12.
26. Bennett, *Thoreau's Nature*, 52.
27. Walls, "Greening Darwin's Century," 92–103.
28. Walls, *Thoreau*, 275.
29. During Dickinson's lifetime, the fossil Cabinet continued to expand and grow. In 1855 the Cabinet was moved into a separate building and renamed the Appleton Cabinet.
30. In 1847 Harriet Kezia Hunt requested permission to attend lectures at Harvard Medical School. The president and fellows of Harvard College refused, explaining that their regulations implied that "students are exclusively of the male sex." In 1850, when Hunt repeated her request, the administration voted to grant her admission along with three Black students, but the students who were already enrolled protested so vehemently that administrators privately persuaded Hunt not to attend. Women would not gain admission to Harvard Medical School until 1945. "History of Women at Harvard Medical School."
31. Mott, "Discourse on Women."
32. Abby Wood to Abiah Root, March 29, 1850, reprinted in Longsworth, "'And Do Not Forget Emily,'" 343.
33. Emily Dickinson to Abiah Root, May 7, 1850, JL 36.
34. Emily Dickinson to Abiah Root, late 1850, JL 39.
35. Emily Dickinson to Thomas Wentworth Higginson, April 25, 1862, JL 261.
36. Hitchcock, *Religion of Geology*, 365.
37. Hitchcock, *Religion of Geology*, 333.
38. Hitchcock, *Religion of Geology*, 334.
39. Hitchcock, *Religion of Geology*, 333.
40. Emily Dickinson to Abiah Root, late 1850, JL 39.

Chapter 8

1. Gopnik, *Angels and Ages*, 86.
2. Dickinson, "Some – keep the Sabbath – going to church –" (Emily Dickinson Archive).
3. Heiligman, *Charles and Emma*, 38.
4. Heiligman, *Charles and Emma*, 38.
5. Charles Darwin to William Erasmus Darwin, October 3, 1851, Darwin Correspondence Project, "Letter no. 1456."
6. Heiligman, *Charles and Emma*, 42–46.
7. "Darwin and the Church."
8. Emma Wedgwood to Charles Darwin, November 21 to November 22, 1838, Darwin Correspondence Project, "Letter no. 441."

9. Emily Dickinson to Mary Bowles, after Christmas 1859, JL 213.

10. *Oxford English Dictionary*, s.v. "evolution," July 2023, https://doi.org/10.1093/OED/3466062978. The entry cites Lyell's *Principles of Geology* (1832) and Martineau's translation of *The Positive Philosophy of Comte* (1853).

11. Charles Darwin to Emma Darwin, July 5, 1844, Darwin Correspondence Project, "Letter no. 761."

12. Chambers, *Vestiges*, 154.

13. Quoted in Browne, *Charles Darwin: Voyaging*, 468.

14. Tennyson, "In Memoriam, AHH."

15. Tennyson, "In Memoriam, AHH."

16. Charles Darwin to Joseph Dalton Hooker, [January 11, 1844], Darwin Correspondence Project, "Letter no. 729."

17. Charles Darwin to Leonard Jenyns, October 12, [1844], Darwin Correspondence Project, "Letter no. 782."

18. Sedgwick, "Vestiges of the Natural History of Creation," 2.

19. "Nevertheless it a grand piece of argument against mutability of species; & I read it with fear & trembling, but was well pleased to find, that I had not overlooked any of the arguments, though I had put them to myself as feebly as milk & water." Charles Darwin to Charles Lyell, October 8, 1845, Darwin Correspondence Project, "Letter no. 919."

20. Darwin, *Origin of Species* (1861), xvi.

21. Charles Darwin to William Darwin Fox, April 24, 1845, Darwin Correspondence Project, "Letter no. 859."

22. Cited in Browne, *Charles Darwin: Voyaging*, 470.

23. Darwin, *Charles Darwin's Beagle Diary*, December 25, 1834, 271–272.

24. Stott, *Darwin and the Barnacle*, 62.

25. Emily Dickinson, "We see – Comparatively –" (Emily Dickinson Archive).

26. Stott, *Darwin and the Barnacle*, 82.

27. Charles Darwin to Robert FitzRoy, October 1, 1846, Darwin Correspondence Project, "Letter no. 1002."

28. Charles Darwin to Robert FitzRoy, October 28, [1846], Darwin Correspondence Project, "Letter no. 1014."

29. Browne, *Charles Darwin: Voyaging*, 473.

30. Charles Darwin to Edward Cresy, May 1848, Darwin Correspondence Project, "Letter no. 1171."

31. Charles Darwin to Joseph Dalton Hooker, February 3, [1850], Darwin Correspondence Project, "Letter no. 1300."

32. Charles Darwin to William Darwin Fox, [October 24, 1852], Darwin Correspondence Project, "Letter no. 1489."

33. Darwin, "Geology," 195.

34. Charles Darwin to Alfred Russel Wallace, December 22, 1857, Darwin Correspondence Project, "Letter no. 2192."

35. See Cowles, *Scientific Method*.

36. Browne, *Charles Darwin: Voyaging*, 499.

37. Darwin, *Autobiography*, 97.

38. Darwin, *Life and Letters*, 134.

39. Charles Darwin to John Fordyce, May 7, 1879, Darwin Correspondence Project, "Letter no. 12041."

40. Charles Darwin to William Darwin Fox, [October 24, 1852], Darwin Correspondence Project, "Letter no. 1489."

Chapter 9

1. Emily Dickinson to Austin Dickinson, October 2, 1851, JL 53.

2. Olmsted, *On the Recent Secular Period*, 3. See also "Aurora Borealis," 740, and Quinn, "Dickinson, Telegraphy, and the Aurora Borealis," 58.

3. Somerville, *On the Connexion of the Physical Sciences*, 324.

4. Olmsted, *On the Recent Secular Period*, 26.

5. "Sleep is supposed to be" is from 1858; other significant poems that use the term *aurora* include "These saw visions" (1863), "Struck was I, nor yet by Lightning –" (1864), and "Aurora is the effort" (1865). Many other poems describe the northern lights without using the term "aurora." See, for example, "Of Bronze – and Blaze" (1862) and "I saw no Way – the Heavens were stitched –" (1863). All poems are available in the Emily Dickinson Archive.

6. Emily Dickinson to Austin Dickinson, October 2, 1851, JL 53.

7. Hitchcock, *Religion of Geology*.

8. Byron, "Childe Harold's Pilgrimage." In Lydia Maria Child's *Hobomok*, the main character sees a ship apparition in the sky, quotes Byron's electric chain passage, and is convinced that her fiancé has been shipwrecked. Nathaniel Hawthorne's *Scarlet Letter* features an aurora in the shape of the letter *A*, which characters in the novel interpret in a variety of ways.

9. Browning, *Aurora Leigh*.

10. Brontë, *Jane Eyre*, 161.

11. Herschel, *Preliminary Discourse*, 330.

12. Sewall, *Life of Emily Dickinson*, 109.

13. Sewall, *Life of Emily Dickinson*, 109.

14. Austin Dickinson to Susan Gilbert, December 6, 1851, in Leyda, *Years and Hours of Emily Dickinson*, 226.

15. Emily Dickinson to Susan Gilbert, October 9, 1851, JL 56.

16. There are also darker stains nearby on the page that look like an editor's coffee stains. See the manuscript image at Dickinson Electronic Archives, Emily Dickinson's Correspondences, Correspondence with Susan Dickinson, http://archive.emilydickinson.org/working/zhl5a.htm.

17. Emily Dickinson to Susan Gilbert, October 9, 1851, JL 56.

18. Dickinson, "Why Susie!," 6.

19. Austin Dickinson to Susan Gilbert, April 19, 1852, in Leyda, *Years and Hours of Emily Dickinson*, 226.

20. Emily Dickinson to Austin Dickinson, October 2, 1851, JL 53.

21. Emily Dickinson to Austin Dickinson, February 6, 1852, JL 72.

22. *Springfield Daily Republican*, February 20, 1852, in Ackmann, *These Fevered Days*, 69.

23. *Springfield Daily Republican*, February 20, 1852, in Ackmann, *These Fevered Days*, 69.

24. Leyda, *Years and Hours of Emily Dickinson*, 303.

25. Leyda, *Years and Hours of Emily Dickinson*, 302.

26. Austin Dickinson to Susan Gilbert, September 23, [1851 or 1854], quoted in Sewall, *Life of Emily Dickinson*, 167.

27. Emily Dickinson to Susan Gilbert, January 1855, JL 177. In the letter to her cousin, she wrote: "Ah John – Gone? Then I lift the lid to my box of Phantoms, and lay another in, unto the Resurrection – Then will I gather in Paradise, the blossoms fallen here, and on the shores of the sea of Light, seek my missing sands. Your Coz – Emilie." Emily Dickinson to John Graves, [about 1856], JL 186.

28. Emily Dickinson, letter to Susan Gilbert, January 1855, JL 177.

29. Bingham, *Emily Dickinson's Home*, 375.

30. Hitchcock, "Description of a Large Bowlder."

31. Hitchcock, *Reminiscences of Amherst College*, 6, cited in Thrailkill, "Fables of Extinction," 222.

32. Emily Dickinson to Elizabeth Holland, January 1856, JL 182.

33. Dickinson claimed that she never read Whitman. In a letter to Thomas Wentworth Higginson, she wrote, "You speak of Mr. Whitman—I never read his Book—but was told that he was disgraceful." Emily Dickinson to Thomas Wentworth Higginson, April 25, 1862, JL 261.

34. Dickinson, "The Grass so little has to do" (Emily Dickinson Archive).

35. Emily Dickinson to Elizabeth Holland, early January 1871, JL 359.

36. See Schweitzer, "June 11–17, 1862: Elizabeth Barrett Browning."

37. Elizabeth Barrett to Robert Browning, August 20, 1845, Brownings' Correspondence, letter 2007.

38. Elizabeth Barrett to Robert Browning, July 31, 1845, Brownings' Correspondence, letter 1988.

39. Field, "Elizabeth Barrett Browning," 371.

40. Browning, "The Dead Pan." See also Browning, "A Musical Instrument." These poems are discussed in Morlier, "Death of Pan," 131–155.

41. Browning, *Aurora Leigh*, 1.835–838.

42. Quoted in Finnerty, "On the Wall," 345.

43. Browning, *Aurora Leigh*, 2.294–298.

44. Browning, *Aurora Leigh*, 5.1115–1119, discussed in Cameron, "Renegotiating Science."

45. Finnerty, "On the Wall," 345.

46. Dickinson, "Volcanoes be in Sicily" (Emily Dickinson Archive).

47. Dickinson, "The Admirations – and Contempts – of time –" (Emily Dickinson Archive).

48. Dickinson, "I think I was enchanted" (Emily Dickinson Archive).

Chapter 10

1. Charles Darwin to Joseph Dalton Hooker, September 7, [1854], Darwin Correspondence Project, "Letter no. 1588."

2. Charles Darwin to William Darwin Fox, March 19, [1855], Darwin Correspondence Project, "Letter no. 1651."

3. Charles Darwin to Alfred Russel Wallace, December 22, 1857, Darwin Correspondence Project, "Letter no. 2192."

4. Charles Darwin to Charles Lyell, [June 18, 1858], Darwin Correspondence Project, "Letter no. 2285."

5. Charles Darwin to Charles Lyell, [June 18, 1858], Darwin Correspondence Project, "Letter no. 2285."

6. Charles Darwin to Charles Lyell, [June 25, 1858], Darwin Correspondence Project, "Letter no. 2294."

7. Charles Darwin to Charles Lyell, [June 26, 1858], Darwin Correspondence Project, "Letter no. 2295."

8. Charles Darwin to Joseph Dalton Hooker, [June 29, 1858], Darwin Correspondence Project, "Letter no. 2297."

9. Darwin, *Origin of Species* (1859), 129.

10. Darwin, *Origin of Species* (1859), 130.

11. Darwin, *Origin of Species* (1859), ii.

12. Darwin, *Origin of Species* (1859), 1.

13. Cannon, "Impact of Uniformitarianism," 305.

14. Darwin, *Origin of Species* (1859), 2.

15. Darwin, *Origin of Species* (1859), 490.

16. Darwin, *Origin of Species* (1860), 490.

17. Gopnik, *Angels and Ages*, 71.

18. Gopnik, *Angels and Ages*, 71.

19. Darwin, *Power of Movement in Plants*, 574.

20. Huxley, "Agnosticism," 237–238.

21. Huxley, "Darwin on The Origin of Species," 9.

22. Emma Darwin to William Darwin, November 30–December 4, 1859, GBR/0012/MS DAR 210.6: 52, Cambridge University Library, https://archivesearch.lib.cam.ac.uk/repositories/2/archival_objects/784653, accessed September 8, 2023.

23. Charles Darwin to Charles Lyell, January 14, 1860, Darwin Correspondence Project, "Letter no. 2650."

24. Charles Darwin to Charles Lyell, April 10, [1860], Darwin Correspondence Project, "Letter no. 2754."

25. Charles Darwin to Charles Lyell, April 10, [1860], Darwin Correspondence Project, "Letter no. 2734."

26. Charles Darwin to Charles Lyell, April 10, [1860], Darwin Correspondence Project, "Letter no. 2754."

27. Charles Darwin to John Fordyce, May 7, 1879, Darwin Correspondence Project, "Letter no. 12041."

28. Darwin, *Origin of Species* (1859), 490.

29. Charles Darwin to Joseph Dalton Hooker, [June 26, 1860], Darwin Correspondence Project, "Letter no. 2846."

30. For full discussion of the Oxford debate, see Jensen, "Return to the Wilberforce–Huxley Debate," and Hesketh, *Of Apes and Ancestors*, 76–87.

31. See Browne, "Darwin in Caricature."
32. Browne, *Darwin: Power of Place*, 185.
33. Darwin, *Origin of Species* (1859), iii.
34. Darwin, *Expression of the Emotions*, 330.
35. Darwin, *Origin of Species* (1859), 109.

Chapter 11

1. Habegger, *My Wars Are Laid Away in Books*, 341.
2. Habegger, *My Wars Are Laid Away in Books*, 347.
3. See Susan Dickinson, "Annals of the Evergreens."
4. Habegger, *My Wars Are Laid Away in Books*, 347.
5. Lincoln, *Collected Works*, 3:95, quoted in Wald, *Constituting Americans*, 62.
6. Wald, *Constituting Americans*, 72.
7. Hale, *James Russell Lowell and His Friends*, 158.
8. Hale, *James Russell Lowell and His Friends*, 158.
9. When it was founded in 1857, the *Atlantic Monthly*'s masthead proclaimed it to be a magazine of "Literature, Art and Politics." In July 1865 the publishers would add "Science" to the list—"Literature, Science, Art and Politics."
10. Socarides, *Dickinson Unbound*, 26–35.
11. Werner, *Writing in Time*, 27–29.
12. Werner, *Writing in Time*, 83.
13. Charles Darwin to Asa Gray, October 16, [1862], Darwin Correspondence Project, "Letter no. 3766."
14. Fuller, *Book That Changed America*, 59.
15. Gray, "Darwin on the Origin of Species."
16. Gray, "Darwin on the Origin of Species." The quotation is from Shakespeare's *Troilus and Cressida*.
17. Gray, "Darwin on the Origin of Species."
18. Gray, "Darwin on the Origin of Species."
19. Gosden, *Magic*, 13.
20. Weber, *Charisma and Disenchantment*, 18.
21. Werner, *Writing in Time*, 59.
22. Werner, *Writing in Time*, 63–65.
23. Dickinson, "Master. / If you saw a bullet," in Werner, *Writing in Time*, 75–81. See also JL 233.
24. Dickinson, "Master. / If you saw a bullet," in Werner, *Writing in Time*, 75–81. See also JL 233.
25. Emily Dickinson to Thomas Wentworth Higginson, April 25, 1862, JL 261.
26. Guthrie, "Darwinian Dickinson," 88.
27. Emily Dickinson to Samuel Bowles, early summer 1862, JL 266.
28. Dickinson, "Her – last Poems –" (sent to Susan Dickinson, 1863; Emily Dickinson Archive).
29. Dickinson, "I think I was enchanted" (Emily Dickinson Archive).
30. Gosden, *Magic*, 13.

31. Norton, "The Advantages of Defeat," *Atlantic Monthly*, September 1861, 360.
32. Holmes, "Bread and the Newspaper," *Atlantic Monthly*, September 1861, 346.
33. Holmes, "Bread and the Newspaper," 347.
34. Dickinson, "I felt a Funeral, in my Brain" (Emily Dickinson Archive).
35. Finnerty, "'On the Wall,'" 346.
36. Higginson, "My Out-Door Study," *Atlantic Monthly*, September 1861, 302.
37. Dickinson, "In the name of the Bee –" (Emily Dickinson Archive).
38. Higginson, "My Out-Door Study," 304.
39. Higginson, "My Out-Door Study," 304.

Chapter 12

1. Darwin, *Insectivorous Plants*, 1–2. For a discussion of *Drosera* in Massachusetts, see Grima, "Natural Hybrid."
2. Charles Darwin to Charles Lyell, November 24, [1860], Darwin Correspondence Project, "Letter no. 2996."
3. Charles Darwin to Charles Lyell, November 24, [1860], Darwin Correspondence Project, "Letter no. 2996."
4. Erasmus Darwin, *Botanic Garden*, 24–25, cited in Schaefer, "From Poetry to Pulp Fiction."
5. Erasmus Darwin, *Botanic Garden*, 24, cited in Schaefer, "From Poetry to Pulp Fiction."
6. Charles Darwin to Daniel Oliver November 16, 1860, Darwin Correspondence Project, "Letter no. 2985."
7. Costa, *Darwin's Backyard*, 269.
8. Emma Darwin to Mary Elizabeth Lyell, August 28, 1860, in Litchfield, *Emma Darwin*, 2:177.
9. Charles Darwin to Joseph Dalton Hooker, [August 31, 1860], Letter no. 2886.
10. Charles Darwin to Charles Lyell, November 24, [1860], Darwin Correspondence Project, "Letter no. 2996."
11. Charles Darwin to Charles Lyell, November 24, [1860], Darwin Correspondence Project, "Letter no. 2996"; Charles Darwin to John Lubbock, [November 18, 1860], Darwin Correspondence Project, "Letter no. 2988"; Charles Darwin to Asa Gray, November 26, [1860], Darwin Correspondence Project, "Letter no. 2998."
12. Schaefer, "From Poetry to Pulp Fiction."
13. Charles Darwin to Charles Lyell, April 10, [1860], Darwin Correspondence Project, "Letter no. 2754."
14. Charles Darwin to Daniel Oliver, November 16, 1860, Darwin Correspondence Project, "Letter no. 2985."
15. Costa, *Darwin's Backyard*, xi.
16. Costa, *Darwin's Backyard*, xi–xii.
17. Charles Darwin to Joseph Dalton Hooker, March 26, 1863, Darwin Correspondence Project, "Letter no. 4061."
18. Dickinson, "A little madness in the Spring" (Emily Dickinson Archive).

19. Franklin, *Poems of Emily Dickinson*, 3:1176–1177.
20. Miller, *Emily Dickinson's Poems*, 586.
21. Genesis 1:26, King James Version. See also Milton, *Paradise Lost*, 4:430–432.
22. Darwin, *Descent of Man*, 1:213.
23. Charles Darwin to Charles Lyell, November 24, [1860], Darwin Correspondence Project, "Letter no. 2996; Emma Darwin to Mary Elizabeth Lyell, August 28, 1860, in Litchfield, *Emma Darwin*, 2:177.
24. Darwin, *Power of Movement in Plants*, 574.
25. Wordsworth, "Observations Prefixed to *Lyrical Ballads*."
26. Levine, *Darwin Loves You*, 250.
27. Beer, "Darwin's Reading," 546.
28. Levine, *Darwin Loves You*, 251.
29. Dickinson, "Pink – small – and punctual –" (Emily Dickinson Archive); "There is a flower that Bees prefer –" (Emily Dickinson Archive).
30. Gerhardt, *Place for Humility*, 39.
31. Dickinson, "There is a flower that Bees prefer –" (Emily Dickinson Archive).
32. Dickinson, "Of Bronze – and Blaze –" (Emily Dickinson Archive).
33. Dickinson, "Of Bronze – and Blaze –" (Emily Dickinson Archive).
34. Costa, *Darwin's Backyard*, xii.
35. Dickinson, "Experiment escorts us last –" (Emily Dickinson Archive).
36. Emily Dickinson to Thomas Wentworth Higginson, February 1863, JL 280.
37. Hitchcock, *Religion of Geology*.
38. Holmes, *Darwin's Bards*, 60–61.
39. Holmes, *Darwin's Bards*, 62.
40. Holmes, *Darwin's Bards*, 22.
41. Meredith, "Modern Love, XXX."
42. Quoted in Holmes, *Darwin's Bards*, 249.
43. Quoted in Holmes, *Darwin's Bards*, 80.
44. Hardy, "Hap."
45. Levine, *Reading Thomas Hardy*, 29.
46. Kingsley, *Water-Babies*, 156.
47. Dickinson, "The Brain – is wider than the Sky –" (Emily Dickinson Archive).

Chapter 13

1. Emily Dickinson to Louisa Norcross, December 31, 1861, JL 245.
2. After the war, William S. Clark would become president of the Massachusetts Agricultural College in Amherst (now the University of Massachusetts).
3. Emily Dickinson to Louisa Norcross, December 31, 1861, JL 245. Dickinson consistently spelled *Frazar* incorrectly.
4. Emily Dickinson to Samuel Bowles, late March 1862, JL 256.
5. Emily Dickinson to Frances and Louisa Norcross, late March 1862, JL 255.
6. Tennyson, "Charge of the Light Brigade."

7. Emily Dickinson to Frances and Louisa Norcross, late March 1862, JL 255.
8. Higginson, "Letter to a Young Contributor."
9. Higginson, "Ought Women to Learn the Alphabet?"
10. Higginson, "Emily Dickinson's Letters."
11. Emily Dickinson to Thomas Wentworth Higginson, April 15, 1862, JL 260.
12. See Thomas Johnson's note to JL 260.
13. Emily Dickinson to Thomas Wentworth Higginson, April 25, 1862, JL 261.
14. Emily Dickinson to Thomas Wentworth Higginson, April 25, 1862, JL 261, discussed in Boggs, *Animalia Americana*, 143–155.
15. Emily Dickinson to Thomas Wentworth Higginson, April 25, 1862, JL 261.
16. In his note to JL 261, Thomas Johnson explains that Higginson might be wrong about when he received this poem—the folds of the poem manuscript do not line up with the creases in the letter.
17. Darwin, *Origin of Species* (1859), 62.
18. Darwin, *Origin of Species* (1859), 489.
19. Shakespeare, *A Midsummer Night's Dream*, act 2, scene 1.
20. Costa, *Darwin's Backyard*, 84, 85.
21. Darwin, *Origin of Species* (1859), 490.
22. Dickinson, "A Bird, came down the Walk" (Emily Dickinson Archive).
23. Mondello, "Of Toads and Men," 9.
24. Darwin, *Voyage of the Beagle*, April 8, 1832, 19.
25. Darwin, *Voyage of the Beagle*, December 6, 1833, 132.
26. Dickinson, "There are Two Ripenings –" (Emily Dickinson Archive).
27. Emily Dickinson to Thomas Wentworth Higginson, August 1862, JL 271.
28. Higginson, *Cheerful Yesterdays*, 248–249, quoted in Wineapple, *White Heat*, 121.
29. Higginson's regiment preceded Robert Gould Shaw's Massachusetts Fifty-Fourth by five months. See Wineapple, *White Heat*, 123.
30. Emily Dickinson to Thomas Wentworth Higginson, February 1863, JL 280.
31. Higginson, *Army Life in a Black Regiment*, 134, quoted in Shoptaw, "Dickinson's Civil War Poetics," 18.
32. Higginson, *Army Life in a Black Regiment*, 208.
33. Higginson, "Procession of the Flowers," 319. The essay was first published in the *Atlantic Monthly*, December 1862.
34. Higginson, "Procession of the Flowers," 325.
35. Higginson, "Procession of the Flowers," 325. Higginson was right about this. Nineteenth-century herbaria and notebooks (particularly Thoreau's notebooks) have provided important data for twenty-first-century climate scientists.
36. Emily Dickinson to Thomas Wentworth Higginson, February 1863, JL 280.
37. Dickinson to Thomas Wentworth Higginson, spring 1876, JL 458.
38. Wineapple, *White Heat*, 159.
39. Dickinson, "The Tint I cannot take – is best –" (Emily Dickinson Archive), discussed in Mondello, "Of Toads and Men," 9.
40. Emily Dickinson to Abiah Root, late 1850, JL 39.

41. Emily Dickinson to Thomas Wentworth Higginson, February 1863, JL 280.
42. Pollak, "Dickinson and the Poetics of Whiteness," 92, 91. See also Richards, "'How News Must Feel,'" and Fretwell, "Emily Dickinson in Domingo."
43. Wineapple, *White Heat*, 148.
44. Boggs, *Patriotism by Proxy*, 13, 79, 85.
45. Wineapple, *White Heat*, 144.
46. Boggs, *Patriotism by Proxy*, 13.
47. Boggs, *Patriotism by Proxy*, 76.
48. Emily Dickinson to Louisa and Frances Norcross, [1864], JL 298.
49. Emily Dickinson to Louisa and Frances Norcross, [1864], JL 298.
50. Emily Dickinson to Louisa and Frances Norcross, [1864], JL 298.
51. Dandurand, "New Dickinson Civil War Publications," 17.
52. Kelly, "Emily Dickinson and the New York Press."
53. Dandurand, "New Dickinson Civil War Publications," 17.
54. Dickinson, "Flowers – Well – if anybody" (Emily Dickinson Archive).
55. Dickinson, "Flowers – Well – if anybody" (Emily Dickinson Archive).
56. Emily Dickinson to Thomas Wentworth Higginson, February 1863, JL 280.
57. Emily Dickinson to Thomas Wentworth Higginson, June 9, 1866, JL 319.
58. Darwin, *Expression of the Emotions*, 52–55.
59. Charles Darwin to unknown correspondent, May 19, [1871], Darwin Correspondence Project, "Letter no. 13889."
60. Darwin, *Expression of the Emotions*, 57–60.
61. Emma Darwin to Henrietta Litchfield, March 19, [1870], in Litchfield, *Emma Darwin*, 197.
62. Darwin, *Expression of the Emotions*, 53.
63. Emily Dickinson to Thomas Wentworth Higginson, June 1864, JL 290; Emily Dickinson to Joseph Lyman, in Sewall, "Lyman Letters," 772.
64. Emily Dickinson to Thomas Wentworth Higginson, January 1866, JL 314.
65. Dickinson, "Further in Summer than the Birds" (Emily Dickinson Archive).
66. Dickinson, "Further in Summer than the Birds" (Emily Dickinson Archive).
67. Emily Dickinson to Thomas Wentworth Higginson, June 9, 1866, JL 319.
68. Emily Dickinson to Thomas Wentworth Higginson, June 9, 1866, JL 319.

Chapter 14

1. Darwin, *Autobiography*, 43.
2. Darwin, *Autobiography*, 44.
3. Darwin, *Autobiography*, 138.
4. Darwin, *Autobiography*, 138–139.
5. Darwin, *Autobiography*, 138–139.
6. Eliot, "John Ruskin," 1856.
7. James, "Middlemarch," 424.
8. Beer, *Darwin's Plots*, 161.
9. Alcott, *Work*, 442.

10. Bianchi, *Emily Dickinson Face to Face*, 66.

11. Bianchi, *Emily Dickinson Face to Face*, 66.

12. Charles Darwin to Joseph Dalton Hooker, March 30, [1875], Darwin Correspondence Project, "Letter no. 9905."

13. Charles Darwin to Frances Powers Cobbe, March 18, [1873], Darwin Correspondence Project, "Letter no. 8814."

14. Charles Darwin to Frances Powers Cobbe, [January 14, 1875], Darwin Correspondence Project, "Letter no. 9814F."

15. Jane Gray to Susan Loring, October 28–November 2, 1868, Archives of the Gray Herbarium, box G AG-B10: 8, Darwin Correspondence Project.

16. Jane Gray to Susan Loring, October 28–November 2, 1868, describing her visit to Down, Archives of the Gray Herbarium, box G AG-B10: 8, Darwin Correspondence Project.

17. Thomas Wentworth Higginson to his sisters, December 9, 1873. See note on JL 405.

18. Charles Darwin to Thomas Wentworth Higginson, February 27, [1873], Darwin Correspondence Project, "Letter no. 8790."

19. Thomas Wentworth Higginson to Charles Darwin, March 30, 1873, Darwin Correspondence Project, "Letter no. 8830."

20. Thomas Wentworth Higginson to Emily Dickinson, December 31, 1873, JL 405a.

21. The flowers Dickinson presented to Higginson were probably a late-blooming, fragrant Hosta, known as day-lilies in the nineteenth century. See McDowell, *Emily Dickinson's Gardening Life*, 90.

22. Thomas Wentworth Higginson to his wife, August 17, 1870, JL 342B.

23. Thomas Wentworth Higginson to Emily Dickinson, December 31, 1873, JL 405a.

24. Higginson, *Cheerful Yesterdays*, 284.

25. Higginson, *Cheerful Yesterdays*, 284.

26. Thomas Wentworth Higginson to his sisters, December 9, 1873. See note on JL 405.

27. Dickinson, "The Wind begun to rock the Grass" (Emily Dickinson Archive).

28. Thomas Wentworth Higginson to Emily Dickinson, December 31, 1873. See note on JL 405.

29. Higginson, *Cheerful Yesterdays*, 286.

30. Higginson, *Cheerful Yesterdays*, 286.

31. Higginson, *Cheerful Yesterdays*, 286.

32. Higginson, *Cheerful Yesterdays*, 295.

33. Higginson, *Cheerful Yesterdays*, 296.

34. Dickinson, "What mystery pervades a well!" (Emily Dickinson Archive).

35. Dickinson, "What mystery pervades a well!" (Emily Dickinson Archive). See also Franklin, *Variorum*, 1254–1255.

36. Dickinson, "What mystery pervades a well!" (Emily Dickinson Archive).

Chapter 15

1. Wilson, *From so Simple a Beginning*, 11.

2. Fiske, "Charles Darwin."

3. Jane Gray to Susan Loring, October 28–November 2, 1868, Archives of the Gray Herbarium, box G AG-B10: 8, Darwin Correspondence Project.

4. Charles Darwin to Francis E. Abbot, January 8, 1872, Darwin Correspondence Project, letter no. 8151.

5. Gould, *Rocks of Ages*, 5.

6. Charles Darwin to Francis E. Abbot, September 6, 1871, Darwin Correspondence Project, "Letter no. 7924."

7. Keane, *Emily Dickinson's Approving God*, 50–51.

8. Shaw, *Back to Methuselah*, li.

9. Satter, *Each Mind a Kingdom*, 17.

10. Darwin, *Descent of Man*, 111.

11. Darwin, *Life of Erasmus Darwin*, 46.

12. Shaw, *Back to Methuselah*, li.

13. Shaw, *Back to Methuselah*, lxv.

14. Browne, "Presidential Address," 272.

15. Browne, "Presidential Address," 273.

16. Levine, *Darwin Loves You*, 56, 57.

17. Browne, *Charles Darwin: Power of Place*, 495.

18. Harel, "It's Dogged as Does It," 368.

19. Allen, "Obituary: Charles Darwin."

20. Browne, *Charles Darwin: Power of Place*, 495.

21. "Death of Chas. Darwin," *New York Times*, April 21, 1882, 5, PDF A1211.

22. Fiske, "Charles Darwin," 845.

23. Fiske, "Charles Darwin," 845.

24. "Editors Table," *Popular Science Monthly* 21 (July 1882), 266.

25. Browne, *Charles Darwin: Power of Place*, 496.

26. Colp, "Notes and Events."

27. "The Funeral of Mr. Darwin," *Times* (London), April 27, 1882, 5.

28. James Russell Lowell to Sara Sedgwick Darwin (spouse of William E. Darwin), September 1, 1878, in Lowell, *Letters of James Russell Lowell*, 2:259.

29. Browne, "Making Darwin," 365.

30. See Skrupskelis, "Evolution and Pragmatism."

31. Higginson, *Cheerful Yesterdays*, 29.

32. Darwin, *Voyage of the Beagle*, 1.

33. Darwin, *Formation of Vegetable Mould*, 313.

34. I discuss Dickinson's first letter mentioning Darwin in chapter 9 (Emily Dickinson to Elizabeth Holland, early January 1871, JL 359).

35. See Finnerty, *Emily Dickinson's Shakespeare*.

36. Emily Dickinson to Catharine Sweetser (incorrectly dated October 1876; probably September 1880), JL 478. It is likely that Dickinson exaggerated the closeness of the relationship between her father and Lord; see Habegger, *My Wars Are Laid Away in Books*, 587.

37. Martha Nell Smith, "HB103 Note," 1996 Commentary on "Excuse / Emily and / her Atoms," JL 774, Dickinson Electronic Archive, http://archive.emilydickinson.org/working/nhb103.htm.

38. Dickinson, "Cosmopolites without a plea" (Emily Dickinson Archive; see Miller, *Emily Dickinson's Poems*, 640); Dickinson to Susan Dickinson, October 1882, JL 774. See also "Amherst

Manuscript # 160" in the Amherst College Digital Collection, https://acdc.amherst.edu/view/EmilyDickinson/ed0160-0001.

39. Dickinson, "Excuse me for disturbing Susan" [verso of "Cosmopolites without a plea"], draft of letter to Susan Dickinson, October 1882, JL 774. See also "Amherst Manuscript # 160."

Chapter 16

1. Emily Dickinson to Thomas Wentworth Higginson, July 1874, JL 418.
2. Habegger, *My Wars Are Laid Away in Books*, 560–562.
3. *Boston Evening Journal*, November 7, 1873.
4. Emily Dickinson to Thomas Wentworth Higginson, July 1875, JL 441.
5. Emily Dickinson to Thomas Wentworth Higginson, August 1877, JL 513.
6. Emily Dickinson to Thomas Wentworth Higginson, June 1877, JL 503.
7. Emily Dickinson to Susan Gilbert Dickinson, early October 1883, JL 871.
8. Emily Dickinson, JL undated prose fragment 50, 867.
9. Dickinson, "Exhiliration is the Breeze" (Emily Dickinson Archive).
10. Dickinson, "Best Witchcraft is Geometry" (Emily Dickinson Archive). The editors of *Open Me Carefully* state that the poet sent this leaf of poems to Susan Dickinson in 1869. Ralph W. Franklin states that Susan acquired it after Emily's death. See Miller, *Emily Dickinson's Poems*, 777n31.
11. Emily Dickinson to Mr. and Mrs. E. J. Loomis, autumn 1884, JL 945.
12. Emily Dickinson to Samuel Bowles, early summer 1862, JL 266.
13. Emily Dickinson to Elizabeth Holland, late autumn 1884, JL 950.
14. Emily Dickinson to Susan Gilbert Dickinson, October 1883, JL 868.
15. Emily Dickinson to Susan Gilbert Dickinson, October 1883, JL 868. She sent the same lines to Higginson in reference to Helen Hunt Jackson's death, February 1885. JL 972.
16. Emily Dickinson to Samuel Bowles (the younger), mid-August 1885, JL 1008.
17. Emily Dickinson to Frances and Louisa Norcross, January 14, 1885, JL 962.
18. Emily Dickinson to Frances and Louisa Norcross, January 14, 1885, JL 962.
19. Emily Dickinson to Thomas Wentworth Higginson, August 6, 1885, JL 1007.
20. Dickinson, "The Immortality she gave" (Emily Dickinson Archive).
21. Emily Dickinson to Thomas Wentworth Higginson, spring 1886, JL 1042.
22. Emily Dickinson to Louisa and Frances Norcross, around March 1886, JL 1034.
23. Emily Dickinson to Mrs. George S. Dickerman, early spring 1886, JL 1037.
24. Emily Dickinson to Thomas Wentworth Higginson, early May 1886, JL 1045.
25. Emily Dickinson to Louisa and Frances Norcross, May 1886, JL 1046.
26. Leyda, *Years and Hours of Emily Dickinson*, 2:475. See the description of Dickinson's funeral in Dobrow, *After Emily*, 1–7.
27. Leyda, *Years and Hours of Emily Dickinson*, 2:475.
28. Dickinson, "Of God we ask one favor, that we may be forgiven –" (Emily Dickinson Archive).
29. Dickinson, "Somewhere upon the general Earth" (Emily Dickinson Archive), 1871.
30. Dickinson, "Somewhere upon the general Earth" (Emily Dickinson Archive), 1871.

31. Susan Dickinson, "Obituary of Emily Dickinson."

32. Emily Dickinson to Susan Gilbert (Dickinson), October 9, 1851, JL 56.

33. Susan Dickinson, "Obituary of Emily Dickinson."

34. Dickinson's poem "The Admirations – and Contempts – of time" describes dying as a height. "Because I could not Stop for Death –" depicts death as a seducer. There are many posthumous narrative poems. Some of the best known are "Twas just this time, last year, I died"; "I died for Beauty – but was scarce"; and "I heard a Fly buzz – when I died –" (Emily Dickinson Archive).

35. Dickinson, "I died for Beauty – but was scarce" (Emily Dickinson Archive).

36. Dickinson, "I died for Beauty – but was scarce" (Emily Dickinson Archive).

37. James Russell Lowell to Sara Sedgwick Darwin (spouse of William E. Darwin), September 1, 1878, in *Letters of James Russell Lowell*, 2:259.

38. Darwin, *Origin of Species* (1859), 491.

39. Emily Dickinson to Thomas Wentworth Higginson, Spring 1876, JL 459a.

40. Dickinson, "Whoever disenchants" (Emily Dickinson Archive). Variant reading for line 7.

41. Emily Dickinson to Abiah Root, May 16, 1848, JL 23.

42. Dickinson, "A Moth the hue of this" (Emily Dickinson Archive).

43. Charles Darwin to Joseph Dalton Hooker, January 18, 1874, Darwin Correspondence Project, "Letter no. 9247."

44. Emma Wedgwood Darwin to Charles Darwin, about February 1839, Darwin Correspondence Project, "Letter no. 471."

45. Wyhe, [Emma Darwin's memo about Darwin's religious doubts] (c. 2.1839), CUL-DAR210.8.14, Darwin Online.

46. Allen, "Obituary: Charles Darwin."

47. Dickinson, "The Brain – is wider than the Sky –" (Emily Dickinson Archive).

48. Rose, *Darwin's Spectre*, 3.

49. Levine, *Darwin Loves You*, 273.

50. Thomas Wentworth Higginson to his sisters, December 9, 1873, Johnson note to JL 405.

51. Thomas Wentworth Higginson to Anna (Higginson), December 28, 1876, Johnson note to JL 481.

52. Thomas Wentworth Higginson to his sisters, December 9, 1873. See note on JL 405.

53. Higginson, "Emily Dickinson's Letters," 453.

54. Lowell, "The Sisters."

55. Dickinson, "I died for Beauty – but was scarce" (Emily Dickinson Archive).

Afterword

1. "American Antiquity," *Atlantic Monthly* 1, no. 7 (May 1858): 769.

2. Fiske, "Unseen World," 453.

3. Joseph Dalton Hooker to Charles Darwin, February 3, 1865, Darwin Correspondence Project, "Letter no. 4765."

4. Charles Darwin to Joseph Dalton Hooker, February 9, 1865, Darwin Correspondence Project, "Letter no. 4769."

5. Dickinson, "This was a Poet –" (Emily Dickinson Archive).

6. Dickinson, "This was a Poet –" (Emily Dickinson Archive).

7. Higginson, "Letter to a Young Contributor," 410. See Miller, *Emily Dickinson's Poems*, 757n203.

8. Gray, "Darwin on the Origin of Species," 110.

9. Emerson, "The Poet," 211.

10. Dickinson, "'Faith' is a fine invention" (Emily Dickinson Archive).

11. Dickinson, "Hope is a strange invention –" (Emily Dickinson Archive).

BIBLIOGRAPHIC NOTE

When I started this book, a long line of footsteps stretched before me. I began by reading about the intertwined history of magic and science. For me, as for many others, *Religion and the Decline of Magic* by Keith Thomas was the starting point. I consulted Robert Bartlett's *The Natural and the Supernatural in the Middle Ages*, Brian Copenhaver's *Magic in Western Culture*, David Wootton's *The Invention of Science*, Henry Cowles's *The Scientific Method*, and Michael Hunt's *The Decline of Magic*. Everything began to make sense for me with *Magic: A History* by Chris Gosden. His definition of magic has become a touchstone for me. As I tried to understand enchantment, I turned most often to works by Max Weber, Charles Taylor, and Jane Bennett. Bennett's *The Enchantment of Modern Life* lies at the roots of this project.

Quite a few scholars (most notably Jane Donahue Eberwein, Joan Kirkby, and James Guthrie) have written essays about Dickinson and Darwin. Kaitlin Mondello's article on Dickinson and Darwin's "Brutal Kinship" was published in 2020, anticipating and reinforcing many of the parallels I had been observing. There is a larger group of books and essays about Dickinson and science and/or Dickinson and philosophy. For science, I followed the tracks of James Sweet, Nina Baym, Hiroko Uno, Robin Peel, and Richard Brantley. For philosophy, I looked to Jed Deppman, Marianne Noble, and Christian Haines most frequently. The recent ecological turn in Dickinson studies thrills me. I particularly appreciate works by Christine Gerhardt, Gillian Osborne, Paul Crumbley, Cody Marrs, and Colleen Glenney Boggs. Michelle Kohler's *The New*

Dickinson Studies contains many scholarly approaches to Dickinson that I have found inspiring.

Along a parallel track, I often reached for writers about Thoreau's approach to science, philosophy, and ecology—most significantly Laura Dassow Walls, whose magisterial biography of Thoreau built on previous books on Humboldt and Emerson. Branka Arsic's *Bird Relics* brought Thoreau's vital materialism alive. Jane Bennett's readings of Thoreau provided waymarks for me, as Bennett's works have for so many other scholars, including George Levine.

There is a rich tradition of writing about Darwin and literature. I built on *Darwin's Plots* by Gillian Beers, *The Age of Wonder* by Richard Holmes, and *Darwin's Bards* by John Holmes. Many writers have offered their (often wildly varying) insights into Darwin and religion. For me, *Darwin's Ghosts* by Rebecca Stott, *Charles and Emma* by Deborah Heiligman, and *Angels and Ages* by Adam Gopnik were particularly helpful. George Levine's *Darwin Loves You* is my favorite book on this shelf, since he describes Darwin as a secular enchanter. I couldn't agree more. I also particularly like Banu Subramaniam's *Ghost Stories for Darwin* and James T. Costa's *Darwin's Backyard* for their accounts of the ecological implications of Darwin today. *Darwin's Fossils* by Adrian Lister and *Curious Footprints* by Nancy Pick helped me to make sense of the *Megatherium* bones and the strange fossil footprints that crisscross these pages.

The definitive biography of Charles Darwin is Janet Browne's two-volume work *Charles Darwin: Voyaging* and *Charles Darwin: The Power of Place*. More than thirty years ago, Janet Browne joked that her magisterial biography of Charles Darwin was just "another Darwin biography." Since then, Darwin biographies have continued to appear with surprising frequency. From this shelf, I particularly recommend *Young Charles Darwin* by Richard Thomson, *Darwin and the Barnacle* by Rebecca Stott, and *The Ghost in the Garden* by Jude Piesse.

A surprising number of Dickinson biographies are also available. My library includes biographical works by Jay Leyda, Richard Sewall, Polly Longworth, Martha Nell Smith, Alfred Habegger, Lyndall Gordon,

Brenda Wineapple, Christopher Benfey, Julie Dobrow, and, most recently, Martha Ackmann.

Sometimes fiction offers insights that more factual history cannot provide. For me, Patrick O'Brian's fictional Stephen Maturin character in the Master and Commander series of novels brought Darwin's years as a naval naturalist to life. By the same token, the wholly fictionalized depiction of cross-dressing Dickinson at an Amherst geology lecture in the Apple TV series *Dickinson* captured the essence of twenty-first-century imaginings of Dickinson as a science student.

And then there are audiobooks. Making the long drive from Hanover, New Hampshire, to Boston, I have particularly enjoyed Richard Dawkins's narration of Darwin's own words. I wish there was a similar audio of Dickinson. Along the way I listened to Janet Browne's account of the publication and reception of Darwin's *Origin of Species*, narrated by Josephine Bailey; Deborah Heiligman's *Charles and Emma*, narrated by Rosalyn Landor; Adam Gopnik's own narration of *Angels and Ages*; Lyndall Gordon's *Lives like Loaded Guns*, narrated by Wanda McCaddon; and Martha Ackmann's very appealing rendition of her book *These Fevered Days*. The accident of availability meant that these books permeated my understanding in ways that may not always be footnoted.

During the COVID-19 pandemic, when libraries were closed and travel forbidden, archival websites took on new importance. I relied heavily on the Emily Dickinson Archive: An Open-Access Website for the Manuscripts of Emily Dickinson, sponsored by Harvard University and Amherst College, the Darwin Correspondence Project website hosted by Cambridge University, and the Darwin Online website edited by John van Wyhe. I also turned frequently to Martha Nell Smith's Dickinson Electronic Archives and Ivy Schweitzer's White Heat. Although I usually started with the websites, I verified Emily Dickinson poems with Cristanne Miller's *Emily Dickinson's Poems: As She Preserved Them*. For Emily Dickinson's letters, I referred to Thomas Johnson's edition of *The Letters of Emily Dickinson*. To verify Darwin's published writings and diaries, I turned to John van Wyhe's digital versions of Darwin's writings, collected at Darwin Online. I verified Darwin's letters using the Darwin Correspondence Project website.

For general readers, I recommend R. W. Franklin's reading edition of *The Poems of Emily Dickinson* and Edward O. Wilson's edition of Darwin's major works, *From so Simple a Beginning: The Four Great Books of Charles Darwin*.

BIBLIOGRAPHY

Note: Details of publication have been omitted for works that are widely available in many different editions/printings.

Abrams, M. H. (Meyer Howard). *Natural Supernaturalism: Tradition and Revolution in Romantic Literature.* New York: Norton, 1971.
Ackmann, Martha. *These Fevered Days: Ten Pivotal Moments in the Making of Emily Dickinson.* New York: W. W. Norton, 2020.
Alcock, John. *An Enthusiasm for Orchids: Sex and Deception in Plant Evolution.* New York: Oxford University Press, 2006.
Alcott, Louisa May. *Little Women.*
Alcott, Louisa May. *Work: A Story of Experience.* Boston: Robert Brothers, 1873.
Allen, Grant. "Obituary: Charles Darwin." *The Academy* 21, no. 521 (April 29, 1882): 306–307.
"Aurora Borealis." *Atlantic Monthly* 4, no. 26 (1859): 740–750.
Bartlett, Robert. *The Natural and the Supernatural in the Middle Ages: The Wiles Lecture Given at the Queen's University of Belfast, 2006.* Cambridge: Cambridge University Press, 2008.
Beer, Gillian. *Darwin's Plots: Evolutionary Narrative in Darwin, George Eliot and Nineteenth-Century Fiction.* Cambridge: Cambridge University Press, 2000.
Bennett, Jane. *The Enchantment of Modern Life: Attachments, Crossings, and Ethics.* Princeton, NJ: Princeton University Press, 2001.
Bennett, Jane. *Thoreau's Nature: Ethics, Politics, and the Wild.* 2nd ed. Washington, DC: Rowman and Littlefield, 2002.
Bernhard, Mary Elizabeth Kromer. "Portrait of a Family: Emily Dickinson's Norcross Connection." *New England Quarterly* 60, no. 3 (September 1987): 363–381.
Bianchi, Martha Dickinson. *Emily Dickinson Face to Face: Unpublished Letters, with Notes and Reminiscences.* Boston: Houghton Mifflin, 1932.
Bingham, Millicent Todd. *Emily Dickinson's Home: Letters of Edward Dickinson and His Family.* New York: Harper, 1955.
Boggs, Colleen Glenney. *Animalia Americana: Animal Representations and Biopolitical Subjectivity.* New York: Columbia University Press, 2013.

Boggs, Colleen Glenney. *Patriotism by Proxy: The Civil War Draft and the Cultural Formation of Citizen-Soldiers, 1863–1865*. New York: Oxford University Press, 2020.

Brewster, David. *Letters on Natural Magic: Addressed to Sir Walter Scott*. 1832. Reprint, Cambridge: Cambridge University Press, 2011. doi:10.1017/CBO9780511792663.

Brontë, Charlotte. *Jane Eyre: An Autobiography*. New York: Harper & Brothers, 1848.

Browne, Janet. "Botany for Gentlemen: Erasmus Darwin and 'The Loves of the Plants.'" *Isis* 80, no. 4 (1989): 592–621. https://www.jstor.org/stable/234174.

Browne, Janet. *Charles Darwin: Power of Place*. New York: Alfred A. Knopf, 2002.

Browne, Janet. *Charles Darwin: Voyaging*. New York: Alfred A. Knopf, 1995.

Browne, Janet. "Darwin in Caricature: A Study in the Popularisation and Dissemination of Evolution." *Proceedings of the American Philosophical Society* 145, no. 4 (2001): 496–509.

Browne, Janet. "Making Darwin: Biography and the Changing Representations of Charles Darwin." *Journal of Interdisciplinary History* 40, no. 3 (2010): 347–373.

Browne, Janet. "Presidential Address Commemorating Darwin." *British Journal for the History of Science* 38, no. 3 (2005): 251–274.

Browning, Elizabeth Barrett. *Aurora Leigh*.

Browning, Elizabeth Barrett. "The Dead Pan."

Browning, Elizabeth Barrett. "A Musical Instrument."

Browning, Robert. "Caliban upon Setebos, or Natural Theology in the Island." In *Dramatis Personae*. London: Chapman and Hall, 1864. Google Books.

Brownings' Correspondence. https://www.browningscorrespondence.com/.

Buell, Lawrence. *Emerson*. Cambridge, MA: Harvard University Press, 2009.

Byron, George Gordon. "Childe Harold's Pilgrimage."

Cameron, Lauren N. "Renegotiating Science: British Women Novelists and Evolution Controversies, 1826–1876." Carolina Digital Repository, 2013. https://doi.org/10.17615/eyfh-3126.

Cannon, Walter F. "The Impact of Uniformitarianism: Two Letters from John Herschel to Charles Lyell, 1836–1837." *Proceedings of the American Philosophical Society* 105, no. 3 (1961): 301–314. http://www.jstor.org/stable/985457.

Carlyle, Thomas. "Signs of the Times" (1829). In *The Spirit of the Age: Victorian Essays*, 31–49. New Haven, CT: Yale University Press, 2007.

[Chambers, Robert]. *Vestiges of the Natural History of Creation*.

Child, Lydia Maria. *Hobomok*, 1824.

Coleridge, Samuel Taylor. *Anima Poetae*. London: Heinemann, 1895. Originally published 1805.

Coleridge, Samuel Taylor. *Collected Letters*. Edited by Earl Leslie Griggs. 6 vols. Oxford: Clarendon Press, 1956–1971.

Coleridge, Samuel Taylor. *The Collected Works of Samuel Taylor Coleridge*. Vol. 9, *Aids to Reflection*. Edited by John Beer. Princeton, NJ: Princeton University Press, 1993.

Colp, Ralph. "Notes and Events." *Journal of the History of Medicine and Allied Sciences* 35, no. 1 (1980): 59–74.

Copenhaver, Brian P. *The Book of Magic: From Antiquity to the Enlightenment*. London: Penguin Classics, 2016.

Copenhaver, Brian P. *Magic in Western Culture: From Antiquity to the Enlightenment*. Cambridge: Cambridge University Press, 2015.

Costa, James T. *Darwin's Backyard: How Small Experiments Led to a Big Theory*. New York: W. W. Norton, 2017.

Cowburn, Philip, ed. *Salopian Anthology: Some Impressions of Shrewsbury School during Four Centuries*. London: Macmillan, 1964.

Cowles, Henry M. *The Scientific Method: An Evolution of Thinking from Darwin to Dewey*. Cambridge, MA: Harvard University Press, 2020.

Dandurand, Karen. "New Dickinson Civil War Publications." *American Literature* 56, no. 1 (1984): 17-27.

Darwin, Charles. *The Autobiography of Charles Darwin, 1809–1882*. Edited by Nora Barlow. London: Collins, 1958. Darwin Online.

Darwin, Charles. *Charles Darwin's Beagle Diary*. Edited by R. D. Keynes. Cambridge: Cambridge University Press, 1988.

Darwin, Charles. *Charles Darwin's "The Life of Erasmus Darwin."* Cambridge: Cambridge University Press, 2002.

Darwin, Charles. *The Descent of Man, and Selection in Relation to Sex*. 2 vols. London: John Murray, 1871. Darwin Online.

Darwin, Charles. *The Different Forms of Flowers on Plants of the Same Species*. 2 vols. London: John Murray, 1877. Darwin Online.

Darwin, Charles. *The Expression of the Emotions in Man and Animals*. London: John Murray, 1872. Darwin Online.

Darwin, Charles. *The Fertilisation of Orchids*. London: John Murray, 1862. Darwin Online.

Darwin, Charles. *The Formation of Vegetable Mould, through the Action of Worms, With Observations on Their Habits*. London: John Murray, 1881. Darwin Online.

Darwin, Charles. "Geology." In *A Manual of Scientific Enquiry; Prepared for the Use of Her Majesty's Navy: And Adapted for Travellers in General*, edited by John Herschel, 156–195. London: John Murray, 1849.

Darwin, Charles. "Historical Sketch." In *On the Origin of Species*, 3rd ed., xiii–xix. London: John Murray, 1861. Darwin Online.

Darwin, Charles. *Insectivorous Plants*. London: John Murray, 1875. Darwin Online.

Darwin, Charles. *Life and Letters of Charles Darwin*. Edited by Francis Darwin. London: John Murray, 1887. Darwin Online.

Darwin, Charles. *On the Movement and Habits of Climbing Plants*. London: John Murray, 1865. Darwin Online.

Darwin, Charles. *On the Origin of Species*. London: John Murray, 1859. Darwin Online.

Darwin, Charles. *On the Origin of Species*. 2nd ed. London: John Murray, 1860.

Darwin, Charles. *On the Origin of Species*. 3rd ed. London: John Murray, 1861. Darwin Online.

Darwin, Charles. *The Power of Movement in Plants*. London: John Murray, 1880. Darwin Online.

Darwin, Charles. *The Voyage of the Beagle*. Published as *Journal of Researches into the Natural History and Geology of the Countries Visited during the Voyage of H.M.S. Beagle round the World, under the Command of Capt. Fitz Roy, R.N.* 2nd ed. London: John Murray, 1845. Darwin Online.

Darwin, Erasmus. *The Botanic Garden*. Edited by Adam Komisaruk and Allison Leigh Dushane. 2 vols. Abingdon, UK: Routledge, 2017.

Darwin, Erasmus. *The Temple of Nature Or, the Origin of Society: A Poem, With Philosophical Notes*. London: Printed for J. Johnson ... by T. Bensley, 1803.

Darwin, Erasmus. *Zoonomia; or, the laws of organic life ... by Erasmus Darwin, M.D. F.R.S. Author of the Botanic Garden*. Vol. 1. London: Printed for J. Johnson, in St. Paul's Church-Yard, [1794]–96. Eighteenth Century Collections Online. Accessed January 26, 2022.

"Darwin and Dogs." Commentary, Darwin Correspondence Project. https://www.darwinproject.ac.uk/commentary/curious/darwin-and-dogs.

"Darwin and the Church." Commentary, Darwin Correspondence Project. https://www.darwinproject.ac.uk/commentary/religion/darwin-and-church.

Darwin Correspondence Project.

Darwin Online.

Defoe, Daniel. *A System of Magick; or, a History of the Black Art. Being an Historical Account of Mankind's Most Early Dealing with the Devil; and How the Acquaintance on Both Sides First Begun*. Eighteenth Century Collections Online. London: printed and sold by Andrew Millar, at Buchanan's Head, against St. Clement's Church in the Strand, 1728.

Dickens, Charles. *David Copperfield*.

Dickinson, Emily. *Herbarium*. Harvard Library. https://iiif.lib.harvard.edu/manifests/view/drs:4184689$1i.

Dickinson, Emily. *Letters of Emily Dickinson*. Edited by Mabel Loomis Todd. Boston: Roberts Brothers, 1894.

Dickinson, Emily. *The Letters of Emily Dickinson*. Edited by Thomas H. Johnson. Cambridge, MA: Belknap Press of Harvard University Press, 1958.

Dickinson, Susan. "Obituary of Emily Dickinson." Digital Amherst. https://www.digitalamherst.org/items/show/904.

Dickinson, Susan Huntington. "Annals of the Evergreens." Typescript, [n.d.]. Dickinson family papers, MS Am 1118.95 (269), box 12, Houghton Library.

"The Dickinson Daguerreotype." Amherst College. https://www.amherst.edu/library/archives/holdings/edickinson/dickinsondag.

Dobrow, Julie. *After Emily: Two Remarkable Women and the Legacy of America's Greatest Poet*. New York: W. W. Norton, 2018.

Edwards, Jonathan, and Perry Miller. *Images or Shadows of Divine Things*. New Haven, CT: Yale University Press, 1948.

Eliot, George [Mary Ann Evans]. "John Ruskin." *Westminster Review*, 1856.

Eliot, George [Mary Ann Evans]. *Middlemarch*. 1871.

Emerson, Ralph Waldo. *English Traits*. Boston: Phillips Sampson, 1856. Google Books.

Emerson, Ralph Waldo. "Humboldt." In *The Complete Works of Ralph Waldo Emerson*, vol. 11, *Miscellanies*, 457–459. Boston: Houghton Mifflin, [1903–1904]. http://name.umdl.umich.edu/4957107.0011.001.

Emerson, Ralph Waldo. "Nature."

Emerson, Ralph Waldo. "The Poet."

Emerson, Ralph Waldo. *Ralph Waldo Emerson: The Major Prose*. Edited by Ronald A. Bosco and Joel Myerson. Cambridge, MA: Harvard University Press, 2015.

Emily Dickinson Archive.

Field, Kate. "Elizabeth Barrett Browning." *Atlantic Monthly*, September 1861.

Finnerty, Páraic. *Emily Dickinson's Shakespeare*. Amherst: University of Massachusetts Press, 2006.

Finnerty, Páraic. "'On the Wall of Her Own Room Hung Framed Portraits of Mrs. Browning, George Eliot, and Carlyle': Dickinson's Heroes and Hero-Worship." In *Oxford Handbook of Emily Dickinson*. Edited by Cristanne Miller and Karen Sanchez-Eppler, 336–352. Oxford: Oxford University Press, 2022.

Fiske, John. "Charles Darwin." *Atlantic Monthly*, June 1882, 835–845.

Fiske, John. "The Unseen World." *Atlantic Monthly*, 1876, 453.

Franklin, R. W., ed. *The Poems of Emily Dickinson*. Variorum Edition. 3 vols. Cambridge, MA: Belknap Press of Harvard University Press, 1998.

Fretwell, Erica. "Emily Dickinson in Domingo." *J19: The Journal of Nineteenth-Century Americanists* 1, no. 1 (2013): 71–96.

Fuller, Randall. *The Book That Changed America: How Darwin's Theory of Evolution Ignited a Nation*. New York: Viking, 2017.

Gallenga, Antonio Carlo Napoleone. *Episodes of My Second Life: (American and English Experiences)*. New York: J. B. Lippincott, 1885.

Gerhardt, Christine. *A Place for Humility: Whitman, Dickinson, and the Natural World*. Iowa City: University of Iowa Press, 2014.

Goldman, Anne. "Dickinson and Darwin among the Heliotropes." *Georgia Review* 72, no. 3 (Fall 2018): 753–771.

Gopnik, Adam. *Angels and Ages: A Short Book about Darwin, Lincoln, and Modern Life*. New York: Vintage, 2010.

Gosden, Chris. *Magic: A History: From Alchemy to Witchcraft, from the Ice Age to the Present*. New York: Farrar, Straus, and Giroux, 2020.

Gould, Stephen Jay. *Rocks of Ages: Science and Religion in the Fullness of Life*. New York: Ballantine, 1999.

Gray, Asa. "Darwin on the Origin of Species: A Book Review." *Atlantic*, July 1860. https://www.theatlantic.com/magazine/archive/1860/07/darwin-on-the-origin-of-species/304152/.

Grima, Peter P. "The Natural Hybrid between *Drosera intermedia* and *Drosera rotundifolia* in Massachusetts." *Rhodora* 122, no. 989 (2020): 23–36.

Guthrie, James R. "Darwinian Dickinson: The Scandalous Rise and Noble Fall of the Common Clover." *The Emily Dickinson Journal* 16, no. 1 (2007): 73–91.

Habegger, Alfred. *My Wars Are Laid Away in Books: The Life of Emily Dickinson*. New York: Random House, 2001.

Hale, Edward Everett. *James Russell Lowell and His Friends*. Boston: Houghton Mifflin, 1901.

Hardy, Thomas. "Hap."

Harel, Kay. "'It's Dogged as Does It': A Biography of the Everpresent Canine in Charles Darwin's Days." *Southwest Review* 93, no. 3 (2008): 368–378, 462.

Hart, Ellen Louise, and Martha Nell Smith. "Why Susie!" In *Open Me Carefully: Emily Dickinson's Intimate Letters to Susan Huntington Dickinson*, edited by Ellen Louise Hart and Martha Nell Smith, 1–6. Ashfield, MA: Paris Press, 1998.

Hawthorne, Nathaniel. *The Scarlet Letter*. 1850.

Heiligman, Deborah. *Charles and Emma: The Darwins' Leap of Faith.* New York: Square Fish, 2011.

Herschel, John F. W. *A Preliminary Discourse on the Study of Natural Philosophy.* Philadelphia: Carey and Lea, 1831.

Hesketh, Ian. *Of Apes and Ancestors: Evolution, Christianity, and the Oxford Debate.* Toronto: University of Toronto Press, 2018.

Higginson, Thomas Wentworth. *Army Life in a Black Regiment.* Boston: Fields, Osgood, 1870.

Higginson, Thomas Wentworth. *Cheerful Yesterdays.* Cambridge, MA: Riverside, 1898.

Higginson, Thomas Wentworth. "Emily Dickinson's Letters." *Atlantic Monthly*, October 1891. https://www.theatlantic.com/magazine/archive/1891/10/emily-dickinsons-letters/306524/.

Higginson, Thomas Wentworth. "Letter to a Young Contributor." *Atlantic Monthly*, April 1862. https://www.theatlantic.com/magazine/archive/1862/04/letter-to-a-young-contributor/305164/.

Higginson, Thomas Wentworth. "The Life of a Bird." In *Out-door Papers*, 293–315. Boston: Ticknor and Fields, 1863.

Higginson, Thomas Wentworth. "Ought Women to Learn the Alphabet?" *Atlantic Monthly*, February 1859. https://www.theatlantic.com/magazine/archive/1859/02/ought-women-to-learn-the-alphabet/306366/.

Higginson, Thomas Wentworth. *Out-door Papers.* Boston: Ticknor and Fields, 1863.

Higginson, Thomas Wentworth. "The Procession of the Flowers." In *Out-door Papers*, 317–337. Boston: Ticknor and Fields, 1863.

History of the Town of Amherst, Massachusetts. Amherst, MA: Carpenter & Morehouse, 1896. Emily Dickinson Museum. https://www.emilydickinsonmuseum.org/samuel-fowler-dickinson-1775-1838-and-lucretia-gunn-dickinson-1775-1840-grandparents/.

"History of Women at Harvard Medical School." Harvard University Joint Committee on the Status of Women. https://jcsw.hms.harvard.edu/history.

Hitchcock, Edward. "Description of a Large Bowlder in the Drift of Amherst, Massachusetts, with Parallel Striae upon Four Sides." *American Journal of Science and Arts* (1856): 397–400. https://acdc.amherst.edu/view/asc:701826/asc:701834.

Hitchcock, Edward. *The Power of Christian Benevolence Illustrated in the Life and Labors of Mary Lyon.* New York: American Tract Society, 1858.

Hitchcock, Edward. *The Religion of Geology and Its Related Sciences.* Boston: Phillips, Sampson, 1851.

Hitchcock, Edward. *Reminiscences of Amherst College*, Northampton, MA: Bridgman & Childs, 1863.

Hitchcock, Edward. *Report on the Geology, Mineralogy, Botany, and Zoology of Massachusetts.* Amherst, MA: Press of J. S. and C. Adams, 1833.

Holmes, John. *Darwin's Bards: British and American Poetry in the Age of Evolution.* Edinburgh: Edinburgh University Press, 2009.

Holmes, Richard. *The Age of Wonder: How the Romantic Generation Discovered the Beauty and Terror of Science.* London: Harper, 2008.

Holmes, Richard. "In Retrospect: On the Connexion of the Physical Sciences." *Nature* 514 (2014): 432–433.

Hughes, Thomas McKenny, and John Willis Clark. *The Life and Letters of the Reverend Adam Sedgwick.* Cambridge: University Press, 1890.

Humboldt, Alexander von. *Letters of Alexander von Humboldt, Written between the Years 1827 and 1858, to Varnhagen von Ense: Together with Extracts from Varnhagen's Diaries, and Letters from Varnhagen and Others to Humboldt.* Authorized translation from the German with explanatory notes and a full index. London: Trübner, 1860.

Hunt, Tristram. *The Radical Potter: Josiah Wedgwood and the Transformation of Britain.* London: Allen Lane, an imprint of Penguin Books, 2021.

Hunter, Michael. *The Decline of Magic: Britain in the Enlightenment.* New Haven, CT: Yale University Press, 2020.

Huxley, Thomas H. "Agnosticism." In *Science and Christian Tradition.* London: Macmillan, 1894, 209–262. Google Books.

Huxley, Thomas H. "Darwin on The Origin of Species." *The Times*, December 26, 1859, 8–9. Darwin Online.

James, Henry. "Middlemarch [A Review]." *The Galaxy*, March 1873, 424–428.

James, William. "'Is Life Worth Living?' (An Address to the Young Men's Christian Association at Harvard University, May 1895)." *International Journal of Ethics* 6, no. 1 (October 1895): 1–24.

Jensen, J. Vernon. "Return to the Wilberforce–Huxley Debate." *British Journal for the History of Science* 21, no. 2 (1988): 161–179.

Josephson Storm, Jason Ā. *The Myth of Disenchantment: Magic, Modernity, and the Birth of the Human Sciences.* Chicago: University of Chicago Press, 2019.

Keane, Patrick J. *Emily Dickinson's Approving God: Divine Design and the Problem of Suffering.* Columbia: University of Missouri Press, 2008.

Keats, John. "Ode on a Grecian Urn."

Kelly, Mike. "Emily Dickinson and the New York Press." *The Consecrated Eminence: Archives and Special Collections at Amherst College* (blog), July 15, 2013. https://consecratedeminence.wordpress.com/2013/07/15/emily-dickinson-and-the-new-york-press/.

Kelly, Mike et al. *The Networked Recluse: The Connected Worlds of Emily Dickinson.* Amherst, MA: Amherst College Press, 2017. https://doi.org/10.3998/mpub.9959167.

Kingsley, Charles. *The Water-Babies: A Fairy Tale for a Land-Baby.* Cambridge, MA: Macmillan and Co., 1863.

Kirkby, Joan. "'We Thought Darwin Had Thrown "the Redeemer" Away': Darwinizing with Emily Dickinson." *The Emily Dickinson Journal* 19, no. 1 (2010): 1–29.

Knapp, Krister Dylan. *William James: Psychical Research and the Challenge of Modernity.* Chapel Hill: University of North Carolina Press, 2017.

Kuhn, Mary. "Dickinson and the Politics of Plant Sensibility." *ELH* 85, no. 1 (2018): 141–170.

Leighton, William Allport. [Recollections of Charles Darwin]. CUL-DAR112.B94-B98. c. 1886. Darwin Online.

Levine, George. *Darwin Loves You: Natural Selection and the Re-enchantment of the World.* Princeton, NJ: Princeton University Press, 2008.

Levine, George. *Reading Thomas Hardy.* Cambridge: Cambridge University Press, 2017.

Leyda, Jay. *The Years and Hours of Emily Dickinson.* New Haven, CT: Yale University Press, 1960.

Lincoln, Abraham. *Collected Works of Abraham Lincoln.* Edited by Roy P. Basler. 8 vols. New Brunswick, NJ: Rutgers University Press, 1953.

Lincoln, Almira Hart. *Familiar Lectures on Botany.* Hartford, CT: H. and F. J. Huntington, 1829.

Linné, Carl von, and Carl Linné. *The Families of Plants: with Their Natural Characters, According to the Number, Figure, Situation, and Proportion of All the Parts of Fructification. Translated from the Last Edition, (as Published by Dr. Reichard) of the Genera Plantarum, and of Mantissæ Plantarum of the Elder Linneus; and from the Supplementum Plantarum of the Younger Linneus, with All the New Families of Plants, from Thunberg and L'Heritier. To Which Is Prefix'd an Accented Catalogue of the Names of Plants, with the Adjectives Apply'd to Them, and Other Botanic Terms, for the Purpose of Teaching Their Right Pronunciation. Two Volumes. By a Botanical Society at Lichfield.* Translated by Erasmus Darwin. Lichfield, UK: printed by John Jackson. Sold by J. Johnson, St. Paul's Church-yard, London. T. Byrne, Dublin. and J. Balfour, Edenburgh, 1787.

Lister, Adrian. *Darwin's Fossils: The Collection That Shaped the Theory of Evolution.* Washington, D.C.: Smithsonian, 2018.

Litchfield, Henrietta. *Emma Darwin: A Century of Family Letters, 1792–1896.* 2 vols. London: John Murray, 1915. Darwin Online.

Longsworth, Polly. "'And Do Not Forget Emily': Confidante Abby Wood on Dickinson's Lonely Religious Rebellion." *New England Quarterly* 82, no. 2 (2009): 335–346.

Lowell, Amy. "The Sisters." In *What's O'Clock.* Boston: Houghton Mifflin, 1925.

Lowell, James Russell. *The Letters of James Russell Lowell.* Vol. 2. Edited by Charles Eliot Norton. London: Osgood, McIlvaine, 1894.

Lyell, Charles. *Principles of Geology: Being an Attempt to Explain the Former Changes of the Earth's Surface, by Reference to Causes Now in Operation.* 2nd ed. London: J. Murray, 1832.

Lyell, Charles. *Travels in North America, in the Years 1841–2; with Geological Observations on the United States, Canada, and Nova Scotia.* New York: Wiley and Putnam, 1845.

Macfarlane, Robert. "Badger or Bulbasaur—Have Children Lost Touch with Nature?" *The Guardian,* September 30, 2017. https://www.theguardian.com/books/2017/sep/30/robert-macfarlane-lost-words-children-nature.

Marché, Jordan D. II. "Edward Hitchcock's Poem, *The Sandstone Bird* (1836)." *Earth Sciences History* 10, no. 1 (1991): 5–8.

Martineau, Harriet. *Retrospect of Western Travel.* London: Saunders and Otley, 1838.

McDowell, Marta. *Emily Dickinson's Gardening Life: The Plants and Places That Inspired the Iconic Poet.* Portland, OR: Timber Press, 2019.

Meredith, George. "Modern Love XXX."

Miller, Cristanne, ed. *Emily Dickinson's Poems: As She Preserved Them.* Cambridge, MA: Belknap Press of Harvard University Press, 2016.

Milton, John. *Paradise Lost.*

Mitchell, Donald Grant (pseud. Ik Marvel). *Reveries of a Bachelor.* 1850.

Mondello, Kaitlin. "'Of Toads and Men': Brutal Kinship in Emily Dickinson and Charles Darwin." *Journal of Literature and Science* 13, no. 2 (2020): 1–19.

Morlier, Margaret M. "The Death of Pan: Elizabeth Barrett Browning and the Romantic Ego." *Browning Institute Studies* 18 (1990): 131–55. https://doi:10.1017/S009247250000290X.

Mott, Lucretia. "Discourse on Women—Dec. 17, 1849." Archives of Women's Political Communication, Iowa State University. https://awpc.cattcenter.iastate.edu/2017/03/21/discourse-on-women-dec-17-1849/.

Olmsted, Denison. *On the Recent Secular Period of the Aurora Borealis*. Washington, DC: Smithsonian Institution, 1856.

Pearson, P. N., and C. J. Nicholas. "'Marks of Extreme Violence': Charles Darwin's Geological Observations at St Jago (São Tiago), Cape Verde Islands." In *Four Centuries of Geological Travel: The Search for Knowledge on Foot, Bicycle, Sledge and Camel*, ed. P. N. Wyse Jackson, 239–253. Special Publications, vol. 287. London: Geological Society, 2007. https://doi.org/10.1144/SP287.19.

Piesse, Jude. *The Ghost in the Garden: In Search of Darwin's Lost Garden*. London: Scribe, 2021.

Pollak, Vivian R. "Dickinson and the Poetics of Whiteness." *The Emily Dickinson Journal* 9, no. 2 (2000): 84–95.

Pollak, Vivian R. *A Poet's Parents: The Courtship Letters of Emily Norcross and Edward Dickinson*. Chapel Hill: University of North Carolina Press, 1988.

Porta, Giambattista della. *Natural Magick*. London: Printed for Thomas Young, and Samuel Speed, and are to be sold at the three Pigeons, and at the . . . https://doi.org/10.5479/sil.82926.39088002126779.

Quinn, Carol. "Dickinson, Telegraphy, and the Aurora Borealis." *The Emily Dickinson Journal* 13, no. 2 (2004): 58–78.

Richards, Eliza. "'How News Must Feel When Traveling': Dickinson and Civil War Media." In *A Companion to Emily Dickinson*, edited by Martha Nell Smith and Mary Loeffelholz, 157–179. Malden, MA: Blackwell, 2008.

Richards, Robert J. *The Romantic Conception of Life Science and Philosophy in the Age of Goethe*. Chicago: University of Chicago Press, 2002.

Richardson, Robert D. *Emerson: The Mind on Fire: A Biography*. Berkeley: University of California Press, 1995.

Ridley, Mark. *The Darwin Reader*. New York: W. W. Norton, 1987.

Rose, Michael R. *Darwin's Spectre: Evolutionary Biology in the Modern World*. Princeton, NJ: Princeton University Press, 1998.

Satter, Beryl. *Each Mind a Kingdom: American Women, Sexual Purity, and the New Thought Movement, 1875–1920*. Berkeley: University of California Press, 1999.

Saxton, Martha, ed. *Amherst in the World*. Amherst, MA: Amherst College Press, 2020.

Schaefer, John R. "From Poetry to Pulp Fiction: Carnivorous Plants in Popular Culture." *Biodiversity Heritage Library Blog Reel*, October 28, 2021. https://blog.biodiversitylibrary.org/2021/10/from-poetry-to-pulp-fiction-carnivorous-plants-in-popular-culture.html.

Schweitzer, Ivy. "June 11–17, 1862: Elizabeth Barrett Browning." *White Heat: Emily Dickinson in 1862: A Weekly Blog*, June 17, 2018. https://journeys.dartmouth.edu/whiteheat/2018/06/17/june-11-17-1862-elizabeth-barrett-browning/.

Scott, Walter. *Letters on Demonology and Witchcraft: Addressed to J. G. Lockhart*. Cambridge: Cambridge University Press, 2011. https://doi:10.1017/CBO9780511792915.

Scott, Walter, and George Cruikshank. *Twelve Sketches: Illustrative of Sir Walter Scott's Demonology and Witchcraft*. London: Published for the artist by J. Robins and Co., 1830.

Sedgwick, Adam. "Vestiges of the Natural History of Creation" (Review). *The Edinburgh Review or Critical Journal* 82, no. 165 (July 1845): 1–85.

Sewall, Richard B. *Life of Emily Dickinson*. Cambridge, MA: Harvard University Press, 1994.

Sewall, Richard B. "The Lyman Letters: New Light on Emily Dickinson and Her Family." *The Massachusetts Review* 6, no. 4 (1965): 693–780.

Shakespeare, William. *A Midsummer Night's Dream.*

Shakespeare, William. *Troilus and Cressida.*

Shaw, Bernard. *Back to Methuselah: A Metabiological Pentateuch.* New York: Brentano's, 1929.

Shelley, Mary. *Frankenstein: or, The Modern Prometheus.* Edited by M. K. Joseph. 1831; reprint, Oxford: Oxford University Press, 2008.

Shmurak, Carole B., and Bonnie S. Handler. "'Castle of Science': Mount Holyoke College and the Preparation of Women in Chemistry, 1837–1941." *History of Education Quarterly* 32, no. 3 (1992): 315–342.

Shoptaw, John. "Dickinson's Civil War Poetics from the Enrollment Act to the Lincoln Assassination." *The Emily Dickinson Journal* 19, no. 2 (2010): 1–19.

Skrupskelis, Ignas K. "Evolution and Pragmatism: An Unpublished Letter of William James." *Transactions of the Charles S. Peirce Society* 43, no. 4 (2007): 745–752.

Snyder, Laura J. *The Philosophical Breakfast Club: Four Remarkable Friends Who Transformed Science and Changed the World.* New York: Broadway Books, 2011.

Socarides, Alexandra. *Dickinson Unbound: Paper, Process, Poetics.* New York: Oxford University Press, 2012.

Somerville, Mary. *On the Connexion of the Physical Sciences.* London: Clowes, 1834. Darwin Online.

Stoddart, D. R. "Darwin, Lyell, and the Geological Significance of Coral Reefs." *British Journal for the History of Science* 9, no. 2 (1976): 199–218.

Stott, Rebecca. *Darwin and the Barnacle: The Story of One Tiny Creature and History's Most Spectacular Scientific Breakthrough.* New York: W. W. Norton, 2003.

Stott, Rebecca. *Darwin's Ghosts: The Secret History of Evolution.* New York: Random House, 2012.

Subramaniam, Banu. *Ghost Stories for Darwin: The Science of Variation and the Politics of Diversity.* Urbana-Champaign: University of Illinois Press, 2014.

Temple, Nicola. "A Portrait of a Boy and His Plant." University of Bristol Botanic Garden. https://botanicgarden.blogs.bristol.ac.uk/category/history/.

Tennyson, Alfred. "Charge of the Light Brigade."

Tennyson, Alfred. "In Memoriam, AHH."

Thomson, Keith. *The Young Charles Darwin.* New Haven, CT: Yale University Press, 2009.

Thoreau, Henry David. "Resistance to Civil Government." In *The Writings of Henry D. Thoreau: Reform Papers*, edited by Wendell Glick. Princeton, NJ: Princeton University Press, 1973.

Thoreau, Henry David. "7 June 1851." In *Journal*, vol. 3, edited by Robert Sattelmeyer, Mark R. Patterson, and William Rossi. Princeton, NJ: Princeton University Press, 1991.

Thoreau, Henry David. "28 October 1853." In *Journal*, vol. 7, edited by Nancy Craig Simmons and Ron Thomas. Princeton, NJ: Princeton University Press, 2009.

Thrailkill, Jane F. "Fables of Extinction." In *Amherst in the World*, ed. Martha Saxton. Amherst, MA: Amherst College Press, 2020, 216-234.

Tolley, Kim. *The Science Education of American Girls: A Historical Perspective.* Abingdon, UK: RoutledgeFalmer, 2003.

Uglow, Jennifer S. *The Lunar Men: Five Friends Whose Curiosity Changed the World.* New York: Farrar, Straus, and Giroux, 2002.

van Wyhe, John, ed. *The Complete Work of Charles Darwin Online.* 2002–. http://darwin-online.org.uk/.

Ventriloquism Explained: And Juggler's Tricks, or Legerdemain Exposed: With Remarks on Vulgar Superstitions. In a Series of Letters to an Instructor. Amherst, MA: J. S. and C. Adams, 1834. Pdf. Library of Congress, Harry Houdini Collection, and McManus-Young Collection. https://www.loc.gov/item/34015038/.

Wainwright, William. S.v. "Jonathan Edwards." In *The Stanford Encyclopedia of Philosophy*, edited by Edward N. Zalta, Fall 2020. https://plato.stanford.edu/archives/fall2020/entries/edwards/.

Wald, Priscilla. *Constituting Americans: Cultural Anxiety and Narrative Form.* Durham, NC: Duke University Press, 1995.

Walls, Laura Dassow. *Emerson's Life in Science: The Culture of Truth.* Ithaca, NY: Cornell University Press, 2003.

Walls, Laura Dassow. "Greening Darwin's Century: Humboldt, Thoreau, and the Politics of Hope." *Victorian Review* 36, no. 2 (2010): 92–103.

Walls, Laura Dassow. *Thoreau: A Life.* Chicago: University of Chicago Press, 2017.

Weber, Max. *Charisma and Disenchantment: The Vocation Lectures.* New York: New York Review of Books, 2020.

Werner, Marta L. *Writing in Time: Emily Dickinson's Master Hours.* Amherst, MA: Amherst College Press, 2021. https://doi.org/10.3998/mpub.12023683.

West, Peter. "The City in Frames: Otis Bullard's Moving Panorama of New York." Commonplace. http://commonplace.online/article/the-city-in-frames/.

Whewell, William. *Astronomy and General Physics: Considered with Reference to Natural Theology.* The Bridgewater Treatises, vol. 3. London: William Pickering, 1833.

Whewell, William. *Philosophy of the Inductive Sciences.* 1840; repr., Cambridge: Cambridge University Press, 2014.

Wilson, Edward O. *From So Simple a Beginning: The Four Great Books of Charles Darwin.* New York: W. W. Norton, 2006.

Wineapple, Brenda. *White Heat.* New York: Anchor Books, 2008.

Wingate, Zenobia Grant. "Otis Bullard: Biography." Caldwell Gallery. https://www.caldwellgallery.com/artists/otis-bullard/biography.

Wootton, David. *The Invention of Science: A New History of the Scientific Revolution.* London: Allen Lane, 2015.

Wordsworth, William. "I Wandered Lonely as a Cloud." https://www.poetryfoundation.org/poems/45521/i-wandered-lonely-as-a-cloud.

Wordsworth, William. "Observations Prefixed to *Lyrical Ballads*." https://www.poetryfoundation.org/articles/69383/observations-prefixed-to-lyrical-ballads.

Wordsworth, William. "Tables Turned." https://www.poetryfoundation.org/poems/45557/the-tables-turned.

Wulf, Andrea. *The Invention of Nature: Alexander von Humboldt's New World.* New York: Alfred A. Knopf, 2015.

INDEX

Notes: Page numbers in *italics* indicate figures. ED indicates Emily Dickinson, and CD indicates Charles Darwin. *On the Origin of Species* and Darwin's other works are in the entry "Darwin, Charles, works." Likewise, the poems of Dickinson are in the entry "Dickinson, Emily, poems."

Abbot, Francis, 316, 324
abolitionism/abolitionists: antislavery medallion, 247, *248*; *Atlantic Monthly*'s antislavery stance, 243; and Darwin's theory, 246–248, 249; ED's attitude, 290, 294–295
Ackmann, Martha, 119, 128–129
Agassiz, Louis, 172, 262–263
agnostic as term, 230–231
Alcott, Bronson, 249–250
Alcott, Louisa May, 293, 303
Allen, Grant, 319–320, 344
Amherst Academy: and Amherst College, 163–164; curriculum, 77–78, 122, 156; establishment, 76; women teachers, 131–132
Amherst boulder, 213–214
Amherst College: Appleton Cabinet, 213; establishment, 4–5, 23, 72, 91; influence of Edwards on, 42, 43; natural sciences and classics, 72–73; Octagon Cabinet (museum), 170–172, *171*, 374n29; women at, 78, 125, 172
Amherst General School Committee, 77
Amherst railroad, 206, 208, 241
Anglican Church, 173, 184
anthropomorphism, 267–268
Appleton Cabinet (museum), 213

Argentina, 104
Atlantic Monthly: Fiske's essays, 314, 320; founding of, 243–244, 379n9; Higginson's essays, 279, 287–288, 382n35; review of *Origin*, 246, 250–251, 354; September 1861 issue, 257–258, 259
aurora borealis, 201–204, 207, 376n8; in poetry, 218–219, 268–269, 376n5

BAAS (British Association for the Advancement of Science): attitude regarding amateur scientists, 106; debate about CD's theory, 233–234; dissection of ape brain, 272; divorce between arts and sciences, 107–109, 271; founding, 11–12; rift with natural philosophers, 83
Babbage, Charles, 11–12, 108
Bacon, Francis, 10, 11, 191, 228
Bagehot, Walter, 317
Banks, Sir Joseph, 18, 50
Barad, Karen, 15
Barlow, Nora (granddaughter of CD), 324
barnacles, 193–196, *195*, 221–222
Barrett Browning, Elizabeth. *See* Browning, Elizabeth Barrett

405

Beagle expedition: butterfly flocks, 285; final leg of voyage, 133; fossils, 103–104; geological studies, 94, 98–100, 101; life aboard ship, 1, 46, 70–71, 93, 94–95, 247; specimens from, 104, 106–107, 127; in Uruguay, 101–106. *See also* Darwin, Charles; Darwin, Charles, works, *On the Origin of Species*; Darwin, Charles, works, *The Voyage of the Beagle*; natural selection theory (Darwin)
Beer, Gillian, 69, 267, 300–301, 302
Bell, Thomas, 192–193
Bennett, Jane, 15, 16, 167
Bernhard, Mary, 124–125
Bianchi, Martha Dickinson (niece of ED), 303, 349
Bingham, Millicent Todd, 349
Boggs, Colleen Glenney, 291, 382n14
botany: climate data, 287–288, 382n35; *Drosera* (sundew), 260–263, *261*; as feminine, 39–40; orchid fertilization, xiv–xv; plant consciousness, 266, 276; and Romanticism, 157; as school subject, 37–38; and sexuality, 128–129; as socially disruptive, 23, 24, 31–32, 43–45. *See also* Darwin, Charles, natural studies; Darwin, Erasmus (grandfather of CD)
Boulton, Matthew, 26
Bowles, Samuel, 78, 208, 212, 240, 332–333; correspondence with ED, 256, 278, 355
Brace, Charles Loring, 249–250
brains, 271–272, 275
Brewster, David, 11, 12–13, 108–109
British Association for the Advancement of Science (BAAS). *See* BAAS (British Association for the Advancement of Science)
British Museum, 135
Brontë, Charlotte, 203–204
Brown, John, 248–250
Browne, Janet, 13, 28, 128, 234, 318, 322
Browning, Elizabeth Barrett, *219*; *Aurora Leigh*, 203, 215–219, 257; death of, 253, 258, 292; influence on CD, 340; influence on ED, 215–219, 255–257, 280–281, 286, 310, 311

Browning, Robert, 272, 292–293, 309
Bryant, William Cullen, 293
Buell, Lawrence, 83
Bullard, Otis Allen, *21*, 21–22, 24, 38–39, 124
Burke, William, 59
Burnett, Frances Hodgson, 240
Butler, Samuel, 54–56, 74, 122, 134
butterflies, 285, 294
Byron, Lord, 36, 203, 376n8

Called Back (Hugh Conway), 334–335, 336
"Called Back" as epitaph, 337–338
Cambridge Philosophical Society, 133, 134
Cambridge University, 56–57, 65–66, 74–75, 109–110
Cameron, Julia, 303–304, 309–310, 322
cardinal flowers, 115, *116*, 127–128
Carlyle, Thomas, 108, 310–311, *310*
carnivorous plants, 260–263, *261*
Case, George, 49–50, 51
Case school, 49
Chambers, Robert, 189–192
Child, Lydia Maria, 203, 376n8
Christians/Christianity: influence of Edwards on, 42–43; and magic, 16; and sciences, 172–174; theological differences, 184; Unitarianism, 50–51. *See also* natural magic; natural philosophy; science and religion
Christmas holiday, 160–161
Civil War (U.S.): Amherst residents in, 277–278; *Atlantic Monthly* articles, 257–258; beginning, 252–253; end, 297; 1st South Carolina Volunteers, 250, 286, 382n29; Higginson in, 250, 286–287, 291, 382n29; prewar tensions, 241–242, 248–250; racism during, 291–292; social effects, 303
Clark, William S., 277, 381n2
climate data in botanical observations, 287–288, 382n35
Cobbe, Frances Powers, 304–305, 317
Coleman, Eliza (cousin of ED), 212
Coleridge, Samuel Taylor: criticism of Erasmus Darwin, 51–52; criticism of materialism, 51; criticism of materialism at BAAS meeting,

107, 108–109, 271; criticism of naturalists, 33, 34, 36, 61, 366n25; financial settlement, 49–52; influence of Humboldt on, 99
—, works: *Aids to Reflection*, 61; *Lyrical Ballads* (with Wordsworth), 33, 51, 266
College of Surgeons, 135
Comet Mitchell, 154–156
Congregational Church, 173–174
consciousness in plants, 262–263, 266
Conway, Hugh, 334–335, 336
coral reefs, 136
cosmology, 97–98
Costa, James T., 263, 283
Crane, Stephen, 325–326
Cresy, Edward, 195–196

Dandurand, Karen, 293
Darwin, Anne Elizabeth (daughter of CD), 181, 198–199, *199*
Darwin, Caroline (sister of CD), 47, 55, 145; correspondence with CD, 61, 112–113, 138
Darwin, Catherine (sister of CD), 19, *20*, 24, 47, 54; correspondence with CD, 103, 104, 111
Darwin, Charles: Darwin (the man) vs. Darwinism, 324–326; as intellectual "kinsman" of ED, 340–342, 349–350, 357; portraits and caricatures, *20*, *68*, *147*, *196*, *235*, *310*. See also *Beagle* expedition; Darwin, Charles, works; Darwinism; natural selection theory (Darwin)
—, beliefs and ideas: abolition of slavery, 246–247; death as generative force, 270, 284, 341–342, 351–353; human-nonhuman relationships, 267, 270, 299–300; morality as instinctive, 182–183; natural magic, 345; political activism, 304–305, 319; reading preferences, 300–302; religious feelings, 178–179, 183–184, 198–200, 226–233, 315–317, 344; religious language use, 226–227; social inequalities, 315–319
—, career: career choice, 60, 105–106, 111–112; correspondents and networks, 133–137, 220, 221, 300; elected to Royal Society, 145; exhibition of specimens, 106–108; expertise as biologist established, 193–196, *195*; personal popularity and image curation, 314–316, 324; reactions to reception of *Origin*, 232–236, *235*, 260, 262–263; reluctance to publish, 111, 137, 185–186; writing, editing, and publishing, 113–114, 192–193, 276, 299–300
—, education: at Cambridge, 66–70, 74–75; at Case school, 49; conventional schooling, 36; at Edinburgh, 57–59, 61–62; at Shrewsbury School, 54–56, 74
—, family: background, 5–6, 18–19; childhood, 46–49, *49*, 52–54, *54*; childhood portrait, *19*, *20*, *21*, 23–24; children, 145, 147–148, 181, *181*, 182–183, 225; children's deaths, xi–xii, 7, 198–200, 222–223, 225, 226; death, funeral, and legacy of CD, 319–322, *321*, *323*, 324, 344; death of mother, 52–53; dogs, 295–297, 319; Down House, 148, 180–182, *181*, 200, 263, 305; health, 196–197, 233–234, 309; marriage relationship, 144–146, 179–181; social circles, 305–306, 307; travels, 303–304; work space, 181, *200*
—, influences on: Eliot, 302; Herschel, 81, 112–113; Humboldt, 70, 98–100, 313; Lyell, 94, 95–96, 98, 100–101, 112, 136–137, 142–143; Malthus, 143; Martineau, 137–144, *138*; Romanticism, 114
—, natural studies: approach to, 2–4, 7–8, 16–17, 265–266, 345–346; biology of earthworms, 263, 270–271, 303; biology of marine invertebrates, 62–64, *62*, 193–196, *195*, 221–222; botany, xiv–xv, 36–37; botany and plant consciousness, 266, 276; botany of *Drosera*, 260–263, *261*; chemistry, 56; childhood collections, 52; emotions, 295–297, *296*; entomology, 67, *68*, 269; experiments, 270–271; geology, 64; geology of coral reefs, 136; geology of Wales, 70, 93–94; geology on *Beagle* expedition, 94, 98–100; pigeons, 222; specimen collections on *Beagle* expedition, 104, 105, 106–107, 127; taxidermy, 64–65

Darwin, Charles, works: barnacle books, 221–222; *Geological Observations on South America*, 150; "Geology" in *Manual of Scientific Enquiry*, 197–198; *Journal of Researches*, 145; nontheoretical works, 314; on orchid fertilization, xiv–xv; on plant movement and consciousness, 266, 276; theoretical books, 313–314; various projects in later decades, 299, 303; *The Zoology of the Voyage of H.M.S. Beagle* (ed.), 192–193. See also *Beagle* expedition; Darwin, Charles; natural selection theory (Darwin)

—, *The Descent of Man*: beauty and sexual selection, 3, 235–236, 235; Darwinian arguments for women's rights, 327; publication of, 303, 313; and social Darwinism, 317

—, *The Expression of the Emotions*: dog emotions, 295–297, 296; as foundational text, 304; human–nonhuman relationships, 313–314; images for, 305; mystery and wonder, 235–236

—, *On the Origin of Species*: 1859 edition epigraphs and introduction, 228–229; 1860 second edition (revision), 229; 1861 third edition "Historical Sketch," 191–192; and abolitionism, 246–248, 249, 250; as best seller, 227–228; death as generative force, 341–342; humans omitted from, 313; interconnectedness of life-forms, 282–284; publication, xiv; reactions to, xiv, 233–236, 235, 249, 260; reactions to in United States, 246–248, 250–251; reviews of, 231, 232, 354; wonder for natural world, 3–4

—, *The Voyage of the Beagle*: butterfly flocks, 285; influence on Thoreau, 167; popularity of, 313; publication, 3; Romanticism in, 119–121, 325, 371n12

Darwin, Charles (uncle of CD), 57, 59
Darwin, Charles Waring (son of CD), xi–xii, 222–223, 225, 226
Darwin, Emma Wedgwood (wife of CD), 146; on burial of CD, 321; on CD's work, 262; correspondence with CD, 143, 188, 344; marriage and relationship, xiv, 144–148, 179–182, 198–200; religious feelings, 179, 183, 184–185, 229, 344

Darwin, Erasmus (grandfather of CD), 31; botanical studies, 20, 24–25, 27–30, 32, 261–262, 365n16; criticism of, 33–34, 35–36, 51–52, 59–60, 111, 366n25; egalitarian impulses, 317; Lunar Society, 18, 26; medical studies, 57; other studies, 34–35; as physician, 50; religious beliefs, 184; transmutation, 59–61. See also Lunar Society of Birmingham

—, works: *The Botanic Garden*, 28–30, 32, 261–262; *The Economy of Vegetation*, 30, 32; *The Loves of the Plants*, 28–30, 128, 317; *Phytologia*, 34; *The Temple of Nature*, 34–35, 60; *Zoonomia*, 34, 35, 59–60

Darwin, Erasmus "Ras" (brother of CD): at Edinburgh University, 57–58; and Martineau, 138, 139–142; relationship with CD, 47, 48–49, 56

Darwin, Francis (son of CD), 181, 269, 324
Darwin, Henrietta "Etty" (daughter of CD), 181, 198, 223, 225, 260
Darwin, Horace (son of CD), 181, 269
Darwin, Leonard (son of CD), 181, 269
Darwin, Marianne (sister of CD), 47
Darwin, Robert Waring (father of CD): correspondence with CD, 100; education, 26; education of children, 36, 57, 65; family background, 18, 60; interest in science, 52; personality, 47–48

Darwin, Susanna Wedgwood (mother of CD): death of, 52–53; education, 26, 48; education of children, 36; family background, 18; gardening, 53; spelling of name, 365n1

Darwin, Susan (sister of CD), 47, 139
Darwin, William (son of CD), 181, 222, 223, 324
Darwinian thought: and enchantment, 17, 345–346; evolutionary kinship and activism, 304–305; influence on ED, xv, 17; influence on Eliot, 302; and materialism, 270–271;

and new fields of study, 304. *See also* Darwinism; human–nonhuman relationships; interconnections

Darwinism: and changing conceptions of evolution, 272–274; contradictory interpretations of, 314, 320; vs. Darwin the man, 324–326; and divorce of science from humanities, 231; and naturalism in literature, 325–326; social Darwinism, 234, 304–305, 316–318; as support for varying ideologies, 304–305, 316–318. *See also* Darwinian thought; natural selection theory (Darwin)

Darwinizing, 52

Darwin's ghost as metaphor, 345

Darwin-Wedgwood clan, 6, 48, 49–52, 74

Day, Thomas, 26

death: as generative force, 270, 284, 341–342, 351–353; and shaping of natural world, 214–215, 255, 294–295; as source of creative energy for ED, 339–341, 387n34; as transformative, 15, 251, 331–332. *See also* human–nonhuman relationships; interconnections

Deleuze, Gilles, 15

Dickerman, Mrs. George S., 336

Dickinson, Austin (brother of ED): and Amherst College, 92; childhood portrait, 21, 22, 24; children of, 253; Civil War, 278, 290–292; education, 155; extramarital affair, 328; and Gilbert sisters, 204–206, 210; marriage, 212–213, 238; relationship with ED, 120, 331

—, correspondence with ED, 160, 165, 202

Dickinson, Edward (father of ED), *124*; and Amherst College, 92; and Amherst educational institutions, 77; career, 5, 23, 119, 208–209; church membership, 174; courtship and marriage, 124–125; death of, 330; education, 76; finances, 212–213, 241; politics, 290; readings permitted by, 130; viewing aurora, 201–202

Dickinson, Edward "Ned" (nephew of ED), 253

Dickinson, Emily, 21, *159*; as intellectual "kinsman" of CD, 340–342, 349–350, 357; natural magic and misreadings of, 345–350; posthumous publication, 348–350. *See also* Dickinson, Emily, "Master" documents; Dickinson, Emily, poems

—, beliefs and ideas: approach to natural world, 265–266; biological humility, 294; death and magic, 331–332; death and shaping of natural world, 214–215, 255, 294–295; death as creative force, 352; death as source of creative energy, 339–341, 387n34; evolution and extinction, 354–355; grief, 292–293, 297–298, 308; haunting as metaphor, 341–344; human–nonhuman relationships, 281; ideal poet, 353–354; magic of poetry, 256–257, 338–339; political debates, 174; political upheaval, awareness of, 165; racial justice and abolition, 290–291; religious feelings, 159–163, 168, 174–175, 177, 178–179, 185, 288–290, 336–337, 343; resistance to social expectations, 168; science and emotion, 214–215; science and spiritual views of world, 355; telegraphy and telepathy, 202–204, 205; uncertainty in natural science, 251–252

—, education: at Amherst Academy, 77–78; botany, x, 37–38, 39, 43–44, 127–128; intellect and temperament, 117–118, 121–123, 126–127, 128–130; at Mount Holyoke, 153–158, 163–164; natural sciences, 8, 121; natural sciences and classics, 72–74, 76

—, family: childhood portrait, 21, 21–24; church membership, 174–175; deaths in, 330, 334; *Dickinson Family Silhouette*, 118; garden, 365n8, 366n31; house, 90, 91–92, 212–213, 238; mother's illness, xi, xii–xiii, 214, 238, 328, 330–331; readings permitted in, 119–120, 130, 185, 275; relationship with Austin, 328, 331; relationship with father, 130; relationship with mother, 115, 130; social circles, 4–6, 7, 72–74, 119

—, influences on: Elizabeth Barrett Browning, 215–219, 255–257, 280–281, 286, 310, 311, 340; Robert Browning, 280–281; Darwin, xv, 17, 267–271, 282–285, 289–290, 294–295, 314, 326–329, 343–344; Eliot, 302; Emerson, 89, 92, 170; female role models, 131–132; Herschel, 89–90; Higginson, 5, 79, 279–282, 285–286, 307–308, 334, 335–336, 384n21; Hitchcock, 89; Keats, 280, 340; Lowell, 321–322; Romanticism, 203, 204, 266–267; Tennyson, 278

—, lifestyle and events: adolescence and young adulthood, 129–130; in Amherst with Sue, 209–210; bedroom portraits, 310–311; Civil War effects, 252–253, 257–258, 277–278, 286–287, 290–292, 297; correspondents, 212, 258–259; death and funeral, 335–338, *338*, 339; deaths of parents and others, 330–335; dog Carlo, 295, 297–298, 308; health, 295, 297, 335–336; publishing, 279, 293–295; reading interests, 130–131, 243–244, 257–258, 274–275, 334–335, 379n9; reclusiveness, 237–238, 306–307; social circles, 240–241, 299–300, 303, 306–308, 332–335, 384n21; "time of troubles," 238–243; travel to Washington, DC, 211–212

—, natural studies: botany, x, 43–45, 115–116, 127–130, 157; chemistry, 157–158; geology, 150–151

—, writing practices: ambiguities, 129–130, 355–356; cryptic language, 286; experimental writing, xi, 263–265, *264*, 266–267; experimental writing and magical kinship, 251–252; frame of reference changes, 288–289; human–nonhuman relationships, 267–271; reluctance to choose, 290–291; subtle allusions, 326–327; variant wordings, 264, 269, 312, 328–329, 371n3; writing from personas, 79

Dickinson, Emily, "Master" documents: "Master. / If you saw a bullet" (ED to unknown, summer 1861, JL 233), 253–255, 278–279; "Oh' did I offend it –" (ED to unknown, early 1861, JL 248), 252

Dickinson, Emily, poems: "'Faith' is a fine invention," 355–356; "Ah, Teneriffe – Receding Mountain –," 120–121, 371n12; "Best Witchcraft is Geometry," 332, 386n10; "A Bird, came down the Walk –," 282–285, 382n16; "Blazing in Gold – and," 293; "The Brain – is wider than the Sky –," 275–276; on Civil War death, 278; "Cosmopolites without a plea," 328–329; "Exhiliration is the Breeze," 331–332; fascicles, xii–xiii, 244–245, 259, 279; "Flowers – Well – if anybody," 44, 45, 293, 294–295; "Further in Summer than the Birds," 297–298, 302; "Of God we ask one favor, that we may be forgiven –" (Immured the whole of life), 337–338; "The Grass so little has to do," 215; *Herbarium*, x, 127–128; "Her – last Poems –," 256; "Hope is a strange invention –," 356–357; "How happy is the little Stone," 214; "I died for Beauty – but was scarce," 340–342, 349–350; "I felt a Funeral, in my Brain," 258; "The Immortality she gave," 335; "I think I was enchanted," 218–219, 256–257, 342; "It was given to me by the Gods –," 117; I was thinking to-day – as I noticed, that the "Supernatural," 288–289; "A little madness in the Spring," 264–265, *264*; love letters to Susan Gilbert [Dickinson], 204–206, 210–211; "Master" documents, 244, 245–246, 252, 253–255; "A Moth the hue of this," 343–344; "Mute – thy coronation –," 252; "In the name of the Bee –," xiii, xv, 259; "A narrow Fellow in the Grass," 73, 78–79; "Nature and God – I neither knew," 79–80, 81; "Of Bronze—and Blaze –," 268–269, 376n5; "Of Death I try to think like this" (I do remember when a Child), 115–116, 371n3; "A science – so the Savans say," 151; "Sic transit gloria mundi" comic valentine, 206–208, 353; "Sleep is supposed to be," 376n5; "Some – keep the Sabbath – going to church –," 178–179, 293; "Somewhere upon the general Earth," 338–339; "Success is counted sweetest," 293, 333; "There is a flower that Bees prefer –," 268; "These are the days when the Birds come back –," 293; "This was a Poet –," 353–354; "Volcanoes be in Sicily," 214; "We see – Comparatively –," 194; "What mystery pervades a well!," 311–312; "Whoever disenchants," 342–343; "A wife – at Daybreak – I shall be –," 252; "Wild Nights—Wild nights!," 209; "The Wind begun to rock the Grass," 308–309; "Wonder – is not precisely knowing," 90. *See also* Dickinson, Emily

Dickinson, Emily Norcross (mother of ED), 124; death of, 334; education, 123–124, 126; gardening, 23, 365n8, 366n31; illness, xi, xiii, 214, 238, 328, 331; marriage relationship, 119, 124–125

Dickinson, Lavinia (sister of ED): childhood portrait, 21, 22, 24, 39; church membership, 174; at ED's funeral, 336; posthumous publication of ED's work, 348–349; relationship with Austin, 328; travel to Washington, DC, 211–212; visit with Hollands, 208

Dickinson, Lucretia (aunt of ED), 125

Dickinson, Samuel Fowler (grandfather of ED): and Amherst College, 23, 76, 91–92; on education of girls, 40; move to Ohio, 92

Dickinson, Susan Gilbert (sister-in-law of ED), 206; children of, 253; church membership, 174; death of ED, 337, 339; on ED's poetry, 8–9, 339; family background, 209–210; as hostess, 240–241; marriage relationship, 212–213, 237, 238; posthumous publication of ED's work, 348–349; relationship with ED, 240–241, 328–329

—, correspondence with ED, 204–206, 263–264, 264, 312, 328–329, 332, 334, 356–357, 376n16, 386n10

Dickinson, Thomas Gilbert "Gib" (nephew of ED), 331, 334

Dickinson Homestead, 90, 91, 212–213, 238

dinosaur as term, 150

disenchantment: CD's alternative to, 265–266; ED's attitude, 342–343; and human-nonhuman relationships, 347–348; as loss, 341–343; resistance to, 9; and scientific objectivity, 2, 169–170, 363n2

dogs: in CD's family, 295–297, 319; ED's pet Carlo, 295, 297–298, 308; emotions, 295–297, 296

Dred Scott v. Sandford decision, 241–242

Drosera (sundew), 260–263, 261

Drum Beat (magazine), 293, 294

earthworms, 263, 270–271, 303
ecology and Darwin's work, 271, 282–284
economics and social Darwinism, 317
Edgeworth, Richard Lovell, 26–27
Edinburgh University, 57–59, 61–62
Edmonstone, John, 64–65, 247
education. *See* Amherst College
—, girls and women: in Britain, 77; curriculum, 8, 40, 75–76; post–Civil War curriculum, 78; postsecondary, 124, 125–126, 154–156, 158, 172, 374n30
—, natural sciences: and classics, 72–73; in curriculum, 75–76, 121, 174; hostility to magic, 9–10; social hierarchies, 169
Edwards, Jonathan, 42–43, 366n37
electromagnetism, 201–204
Eliot, George (Mary Ann Evans), 302, 304, 310, 311, 333, 344
Emerson, Ralph Waldo, 82; abolitionism, 247–249; *Atlantic Monthly* contributor, 243; in Europe, 164; guest at Evergreens, 240; influence of Herschel on, 81–84; influence of Humboldt on, 99; natural sciences, 86–87, 170, 281, 355

enchantment in Darwin's work, 17, 345–346. *See also* disenchantment; human–nonhuman relationships; interconnections; natural magic
entomology, 67, 68, 269
environmental change, 351–353, 356–357
Etruria (model industrial village), 25–26, 365n10
Etruscan school, 25–26, 36
eugenics, 317–318
Evans, Mary Ann. *See* Eliot, George (Mary Ann Evans)
evolution: earlier terms for, 35, 63, 186–187, 366n27; pre- and post-Darwinian conceptions of, 272–274; in *Vestiges*, 189–190, 224–225. *See also* Darwin, Charles, works, *On the Origin of Species*; Darwin, Erasmus (grandfather of CD); geology; natural selection theory (Darwin); science and religion

Falconer, Hugh, 352
Faraday, Michael, 135
Field, Kate, 216, 258
Fields, James, 289
fig cultivation, 365n8
Finnerty, Páraic, 311
1st South Carolina Volunteers, 250, 286, 382n29
Fiske, Daniel T., 122
Fiske, John, 314, 320, 352
FitzRoy, Robert, 94–95, 101, 195, 247
Flustra, 62–64, 62
Fordyce, John, 199
fossils: *dinosaur* coined, 150; footprints ("turkey tracks"), 87–89, 87, 148–150; *ichthyolite* (fish), 85–86; *Megatherium*, 103–104, 103, 107–108; and transmutation of species, 186. *See also* geology
Fox, William Darwin (cousin of CD), 67–68; correspondence with CD, 81, 100, 101, 192, 196, 222
Franklin, Benjamin, 27
Franklin, Ralph W., 264

Frazer, J. G., 14
Freud, Sigmund, 304
Fuller, Margaret, 164
Fuller, Randall, 248–249

Gallenga, Antonio, 78
Galton, Francis, 317
gender expectations: and Martineau, 139–142; and misreadings of ED, 346–348; and scientific networks, 152. *See also* education, girls and women; women; women's rights
Geological Society, 133–134, 136–137
geology: CD on *Beagle* expedition, 94, 98–100; CD on expedition with Sedgwick, 70, 93–94; CD's essay on, 197–198; gradualism in geological changes, 95–98, 100–101; survey of Massachusetts, 85–86; timescale of earth, 95–96, 168, 176–177, 186, 352; uniformitarian theory (Lyell), 251. *See also* fossils
Gerhardt, Christine, 268
Gilbert, Martha (Mattie), 204–205
Gilbert, Susan (Sue). *See* Dickinson, Susan Gilbert (sister-in-law of ED)
Godwin, Mary Wollstonecraft, 37
Godwin, William, 32, 35
Gopnik, Adam, 229–230, 302
Gosden, Chris, 15, 251, 257
Gould, John, 192–193
Gould, Stephen Jay, 316
Grant, Robert, 62–64, 187
Graves, John (cousin of ED), 209, 210, 377n27
Gray, Asa: as ally of CD, 315–316; antiracist movements, 317; correspondence with CD, 225, 226, 247, 248, 262; review of *Origin*, 246, 250–251, 354; visits with CD, 305–306
Gray, Jane Loring, 305–306, 315
Grima, Peter, 365n8, 366n31
Guittari, Felix, 15
Guthrie, James, 256

Habegger, Alfred, 117, 122, 161
Hamlin, Kim, 327
Haraway, Donna, 15
Hardy, Thomas, 273–274, 276
Hare, William, 59
Harper's Ferry raid, 248–250
Hart, Ellen Louise, 205
Harvard: admission of women, 125, 374n30; curriculum, 172
Hawthorne, Nathaniel, 39, 203, 376n8
Heiligman, Deborah, 182–183
Henslow, John Stevens: evolutionary thought, 191; geological timescales, 96; influence on CD, 68, 93, 94; influence on school curricula, 75–76; promotion of CD's research, 133, 134
—, correspondence with CD, 70, 100, 101, 107, 112, 113, 127
Herrick School, 124, 126
Herschel, Caroline, 80, 132
Herschel, John: on electricity, 204; experimental method, 11, 80–82; grave of, 320, 322, 323; influence of, 69, 75–76, 106, 108–109, 167; influence on CD, 112–113, 197; on "mystery of mysteries," 228–229; portraits, 75, 322, 323; research in South Africa, 113
Herschel, William, 80
"Herschel's private interest," 79–81
hierarchies: botany as threat to, 44–45; challenged in ED's writing, 251–252, 290, 294; challenged in *Vestiges*, 189–190; Linnaeus's naming system, 28, 40, 186; *Scala Naturae* (Great Chain of Being), 40, *41*, 42, 186; social hierarchies and Darwinism, 315, 324; social hierarchies and education, 169. *See also* human–nonhuman relationships; interconnections; science
Higginson, Thomas Wentworth, *282*; antiracist movements, 317; Civil War service, 286–287, 291, 382n29; on Darwinism, 324; at ED's funeral, 336–337; as literary Darwinian, 5, 281, 289; as member of Secret Six, 250; posthumous publication of ED's work, 348–349; relationship with CD, 300, 307, 308, 309, 310; relationship with ED, 212, 307, 308–309, 346, 384n21; visit with Browning, 309; visit with Tennyson and Cameron, 309–310
—, correspondence with ED, 5, 79, 115, 279–282, 285–286, 295, 297–298, 307–308, 331, 335–336, 336, 342–343, 356–357, 382n16
—, works: "Letter to a Young Contributor," 279, 353–354; "My Out-Door Study," 258–259; *Out-door Papers*, 289–290; "The Procession of the Flowers," 287–288, 382n35
Hitchcock, Edward, *84*; Amherst College museums, 170, 213; correspondence with CD, 149–150; education and career, 84–86; establishment of Mount Holyoke, 158; geology studies, 87–89, *87*, 148–150, 213–214; influence of Herschel on, 81–82; and Lyell, 148; magic as trickery, 13; on Mary Lyon, 132; and materialists, 176–177, 271; science and religion, 43, 175–177; on Samuel F. Dickinson, 92; visit from Martineau, 144
—, works: *Dyspepsy Forestalled and Resisted*, 85; *Peculiar Phenomena in the Four Seasons*, 43; *The Religion of Geology*, 176–177, 202, 214, 271; *Report on the Geology, Mineralogy, Botany, and Zoology of Massachusetts*, 85–86, 148–149; "The Sandstone Bird," 88–89; "The Telegraphic System of the Universe," 176, 202
Hitchcock, Ned, 209–210
Hitchcock, Orra, 85, 88, 158
Holland, Elizabeth, 208, 211; correspondence with ED, 214–215, 263–264, *264*, 333
Holland, Josiah, 208, 211, 333
Holmes, John, 272–273
Holmes, Oliver Wendell, 243, 257–258, 293
Hooker, Joseph, *188*; as ally of CD, 315–316; bequest from CD, 322; encouragement to CD, 188; at funeral of CD, 321; publication of CD's natural selection theory, 227

—, correspondence with CD, xiv–xv, 3, 190, 196, 222, 225–227, 233–234, 262, 263, 352–353
Howland, William, 207
human–nonhuman relationships: as alternatives to human conventions, 168–170; in CD's work, 299–300, 313–314; Darwinism and its interpretations, 314; and disenchantment, 347–348; in ED's work, 259, 267–271, 281, 297–298, 302, 328–329; influence on Eliot, 302; mind and material world interactions, 177; in twenty-first century, 351–352, 356–357. *See also* death; hierarchies; interconnections
Humboldt, Alexander von, 70, 81, 99, 120, 313
Hunt, Caroline Dutch, 73, 126–127, 131
Hunt, Harriet Kezia, 374n30
Hutto, James, 96–98
Huxley, Thomas: *agnostic* coined, 230–231; and CD, 315–316, 321, 322; debate about CD's theory, 233–234; dissection of ape brain, 272

Innes, John Brodie, 184
interconnections: in CD's work, 16–17, 262–263, 266–267, 270–271, 282–284, 349–350; consciousness in plants, 262–263, 266; Darwinian thought and political activism, 304–305; in ED's work, 251–252, 294–295, 349–350; in Edwards's work, 42–43; in Erasmus Darwin's work, 29–30; in Higginson's work, 259; and Romanticism, 99; science, religion, and magic, 15–16. *See also* death; human–nonhuman relationships; natural magic

Jackson, Helen Hunt (Helen Fiske), 212, 307, 333, 335, 337–338
James, Henry, 291, 302, 324
James, Wilkie, 291
James, William, 9, 304, 324, 344
Jameson, Robert, 64, 96
Jefferson, Thomas, 27, 99
Jenyns, Leonard, 67, 190, 192–193
Josephson Storm, Jason Ā., 9, 14

Keats, John, 34, 36, 340
Keir, James, 26
Kelly, Mike, 293
Kingsley, Charles, 275
kinship. *See* human–nonhuman relationships; interconnections
Knickerbocker magazine, 88
Kuhn, Mary, 157

Lamarck, Jean-Baptiste, 63, 186–187
Leighton, William Allport, 37, 49
Levine, George, 3, 267, 274, 300–301, 319, 345
Lincoln, Abraham, 242, 252
Lincoln, Almira Hart. *See* Phelps, Almira (Hart) Lincoln
Linnaeus, Carl, 28, 40, 186
Linnean Society, 227
literary naturalists, 325–326, 342–343
Longfellow, Henry Wadsworth, 203, 243
Lord, Otis Phillips: death of, 334–335; guest at Evergreens, 240–241; relationship with ED, 303, 307, 326–327, 385n36
Lowell, Amy, 346–347
Lowell, James Russell, 243, 321–322
Lubbock, John, 262
Lunar Society of Birmingham: abolition of slavery, 247; founding of, 6, 18; impact of, 42; studies and discussions, 24–27. *See also* Darwin, Erasmus (grandfather of CD)
Lyell, Charles, 97; correspondence with CD, 223–224, 225, 232, 260; evolutionary thought, 191; *evolution* as term, 187; geological timescales, 95–98, 100–101; and Hitchcock, 148; influence on CD, 98, 100–101, 112, 133–134, 135–136; personality, 146; *Principles of Geology*, 94, 96–98, 100–101; publication of CD's natural selection theory, 227; uniformitarian theory, 251
Lyon, Mary, 131; influence on ED, 73, 131–132; and Mount Holyoke, 157, 158, 160–161

magic, natural. *See* natural magic
magic, supernatural vs. natural, 16. *See also* natural magic
magic tricks, 13–14, 105–106, 364n29
Malthus, Thomas, 138, 142
Martineau, Harriet, 78, 137–144, *138*, 148, 187
Massachusetts: common schools, 77; cultural similarities to Britain, 5–6; girls' curriculum, 76; nineteenth-century culture, 39
materialism: and Browning, 216; and CD's approach to evolution, 187; criticism of, 33–34, 35–36, 45; and Darwinian thought, 270–271; and Darwinism, 3–4, 271, 325–326; *naturalist* as term, 61; search for soul in brain, 271–273, 275–276; and split in natural philosophy, 86–87, 108–109
Mather, Cotton, 9
Megatherium fossils, *103*, 103–104, 107–108
Meredith, George, 272–273, 274
Mexican-American War, 164
Miller, Cristanne, 264
Milton, John, 69, 71, 95
Mirandola, Giovanni Pico della, 10
Mitchell, Maria, 154–156, *155*
Mondello, Kaitlin, 284–285
Monson Academy, 124, *125*, 126–127
Mott, Lucretia, 172–173
The Mount, Shrewsbury, 49, 54
Mount Holyoke, *154*; Christmas holiday, 160–161; curriculum, 156–157, 163–164; entrance exams, 153–154; establishment of, 125–126, 158; natural science and Octagon, 170–172; religious mission, 158–161, 161–162, 163
museums: Appleton Cabinet, 213; expansion of, 135; Octagon Cabinet, 170–172, *171*, 374n29

natural hierarchy. *See* hierarchies
naturalist as term, 61, 325–326
naturalists, literary, 325–326, 342–343
natural magic: attitudes toward, 9–10, 13–14; in CD's work, 345–346; in ED's work, 331–332, 338–339, 342–343, 345–350; as empirical observation, 42–43; and hope in twenty-first century, 356–357; magical thinking in twenty-first century, 14–16; as natural forces and transformations, 10–11, 15–16; vs. natural philosophy, 12, 105–106; scientific objectivity and disenchantment, 2, 363n2; spirits animating world, 251, 257; vs. supernatural magic, 16; as transmutation, 29. *See also* human–nonhuman relationships; interconnections; natural philosophy; science; science and religion
Natural Magick (Porta), *xviii*, 10, 42–43
natural philosophy: vs. naturalist studies, 83; vs. natural magic, 12, 105–106; and nature of God, 42–43; poetry and science relationships, xvi, 1, 7, 32–34, 42–45, 86–87; and Romanticism, 32–34; *science* as term for, 105–106; shift toward empirical inquiry, 42–43, 75–76, 80–82, 108–109; shift toward science, 1–4; vitalism, 157–158. *See also* education, natural sciences; natural magic; Romanticism; science; science and religion
natural science. *See* science
natural selection theory (Darwin): beginnings of, 111, 137, 150; concrete application of, 224; controversy recognized in, 185–186, 190–192, 375n19; evidence for sought, 192–193; misrepresentations of, 234–236, *235*; outline and first draft of, 187–188; publication at Linnean Society, 227; transmutation of species, 186–187, 220; writing of, xi–xii, xiii–xiv, 223–225. *See also Beagle* expedition; Darwin, Charles; Darwin, Charles, works, *On the Origin of Species*; Darwinism
Nature's People, 73–74, 79–80
Neptunians, 96
Newman, Francis, 185
Newton, Isaac, 9, 42, 65–66
New York Times, 320
Nicholas, C. J., 98
Niles, Thomas, 302, 333

Norcross, Emily Lavinia (cousin of ED), 161
Norcross, Erasmus (great-uncle of ED), 125
Norcross, Frances and Louisa (cousins of ED), 277, 278, 292, 334, 336
northern lights. *See* aurora borealis
Norton, Charles Eliot, 257

occult as term, 10–11
Octagon Cabinet (museum), 170–172, *171*, 374n29
Oliver, Daniel, 262, 263
Olmsted, Denison, 201
orchids: CD's study of, xiv–xv, 263, 266; pink lady's slipper (*Cypripedium acaule*), *x*, xiii, 337
Orchis Bank, xiv, 283
Ouless, Walter William, 304
Owen, Richard: and CD, 136, 192; coining of *dinosaur*, 150; debate on human brain, 271–272; reaction to *Origin*, 232, 233, 262–263

Paine, Frank, 291
Paley, William, 61, 69, 71, 95, 168, 179
Panic of 1857, 241–242
Parish, Woodbine, 107–108
Parker, Theodore, 185
Pearson, P. N., 98
personality (soul), 271–272, 275
Phelps, Almira (Hart) Lincoln, 39, 128, 132, 371n26
Phillips, Moses Dresser, 243
Piesse, Jude, 53
pigeons, 222
Plinian Natural History Society, 61–62, 64
Plutonians, 96
political questions and science. *See* Darwinian thought; Darwinism; hierarchies; human–nonhuman relationships
political upheaval in Europe, 164
Pollak, Vivian, 290
Popular Science Monthly, 320
Porta, Giambattista della, *xviii*, 10, 42–43

Porter, Hannah, 160–161
Prescott, George B., 258
Priestley, Joseph, 18, 25, 26, 32
Promethean matches, 102–103
psychic phenomena, 344
psychology and Darwin's work, 271
publication: popular nonfiction work, 135; and shift to empiricism, 12–13, 106. *See also* Darwin, Charles, career
public popular culture, 135
Punch (magazine), 247, 249

racism, 291–292, 317–318
Radcliffe College, 172
religion: Christianity and enchantment of natural world, 167; and education, 158–159; revivals, 161, 162, 163; theology split from literature, 51. *See also* science and religion
Ricardo, David, 138
Richards, Robert, 32, 69, 300–301
Richardson, Robert D., 83
Romanticism: and botany, 157; ED's writing different from, 266–267, 340–341; influence of Humboldt on, 99; and naturalists' studies, 108–109, 114; and natural philosophy, 32–34; as reaction against Lunar Society, 36; and science, 204
Root, Abiah, 175; correspondence with ED, 119, 122, 123, 126, 127, 128–129, 130, 157, 161–163, 175, 205
Royal College of Surgeons, 107–108
Royal Institution, 135
Royal Medal for Natural Science, 222
Royal Society, 6, 11, 83

Saint Jago (São Tiago), Cape Verde Islands, 98–101
Sanborn, Franklin Benjamin, 249–250
Satter, Beryl, 317
science: definitions, 319; divorce from arts, 107–108, 114, 151, 271, 354–355; and empiricism, 12–13, 108–109; as imaginative, 151; and materialism, 271–272; and poetry,

258–259, 266, 272–274, 275–276, 300–301, 340–341, 354–355; professionalization of, 7–8, 267; publication and shift to empiricism, 12–13, 106; and Romanticism, 204; as term, 11, 26; women in, 156, 169, 172–173, 346–348, 374n30; and wonder, 167, 168–169, 230, 265–266, 324–326. *See also* botany; education, natural sciences; geology; hierarchies; natural magic; natural philosophy; science and religion; zoology

science and religion: coexistence of, 172–174, 228–234; evolutionary thought, 189–190, 191, 228–234, 250–251, 315–316, 320; geological changes and timescales, 95–98, 100–101, 168, 176–177; insufficient today, 356–357; and magic, 15, 45; separation of, 4, 9–10, 169–170; transmutation of species, 186–187; typology, 43. *See also* Darwin, Charles; Darwin, Charles, works; natural magic; natural philosophy; natural selection theory (Darwin); science

science in literature: aurora borealis, 268–269, 376n5; in Browning, 215–217, 257; electricity and psychic powers, 334–335; telepathy, 203–204, 376n8

scientific networks, 135–137, 146–148, 191

scientists: as amateurs, 106, 267; as cultural figures, 2; as term, 1, 8, 107, 109–111, 319, 355

Secret Six of Harper's Ferry raid, 249–250

secularization. *See* science and religion

Sedgwick, Adam: evolutionary thought, 189, 191; geological expedition, 70, 93–94; geological timescales, 96; promotion of CD's research, 133–134, 137; responses to *Origin*, 232, 262–263

Sedgwick, Catherine Maria, 39, 130

Seneca Falls Convention (1848), 164, 172–173

Seward, Anna, 365n16

Seward, Thomas, 60

Sharples, Ellen, 18–19, *20*, *21*, *22*, 24

Shaw, George Bernard, 316–317, 318

Shaw, Robert Gould, 291, 382n29

Shelley, Mary, 2, 35–36

Shoptaw, John, 287

Shrewsbury School, 54–56, 74

Sismondi, Jessie, 145

slavery, 65, 241–242, 247

Small, Richard, 26, 27

Smith, Martha Nell, 205, 328

Socarides, Alexandra, 245

social Darwinism, 234, 304, 316–318. *See also* Darwinism

Somerville, Mary, 109–110, *110*, 132, 201

soul (personality), 271–272, 275

Spencer, Herbert, 234, 304, 317–318

spiritualists, 344

Springfield Republican, 207, 316

Stearns, Frazar, 277–278

Stearns, Williams, 277

Stott, Rebecca, 195

Stowe, Harriet Beecher, 240, 243, 258

Strachey, Lytton, 322

supernatural and spiritualism, 344

supernatural magic vs. natural magic, 16. *See also* natural magic

survival of the fittest, 234–235, 317–318. *See also* Darwin, Charles, works, *On the Origin of Species*; Darwinian thought; Darwinism; natural selection theory (Darwin)

telepathy, 202–204, 205, 376n8

Tenerife, Canary Islands, 70, 71, 371n12

Tennyson, Alfred, *310*; ED's allusion to, 278; nature as violent, 189–190, 269; and scientific thought, 272; visit with CD, 303–304; visit with Higginson, 309–310

Test Acts (Britain, 1673), 25

Thoreau, Henry David, 164–168, *166*, 170, 249–250, 382n35

Times (London), 321

Todd, Mabel Loomis, 328, 348–349

Tolley, Kim, 8

transformation as term, 63, 186–187

transmutation as term, 35, 63

Tree of Life, 228, 254

Turner, Clara Newman, 163

Uglow, Jenny, 26, 27
Unitarianism/Unitarians, 50–51, 184
Uruguay, 101–106

Valadés, Diego, 40, *41*
Vassar College, 155
Venus flytraps, 263
Vestiges of the Natural History of Creation (anon.; Chambers), 189–192, 224–225, 227–228, 234

Wadsworth, Charles, 212
Wald, Priscilla, 242
Wallace, Alfred Russel: correspondence with CD, 197; psychic phenomena, 344; theory of evolution, xi, 223–224, 225, 227
Walls, Laura Dassow, 167
Waterhouse, George R., 192–193
Way, Albert, 67, *68*
Weber, Max, 2, 252, 265, 363n2
Wedgwood, Emma. *See* Darwin, Emma Wedgwood (wife of CD)
Wedgwood, Fanny, 183
Wedgwood, Josiah (grandfather of CD): antislavery medallion, 247, *248*; business, 25–26, 365n10; on French Revolution, 27; Lunar Society, 18, 26; religious beliefs, 184
Wedgwood, Josiah, II (uncle of CD), 49–51
Wedgwood, Josiah "Joe," III (cousin of CD), 145
Wedgwood, Susanna. *See* Darwin, Susanna Wedgwood (mother of CD)
Wedgwood, Tom (uncle of CD), 49–52
Werner, Marta, 244, 245, 246
Westminster Review, 302

Whewell, William: and CD, 68; coining of *scientist*, 107, 109, 110; epigraph in *Origin*, 228; science in curricula, 75–76
Whitehurst, John, 26
Whitman, Walt, 215, 377n33
Whitney, Maria, 332–333
Wilberforce, Samuel, 233–234, 262–263
Wilson, Edward O., 313
Wineapple, Brenda, 289, 291
witches/witchcraft, 11, 44–45
Withering, William, 26
Wollstonecraft, Mary, 32, 35
women: in science, 156, 169, 172–173, 346–348, 374n30; social effects of Civil War, 303; as teachers, 131–132. *See also* education, girls and women; women's rights
women's rights: CD's support for, 304–305, 317; Darwinian arguments for, 327; Seneca Falls Convention, 164, 172–173; and social Darwinism, 318. *See also* education, girls and women; women
Wood, Abby, 122, 174–175
Woodbridge, R., 123
Wordsworth, William: anthropocentrism, 267; criticism of Erasmus Darwin, 33, 36; influence of, 108; influence of Humboldt on, 99. *See also* Romanticism
—, works: "I Wandered Lonely as a Cloud," 33; *Lyrical Ballads* (with Coleridge), 33, 51, 266; "Tables Turned," 33

Yale University, 125

zoology: barnacles, 193–196, *195*, 221–222; earthworms, 263, 270–271, 303; *Flustra*, 62–64, *62*. *See also* Darwin, Charles, natural studies